T0258278

Cambridge IGCSE™

Physics

Fourth Edition

Cambridge IGCSE™

Physics

Fourth Edition

Heather Kennett
Tom Duncan

AN HACHETTE UK COMPANY

Cambridge International copyright material in this publication is reproduced under licence and remains the intellectual property of Cambridge Assessment International Education. Past paper questions reproduced by permission of Cambridge Assessment International Education.

Exam-style questions [and sample answers] have been written by the authors. In examinations, the way marks are awarded may be different. References to assessment and/or assessment preparation are the publisher's interpretation of the syllabus requirements and may not fully reflect the approach of Cambridge Assessment International Education.

Cambridge Assessment International Education bears no responsibility for the example answers to questions taken from its past question papers which are contained in this publication.

Although every effort has been made to ensure that website addresses are correct at time of going to press, Hodder Education cannot be held responsible for the content of any website mentioned in this book. It is sometimes possible to find a relocated web page by typing in the address of the home page for a website in the URL window of your browser.

Third-party websites and resources referred to in this publication have not been endorsed by Cambridge Assessment International Education.

We have carried out a health and safety check of this text and have attempted to identify all recognised hazards and suggest appropriate cautions. However, the Publishers and the authors accept no legal responsibility on any issue arising from this check; whilst every effort has been made to carefully check the instructions for practical work described in this book, it is still the duty and legal obligation of schools to carry out their own risk assessments for each practical in accordance with local health and safety requirements.

For further health and safety information (e.g. Hazcards) please refer to CLEAPSS at www.cleapss.org.uk.

Hachette UK's policy is to use papers that are natural, renewable and recyclable products and made from wood grown in well-managed forests and other controlled sources. The logging and manufacturing processes are expected to conform to the environmental regulations of the country of origin.

Orders: please contact Hachette UK Distribution, Hely Hutchinson Centre, Milton Road, Didcot, Oxfordshire, OX11 7HH. Telephone: +44 (0)1235 827827. Email education@hachette.co.uk Lines are open from 9 a.m. to 5 p.m., Monday to Friday. You can also order through our website: www.hoddereducation.com

ISBN: 978 1 3983 1054 4

© Tom Duncan and Heather Kennett 2021

First published in 2002

Second edition published in 2009

Third edition published in 2014

This fourth edition published in 2021 by
Hodder Education,
An Hachette UK Company
Carmelite House
50 Victoria Embankment
London EC4Y 0DZ

www.hoddereducation.com

Impression number   10 9 8 7 6 5 4 3 2 1

Year   2024 2023 2022 2021

All rights reserved. Apart from any use permitted under UK copyright law, no part of this publication may be reproduced or transmitted in any form or by any means, electronic or mechanical, including photocopying and recording, or held within any information storage and retrieval system, without permission in writing from the publisher or under licence from the Copyright Licensing Agency Limited. Further details of such licences (for reprographic reproduction) may be obtained from the Copyright Licensing Agency Limited, www.cla.co.uk

Cover photo © Zffoto – stock.adobe.com

Illustrations by Fakenham Prepress Solutions, Wearset and Integra Software Services Pvt. Ltd., Pondicherry, India

Typeset by Integra Software Services Pvt. Ltd., Pondicherry, India

Printed in Slovenia

A catalogue record for this title is available from the British Library.

Contents

How to use this book

To make your study of Physics for Cambridge IGCSE™ as rewarding and successful as possible, this textbook, endorsed by Cambridge Assessment International Education, offers the following important features:

FOCUS POINTS

Each topic starts with a bullet point summary of what you will encounter within each topic.

This is followed by a short outline of the topic so that you know what to expect over the next few pages.

Test yourself

These questions appear regularly throughout the topic so you can check your understanding as you progress.

Revision checklist

At the end of each topic, a revision checklist will allow you to recap what you have learnt in each topic and double check that you understand the key concepts before moving on.

Exam-style questions

Each topic is followed by exam-style questions to help familiarise you with the style of questions you may see in your examinations. These will also prove useful in consolidating your learning. Past paper questions are also provided in the back of the book.

As you read through the book, you will notice that some text is shaded yellow. This indicates that the highlighted material is Supplement content only. Text that is not shaded covers the Core syllabus. If you are studying the Extended syllabus, you should look at both the Core and Supplement sections.

As well as these features, you will also see additional support throughout the topic in the form of:

Key definitions

These provide explanations of the meanings of key words as required by the syllabus.

 ### Practical work

These boxes identify the key practical skills you need to be able to understand and apply as part of completing the course.

 ### Worked example

These boxes give step-by-step guidance on how to approach different sorts of calculations, with follow-up questions so you can practise these skills.

Going further

These boxes take your learning further than is required by the Cambridge syllabus so that you have the opportunity to stretch yourself.

A *Mathematics for Physics* section is provided for reference. This covers many of the key mathematical skills you will need as you progress through your course. If you feel you would benefit from further explanation and practice on a particular mathematical skill within this book, the *Mathematics for Physics* section should be a useful resource.

Answers are provided online with the accompanying *Cambridge IGCSE Physics Teacher's Guide*. A *Practical Skills Workbook* is also available to further support you in developing your practical skills as part of carrying out experiments.

Scientific enquiry

During your course you will have to carry out a few experiments and investigations aimed at encouraging you to develop some of the **skills** and **abilities** that scientists use to solve real-life problems.

Simple experiments may be designed to measure, for example, the temperature of a liquid or the electric current in a circuit. Longer investigations may be designed to establish or verify a relationship between two or more physical quantities.

Investigations may arise from the topic you are currently studying in class, or your teacher may provide you with suggestions to choose from, or you may have your own ideas. However an investigation arises, it will probably require at least one hour of laboratory time, but often longer.

Doing an investigation will involve the following aspects.

1 **Selecting and safely using suitable techniques, apparatus and materials** – your choice of both apparatus and techniques will depend on what you're investigating. However, you must always do a risk assessment of your investigation before proceeding. Your teacher should help ensure that all potential hazards are identified and addressed.

2 **Planning your experiment** – you need to think about how you are going to find answers to the questions regarding the problem posed. This will involve:
   - Making predictions and hypotheses (informed guesses); this may help you to focus on what is required at this stage.
   - Identifying the variables in the investigation and deciding which ones you will try to keep constant (controlled) so that they do not affect the experimental results. The variable you change is the independent variable and the variable you will measure is the dependent variable.
   - Deciding on the range of values you will use for the independent variable, how you will record your results, and the analysis you will do to fulfil the aims of the investigation.
   - The apparatus and materials you choose to use for the investigation need to have enough precision for the required use. For example the smallest division on a metre ruler is 1 mm. A measurement can be read to about half a scale division so will have a precision of about 0.5 mm.
   - Explaining your experimental procedure. A clearly labelled diagram will be helpful here. Any difficulties encountered or precautions taken to achieve accuracy should also be mentioned.

3 **Making and recording observations and measurements** – you need to obtain the necessary experimental data safely and accurately. Before you start taking measurements familiarise yourself with the use of the apparatus.
   - Record your observations and readings in an ordered way. If several measurements are to be made, draw up a table to record your results. Use the column headings, or start of rows, to name the measurement and state its unit; for example 'Mass of load/kg'. Repeat the measurement of each observation; record each value in your table, then calculate an average value. Numerical values should be given to the number of significant figures appropriate to the measuring device.
   - If you have decided to make a graph of your results you will need at least eight data points taken over as large a range as possible; be sure to label each axis of a graph with the name and unit of the quantity being plotted.
   - Do not dismantle the equipment until you have completed your analysis and you are sure you do not need to repeat any of the measurements!

4 **Interpreting and evaluating the observations and data** – this is important as doing it correctly will allow you to establish relationships between quantities.
   - You may need to calculate specific values or plot a graph of your results, then draw a line of best fit and calculate a gradient. Explain any anomalous results you obtained and how you dealt with them. Comment on any graph drawn, its shape and whether the graph points lie on the line; mention any trend you noticed in the data.
   - Draw conclusions from the evidence that are justified by the data. These can take the form of a numerical value (and unit), the statement of a known law, a relationship between two quantities or a statement related to the aim of the experiment (sometimes experiments do not achieve the intended objective).
   - Comment on the quality of the data and whether results are equal within the limits of the accuracy of the experiment. Compare outcomes with those expected.

5 **Evaluating methods and suggesting possible improvements** – not every experiment is flawless and in fact, they rarely are. When looking at evaluation you should:
   - Identify possible sources of error in the experiment which could have affected the

accuracy of your results. These could include random and systematic errors as well as measurement errors.

- Mention any apparatus that turned out to be unsuitable for the experiment.
- Discuss how the experiment might be modified to give more accurate results, for example in an electrical experiment by using an ammeter with a more appropriate scale.
- Suggest possible improvements to the experiment. For example, efforts to reduce thermal energy losses to the environment or changes in a control variable (such as temperature) in an experiment.

Suggestions for investigations

Some suggested investigations for practical work are listed below:

1 Stretching of a rubber band (Topic 1.5.1).
2 Stretching of a copper wire – **wear eye protection** (Topic 1.5.1).
3 Toppling (Topic 1.5.1).
4 Friction – factors affecting (Topic 1.5.1).
5 Model wind turbine design (Topic 1.7.3).
6 Speed of a bicycle and its stopping distance (Topic 1.7.1).
7 Energy transfer using different insulating materials (Topic 2.3.1).
8 Cooling and evaporation (Topic 2.2.3).
9 Pitch of a note from a vibrating wire (Topic 3.4).
10 Variation of the resistance of a thermistor with temperature (Topic 4.2.4).
11 Variation of the resistance of a wire with length (Topic 4.2.4)
12 Heating effect of an electric current (Topic 4.2.2).
13 Strength of an electromagnet (Topic 4.1).
14 Efficiency of an electric motor (Topic 4.2.5).

Ideas and evidence in science

In some of the investigations you perform in the school laboratory, you may find that you do not interpret your data in the same way as your friends do; perhaps you will argue with them as to the best way to explain your results and try to convince them that your interpretation is right. Scientific controversy frequently arises through people interpreting evidence differently.

Observations of the heavens led the ancient Greek philosophers to believe that the Earth was at the centre of the planetary system, but a complex system of rotation was needed to match observations of the apparent movement of the planets across the sky. In 1543, Nicolaus Copernicus made the radical suggestion that all the planets revolved not around the Earth but around the Sun. (His book *On the Revolutions of the Celestial Spheres* gave us the modern usage of the word 'revolution'.) It took time for his ideas to gain acceptance. The careful astronomical observations of planetary motion documented by Tycho Brahe were studied by Johannes Kepler, who realised that the data could be explained if the planets moved in elliptical paths (not circular) with the Sun at one focus. Galileo's observations of the moons of Jupiter with the newly invented telescope led him to support this 'Copernican view' and to be imprisoned by the Catholic Church in 1633 for disseminating heretical views. About 50 years later, Isaac Newton introduced the idea of gravity and was able to explain the motion of all bodies, whether on Earth or in the heavens, which led to full acceptance of the Copernican model. Newton's mechanics were refined further at the beginning of the twentieth century when Einstein developed his theories of relativity. Even today, data from the Hubble Space Telescope is providing new evidence which confirms Einstein's ideas.

Many other scientific theories have had to wait for new data, technological inventions, or time and the right social and intellectual climate for them to become accepted. In the field of health and medicine, for example, because cancer takes a long time to develop it was several years before people recognised that X-rays and radioactive materials could be dangerous (Topic 5.2.5).

At the beginning of the twentieth century scientists were trying to reconcile the wave theory and the particle theory of light by means of the new ideas of quantum mechanics.

Today we are collecting evidence on possible health risks from microwaves used in mobile phone networks. The cheapness and popularity of mobile phones may make the public and manufacturers reluctant to accept adverse findings, even if risks are made widely known in the press and on television. Although scientists can provide evidence and evaluation of that evidence, there may still be room for controversy and a reluctance to accept scientific findings, particularly if there are vested social or economic interests to contend with. This is most clearly shown today in the issue of global warming.

SECTION 1

Motion, forces and energy

Topics

Physical quantities and measurement techniques

FOCUS POINTS

★ Describe how to measure length, volume and time intervals using simple devices.
★ Know how to determine the average value for a small distance and a short time interval.

★ Understand the difference between scalar and vector quantities, and give examples of each.
★ Calculate or determine graphically the resultant of two perpendicular vectors.

This topic introduces the concept of describing space and time in terms of numbers together with some of the basic units used in physics. You will learn how to use simple devices to measure or calculate the quantities of length, area and volume. Accurate measurements of time will be needed frequently in the practical work in later topics and you will discover how to choose the appropriate clock or timer for the measurement of a time interval. Any single measurement will not be entirely accurate and will have an error associated with it. Taking the average of several measurements, or measuring multiples, reduces the size of the error.

Many physical quantities, such as force and velocity, have both magnitude and direction; they are termed vectors. When combining two vectors to find their resultant, as well as their size, you need to take into account any difference in their directions.

▲ **Figure 1.1.1** Aircraft flight deck

Units and basic quantities

Before a measurement can be made, a standard or *unit* must be chosen. The size of the quantity to be measured is then found with an instrument having a scale marked in the unit.

Three basic quantities we measure in physics are **length**, **mass** and **time**. Units for other quantities are based on them. The SI (Système International d'Unités) system is a set of metric units now used in many countries. It is a decimal system in which units are divided or multiplied by 10 to give smaller or larger units.

Measuring instruments on the flight deck of a passenger jet provide the crew with information about the performance of the aircraft (see Figure 1.1.1).

Powers of ten shorthand

This is a neat way of writing numbers, especially if they are large or small. The example below shows how it works.

$$4000 = 4 \times 10 \times 10 \times 10 = 4 \times 10^3$$
$$400 = 4 \times 10 \times 10 \qquad = 4 \times 10^2$$
$$40 = 4 \times 10 \qquad = 4 \times 10^1$$
$$4 = 4 \times 1 \qquad = 4 \times 10^0$$
$$0.4 = 4/10 = 4/10^1 \qquad = 4 \times 10^{-1}$$
$$0.04 = 4/100 = 4/10^2 \qquad = 4 \times 10^{-2}$$
$$0.004 = 4/1000 = 4/10^3 \qquad = 4 \times 10^{-3}$$

The small figures 1, 2, 3, etc. are called **powers of ten**. The power shows how many times the number has to be multiplied by 10 if the power is greater than 0 or divided by 10 if the power is less than 0. Note that 1 is written as 10^0.

This way of writing numbers is called **standard notation**.

Length

The unit of length is the **metre** (m) and is the distance travelled by light in a vacuum during a specific time interval. At one time it was the distance between two marks on a certain metal bar. Submultiples are:

1 decimetre (dm) $= 10^{-1}$ m

1 centimetre (cm) $= 10^{-2}$ m

1 **millimetre** (mm) $= 10^{-3}$ m

1 **micrometre** (μm) $= 10^{-6}$ m

1 **nanometre** (nm) $= 10^{-9}$ m

A multiple for large distances is

1 **kilometre** (km) $= 10^3$ m ($\frac{5}{8}$ mile approx.)

1 **gigametre** (Gm) $= 10^9$ m = 1 billion metres

Many length measurements are made with rulers; the correct way to read one is shown in Figure 1.1.2. The reading is 76 mm or 7.6 cm. Your eye must be directly over the mark on the scale or the thickness of the ruler causes a parallax error.

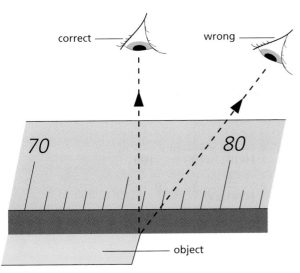

▲ **Figure 1.1.2** The correct way to measure with a ruler

To obtain an average value for a small distance, multiples can be measured. For example, in ripple tank experiments (Topic 3.1), measure the distance occupied by five waves, then divide by 5 to obtain the average wavelength.

Significant figures

Every measurement of a quantity is an attempt to find its true value and is subject to errors arising from limitations of the apparatus and the experimenter. The number of figures, called **significant figures**, given for a measurement indicates how accurate we think it is and more figures should not be given than are justified.

For example, a value of 4.5 for a measurement has two significant figures; 0.0385 has three significant figures, 3 being the most significant and 5 the least, i.e. it is the one we are least sure about since it might be 4 or it might be 6. Perhaps it had to be estimated by the experimenter because the reading was between two marks on a scale.

When doing a calculation your answer should have the same number of significant figures as the measurements used in the calculation. For example, if your calculator gave an answer of 3.4185062, this would be written as 3.4 if the measurements had two significant figures. It would be written as 3.42 for three significant figures. Note that in deciding the least significant figure you look at the next figure to the right. If it is less than 5, you leave the least significant figure as it is (hence 3.41 becomes 3.4), but if it equals or is greater than 5 you increase the least significant figure by 1 (round it up) (hence 3.418 becomes 3.42).

If a number is expressed in standard notation, the number of significant figures is the number of digits before the power of ten. For example, 2.73×10^3 has three significant figures.

> **Test yourself**
>
> 1 How many millimetres are there in these measurements?
>   a 1 cm
>   b 4 cm
>   c 0.5 cm
>   d 6.7 cm
>   e 1 m
> 2 What are these lengths in metres?
>   a 300 cm
>   b 550 cm
>   c 870 cm
>   d 43 cm
>   e 100 mm
> 3 a Write the following as powers of ten with one figure before the decimal point:
>   100 000   3500   428 000 000   504   27 056
>   b Write out the following in full:
> 10^3 2×10^6 6.92×10^4 1.34×10^2 10^9
> 4 a Write these fractions as powers of ten:
>   1/1000   7/100 000   1/10 000 000   3/60 000
>   b Express the following decimals as powers of ten with one figure before the decimal point:
>   0.5   0.084   0.000 36   0.001 04

Area

The **area of the square** in Figure 1.1.3a with sides 1 cm long is 1 square centimetre (1 cm²). In Figure 1.1.3b the rectangle measures 4 cm by 3 cm and has an area of $4 \times 3 = 12$ cm² since it has the same area as twelve squares each of area 1 cm². The area of a square or rectangle is given by

$$\text{area} = \text{length} \times \text{breadth}$$

The SI unit of area is the square metre (m²) which is the area of a square with sides 1 m long. Note that

$$1 \text{ cm}^2 = \frac{1}{100} \text{ m} \times \frac{1}{100} \text{ m} = \frac{1}{10\,000} \text{ m}^2 = 10^{-4} \text{ m}^2$$

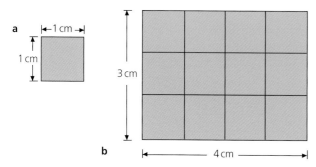

▲ **Figure 1.1.3**

Sometimes we need to know the area of a triangle. It is given by

$$\text{area of triangle} = \frac{1}{2} \times \text{base} \times \text{height}$$

The **area of a circle** of radius r is πr^2 where $\pi = 22/7$ or 3.14; its circumference is $2\pi r$.

> **? Worked example**
>
> Calculate the area of the triangles shown in Figure 1.1.4.
>
> a   area of triangle $= \frac{1}{2} \times$ base $\times$ height
>
>   so area of triangle ABC $= \frac{1}{2} \times$ AB $\times$ AC
>
> $= \frac{1}{2} \times 4 \text{ cm} \times 6 \text{ cm} = 12 \text{ cm}^2$
>
> b   area of triangle PQR $= \frac{1}{2} \times$ PQ $\times$ SR
>
> $= \frac{1}{2} \times 5 \text{ cm} \times 4 \text{ cm} = 10 \text{ cm}^2$

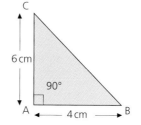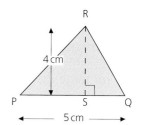

▲ **Figure 1.1.4**

Now put this into practice

1 Calculate the area of a triangle whose base is 8 cm and height is 12 cm.
2 Calculate the circumference of a circle of radius 6 cm.

Volume

Volume is the amount of space occupied. The unit of volume is the **cubic metre** (m³) but as this is rather large, for most purposes the **cubic centimetre** (cm³) is used. The volume of a cube with 1 cm edges is 1 cm³. Note that

$$1 \text{ cm}^3 = \frac{1}{100} \text{ m} \times \frac{1}{100} \text{ m} \times \frac{1}{100} \text{ m}$$

$$= \frac{1}{1\,000\,000} \text{ m}^3 = 10^{-6} \text{ m}^3$$

For a regularly shaped object such as a rectangular block, Figure 1.1.5 shows that

volume = length × breadth × height

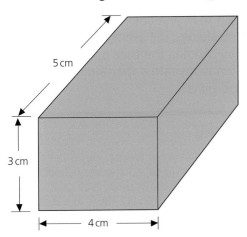

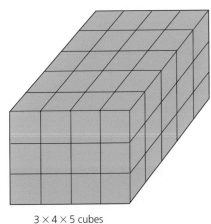

3 × 4 × 5 cubes

▲ **Figure 1.1.5**

The **volume of a cylinder** of radius r and height h is $\pi r^2 h$.

The volume of a liquid may be obtained by pouring it into a measuring cylinder (Figure 1.1.6). When making a reading the cylinder must be upright and your eye must be level with the bottom of the curved liquid surface, i.e. the **meniscus**. The meniscus formed by mercury is curved oppositely to that of other liquids and the top is read.

Measuring cylinders are often marked in millilitres (ml) where $1\,ml = 1\,cm^3$; note that $1000\,cm^3 = 1\,dm^3$ (= 1 litre).

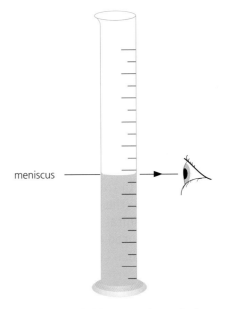

▲ **Figure 1.1.6** A measuring cylinder

? Worked example

a Calculate the volume of a block of wood which is 40 cm long, 12 cm wide and 5 cm high in cubic metres.

volume V = length × breadth × height

$$= 40\,cm \times 12\,cm \times 5\,cm$$
$$= 2400\,cm^3$$
$$= 2400 \times 10^{-6}\,m^3$$
$$= 2.4 \times 10^{-3}\,m^3$$

b Calculate the volume of a cylinder of radius 10 mm and height 5.0 cm in cubic metres.

volume of cylinder $V = \pi r^2 h$

$r = 10\,mm = 1.0\,cm$ and $h = 5.0\,cm$

so $V = \pi r^2 h$

$$= \pi \times (1.0\,cm)^2 \times 5.0\,cm$$
$$= 16\,cm^3 = 16 \times 10^{-6}\,m^3 = 1.6 \times 10^{-5}\,m^3$$

Now put this into practice

1 Calculate the volume of a rectangular box which is 30 cm long, 25 cm wide and 15 cm high in cubic metres.
2 Calculate the volume of a cylinder of radius 50 mm and height 25 cm in cubic metres.

Time

The unit of time is the **second** (s), which used to be based on the length of a day, this being the time for the Earth to revolve once on its axis. However, days are not all of exactly the same duration and the second is now defined as the time interval for a certain number of **energy** changes to occur in the caesium atom.

Time-measuring devices rely on some kind of constantly repeating oscillation. In traditional clocks and watches a small wheel (the balance wheel) oscillates to and fro; in digital clocks and watches the oscillations are produced by a tiny quartz crystal. A swinging pendulum controls a pendulum clock.

To measure an interval of time in an experiment, first choose a timer that is precise enough for the task. A stopwatch is adequate for finding the period in seconds of a pendulum (see Figure 1.1.7 opposite), but to measure the speed of sound (Topic 3.4), a clock that can time in milliseconds is needed. To measure very short time intervals, a digital clock that can be triggered to start and stop by an electronic signal from a microphone, photogate or mechanical switch is useful. Tickertape timers or dataloggers are often used to record short time intervals in motion experiments. Accuracy can be improved by measuring longer time intervals. Several oscillations (rather than just one) are timed to find the period of a pendulum; the average value for the period is found by dividing the time by the number of oscillations. Ten ticks, rather than single ticks, are used in tickertape timers.

> ## Test yourself
>
> 5 The pages of a book are numbered 1 to 200 and each leaf is 0.10 mm thick. If each cover is 0.20 mm thick, what is the thickness of the book?
> 6 How many significant figures are there in a length measurement of
>   a 2.5 cm
>   b 5.32 cm
>   c 7.180 cm
>   d 0.042 cm?
> 7 A rectangular block measures 4.1 cm by 2.8 cm by 2.1 cm. Calculate its volume giving your answer to an appropriate number of significant figures.
> 8 What type of timer would you use to measure the period of a simple pendulum? How many oscillations would you time?

Practical work

Period of a simple pendulum

For safe experiments/demonstrations related to this topic, please refer to the *Cambridge IGCSE Physics Practical Skills Workbook* that is also part of this series.

In this investigation you have to make time measurements using a stopwatch or clock. A motion sensor connected to a datalogger and computer could be used instead of a stopwatch for these investigations.

Attach a small metal ball (called a bob) to a piece of string, and suspend it as shown in Figure 1.1.7 opposite. Pull the bob a small distance to one side, and then release it so that it oscillates to and fro through a small angle.

Find the time for the bob to make several complete oscillations; one oscillation is from A to O to B to O to A (Figure 1.1.7). Repeat the timing a few times for the same number of oscillations and work out the average.

1 The time for one oscillation is the **period** *T*. Determine the period of your pendulum.
2 The **frequency** *f* of the oscillations is the number of complete oscillations per second and equals 1/*T*. Calculate a value for *f* for your pendulum.
3 Comment on how the amplitude of the oscillations changes with time.
4 Plan an investigation into the effect on *T* of (i) a longer string and (ii) a larger bob.
5 What procedure would you use to determine the period of a simple pendulum?
6 In Figure 1.1.7 if the bob is first released at B, give the sequence of letters which corresponds to one complete oscillation.

7 Explain where you would take measurements from to determine the length of the pendulum shown in Figure 1.1.7.

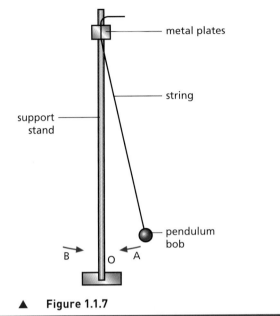

support stand

metal plates

string

pendulum bob

B   O   A

▲ **Figure 1.1.7**

Systematic errors

Figure 1.1.8 shows a part of a ruler used to measure the height of a point P above the bench. The ruler chosen has a space before the zero of the scale. This is shown as the length x. The height of the point P is given by the scale reading added to the value of x. The equation for the height is

$$\text{height} = \text{scale reading} + x$$

$$\text{height} = 5.9 + x$$

By itself the scale reading is not equal to the height. It is too small by the value of x.

This type of error is known as a **systematic error**. The error is introduced by the system. A half-metre ruler has the zero at the *end of the ruler* and so can be used without introducing a systematic error.

When using a ruler to determine a height, the ruler must be held so that it is vertical. If the ruler is at an angle to the vertical, a systematic error is introduced.

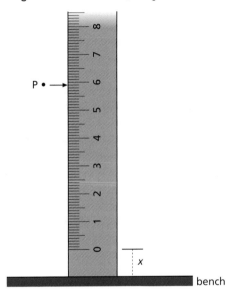

P •→

x

bench

▲ **Figure 1.1.8**

 Going further

Vernier scales and micrometers

Lengths can be measured with a ruler to a precision of about 0.5 mm. Some investigations may need a more precise measurement of length, which can be achieved by using vernier calipers (Figure 1.1.9) or a micrometer screw gauge.

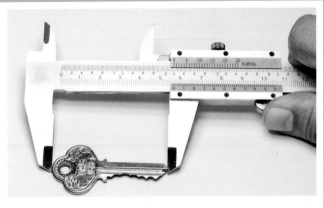

▲ **Figure 1.1.9** Vernier calipers in use

Vernier scale

The calipers shown in Figure 1.1.9 use a vernier scale. The simplest type enables a length to be measured to 0.01 cm. It is a small sliding scale which is 9 mm long but divided into ten equal divisions (Figure 1.1.10a) so

$$1 \text{ vernier division} = \frac{9}{10} \text{ mm}$$

$$= 0.9 \text{ mm}$$

$$= 0.09 \text{ cm}$$

One end of the length to be measured is made to coincide with the zero of the millimetre scale and the other end with the zero of the vernier scale. The length of the object in Figure 1.1.10b is between 1.3 cm and 1.4 cm. The reading to the second place of decimals is obtained by finding the vernier mark which is exactly opposite (or nearest to) a mark on the millimetre scale. In this case it is the 6th mark and the length is 1.36 cm, since

$$OA = OB - AB$$

$$OA = (1.90 \text{ cm}) - (6 \text{ vernier divisions})$$

$$= 1.90 \text{ cm} - 6(0.09) \text{ cm}$$

$$= (1.90 - 0.54) \text{ cm}$$

$$= 1.36 \text{ cm}$$

Vernier scales are also used on barometers, travelling microscopes and spectrometers.

a

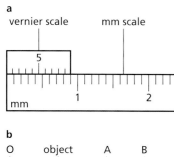

b

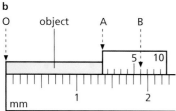

▲ **Figure 1.1.10** Vernier scale

Micrometer screw gauge

This measures very small objects to 0.001 cm. One revolution of the drum opens the flat, parallel jaws by one division on the scale on the shaft of the gauge; this is usually mm, i.e. 0.05 cm. If the drum has a scale of 50 divisions round it, then rotation of the drum by one division opens the jaws by 0.05/50 = 0.001 cm (Figure1.1.11). A friction clutch ensures that the jaws exert the same force when the object is gripped.

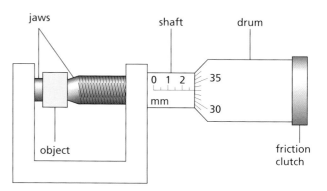

▲ **Figure 1.1.11** Micrometer screw gauge

The object shown in Figure 1.1.11 has a length of

$$2.5 \text{ mm on the shaft scale} + 33 \text{ divisions on the drum scale}$$

$$= 0.25 \text{ cm} + 33(0.001) \text{ cm}$$

$$= 0.283 \text{ cm}$$

Before making a measurement, check to ensure that the reading is zero when the jaws are closed. Otherwise the zero error must be allowed for when the reading is taken.

Scalars and vectors

Length and time can be described by a single number specifying size, but many physical quantities have a directional character.

A **scalar** quantity has magnitude (size) only. Time is a scalar and is completely described when its value is known. Other examples of scalars are distance, speed, time, mass, pressure, energy and **temperature.**

A **vector** quantity is one such as force which is described completely only if both its size (magnitude) and direction are stated. It is not enough to say, for example, a force of 10 N, but rather a force of 10 N acting vertically downwards. Gravitational field strength and electric field strength are vectors, as are weight, velocity, acceleration and momentum.

A vector can be represented by a straight line whose length represents the magnitude of the quantity and whose direction gives its line of action. An arrow on the line shows which way along the line it acts.

Scalars are added by ordinary arithmetic; vectors are added geometrically, taking account of their directions as well as their magnitudes. In the case of two vectors F_X and F_Y acting at right angles to each other at a point, the magnitude of the resultant F, and the angle θ between F_X and F can be calculated from the following equations:

$$F = \sqrt{F_X^{\,2} + F_Y^{\,2}}, \quad \tan\theta = \frac{F_Y}{F_X}$$

The resultant of two vectors acting at right angles to each other can also be obtained graphically.

❓ Worked example

Calculate the resultant of two forces of 3.0 N and 4.0 N acting at right angles to each other.

Let $F_X = 3.0\,\text{N}$ and $F_Y = 4.0\,\text{N}$ as shown in Figure 1.1.12.

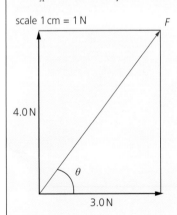

scale 1 cm = 1 N

▲ **Figure 1.1.12** Addition of two **perpendicular** vectors

Then
$$F = \sqrt{F_X^2 + F_Y^2} = \sqrt{3.0^2 + 4.0^2} = \sqrt{9 + 16} = \sqrt{25} = 5.0\,\text{N}$$

and $\tan\theta = \dfrac{F_Y}{F_X} = \dfrac{4.0}{3.0} = 1.3$

so $\theta = 53°$.

The resultant is a force of 5.0 N acting at 53° to the force of 3.0 N.

Graphical method

The values for F and θ can be found graphically by drawing the vectors to scale on a piece of graph paper as shown in Figure 1.1.12.

First choose a scale to represent the size of the vectors (1 cm could be used to represent 1.0 N).

Draw the vectors at right angles to each other. Complete the rectangle as shown in Figure 1.1.12 and draw the diagonal from the origin as shown. The diagonal then represents the resultant force, F. Measure the length of F with a ruler and use the scale you have chosen to determine its size. Measure the angle θ, the direction of the resultant, with a protractor.

Check that the values for F and θ you obtain are the same as those found using the algebraic method.

Now put this into practice

1 Calculate the following square roots.
    a $\sqrt{6^2 + 8^2}$
    b $\sqrt{5^2 + 7^2}$
    c $\sqrt{2^2 + 9^2}$

2 Calculate
    a tan 30°
    b tan 45°
    c tan 60°.

3 Calculate the resultant of two forces of 5.0 N and 7.0 N which are at right angles to each other.

4 At a certain instant a projectile has a horizontal velocity of 6 m/s and a vertical velocity of 8 m/s.
    a Calculate the resultant velocity of the projectile at that instant.
    b Check your answer to a by a graphical method.

Revision checklist

After studying Topic 1.1 you should know and understand the following:

✓ how to make measurements of length and time intervals, minimise the associated errors and use multiple measurements to obtain average values

✓ the difference between scalars and vectors and recall examples of each.

After studying Topic 1.1 you should be able to:

✓ write a number in powers of ten (standard notation) and recall the meaning of standard prefixes

✓ measure and calculate lengths, areas and volumes of regular objects and give a result with the correct units and an appropriate number of significant figures

✓ determine by calculation or graphically the resultant of two vectors at right angles.

Exam-style questions

1 A chocolate bar measures 10 cm long by 2 cm wide and is 2 cm thick.

   a Calculate the volume of one bar. [3]

   b How many bars each 2 cm long, 2 cm wide and 2 cm thick have the same total volume? [3]

   c A pendulum makes 10 complete oscillations in 8 seconds. Calculate the time period of the pendulum. [2]

                                   [Total: 8]

2 **a** A pile of 60 sheets of paper is 6 mm high. Calculate the average thickness of a sheet of the paper. [2]

   b Calculate how many blocks of ice cream each 10 cm long, 10 cm wide and 4 cm thick can be stored in the compartment of a freezer measuring 40 cm deep, 40 cm wide and 20 cm high. [5]

                                   [Total: 7]

3 A Perspex container has a 6 cm square base and contains water to a height of 7 cm (Figure 1.1.13).

   a Calculate the volume of the water. [3]

   b A stone is lowered into the water so as to be completely covered and the water rises to a height of 9 cm. Calculate the volume of the stone. [4]

                                     [Total: 7]

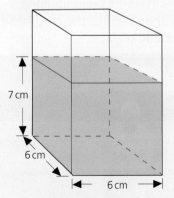

▲ **Figure 1.1.13**

4 **a** State the standard units of length and time. [2]

   b A measurement is stated as 0.0125 mm. State the number of significant figures. [1]

   c Write down expressions for

     i the area of a circle [1]

     ii the circumference of a circle [1]

     iii the volume of a cylinder. [2]

                                     [Total: 7]

Going further

5 What are the readings on the micrometer screw gauges in Figures 1.1.14a and 1.1.14b?

a

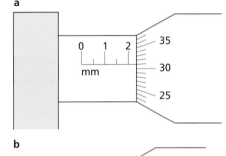

b

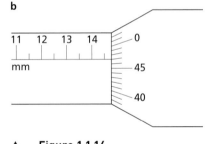

▲ **Figure 1.1.14**

                                     [Total: 4]

6 **a** Select which of the following quantities is a vector.

     A length

     B temperature

     C force

     D time [1]

   b Two forces of 5 N and 12 N act at right angles to each other.
Using a piece of graph paper determine the magnitude and direction of the resultant force graphically. State the scale you use to represent each vector. You will need a protractor to measure the angle the resultant makes with the 5 N force. [7]

                                     [Total: 8]

1.2 Motion

FOCUS POINTS

★ Define speed and velocity and use the appropriate equations to calculate these and average speed.

★ Draw, plot and interpret distance–time or speed–time graphs for objects at different speeds and use the graphs to calculate speed or distance travelled.

★ Define acceleration and use the shape of a speed–time graph to determine constant or changing acceleration and calculate the acceleration from the gradient of the graph.

★ Know the approximate value of the acceleration of freefall, g, for an object close to the Earth's surface.

★ Describe the motion of objects falling with and without air/liquid resistance.

The concepts of speed and acceleration are encountered every day, whether it be television monitoring of the speed of a cricket or tennis ball as it soars towards the opposition or the acceleration achieved by an athlete or racing car. In this topic you will learn how to define speed in terms of distance and time. Graphs of distance against time will enable you to calculate speed and determine how it changes with time; graphs of speed against time allow acceleration to be studied. Acceleration is also experienced by falling objects as a result of gravitational attraction. All objects near the Earth's surface experience the force of gravity, which produces a constant acceleration directed towards the centre of the Earth.

Speed

The **speed** of a body is the distance that it has travelled in unit time. When the distance travelled is s over a short time period t, the speed v is given by

$$v = \frac{s}{t}$$

> **Key definition**
> **Speed** distance travelled per unit time

If a car travels 300 km in five hours, its **average speed** is 300 km/5 h = 60 km/h. The speedometer would certainly not read 60 km/h for the whole journey and might vary considerably from this value. That is why we state the average speed. If a car could travel at a constant speed of 60 km/h for 5 hours, the distance covered would still be 300 km. It is *always* true that

$$\text{average speed} = \frac{\text{total distance travelled}}{\text{total time taken}}$$

To find the actual speed at any instant we would need to know the distance moved in a very short interval of time. This can be done by multiflash photography. In Figure 1.2.1 the golfer is photographed while a flashing lamp illuminates him 100 times a second. The speed of the club-head as it hits the ball is about 200 km/h.

▲ **Figure 1.2.1** Multiflash photograph of a golf swing

Velocity

Speed is the distance travelled in unit time; **velocity** is the distance travelled in unit time in a given direction. If two trains travel due north at 20 m/s, they have the same speed of 20 m/s and the same velocity of 20 m/s *due north*. If one travels north and the other south, their speeds are the same but not their velocities since their directions of motion are different.

$$\text{velocity} = \frac{\text{distance moved in a given direction}}{\text{time taken}}$$
$$= \text{speed in a given direction}$$

> **Key definition**
> **Velocity** change in displacement per unit time

The velocity of a body is **uniform** or constant if it moves with a steady speed in a straight line. It is not uniform if it moves in a curved path. Why?

The units of speed and velocity are the same, km/h, m/s.

$$60 \, \text{km/h} = \frac{60\,000 \, \text{m}}{3600 \, \text{s}} = 17 \, \text{m/s}$$

Distance moved in a stated direction is called the **displacement**. Velocity may also be defined as

$$\text{velocity} = \frac{\text{change in displacement}}{\text{time taken}}$$

Speed is a *scalar* quantity and velocity a *vector* quantity. Displacement is a vector, unlike distance which is a scalar.

Acceleration

When the velocity of an object changes, we say the object *accelerates*. If a car starts from rest and moving due north has velocity 2 m/s after 1 second, its velocity has increased by 2 m/s in 1 s and its acceleration is 2 m/s per second due north. We write this as 2 m/s².

Acceleration is defined as the change of velocity in unit time, or

$$\text{acceleration} = \frac{\text{change of velocity}}{\text{time taken for change}} = \frac{\Delta v}{\Delta t}$$

> **Key definition**
> **Acceleration** change in velocity per unit time

For a steady increase of velocity from 20 m/s to 50 m/s in 5 s

$$\text{acceleration} = \frac{(50 - 20)\,\text{m/s}}{5\,\text{s}} = 6 \, \text{m/s}^2$$

Acceleration is also a vector and both its magnitude and direction should be stated. However, at present we will consider only motion in a straight line and so the magnitude of the velocity will equal the speed, and the magnitude of the acceleration will equal the change of speed in unit time.

The speeds of a car accelerating on a straight road are shown below.

Time/s	0	1	2	3	4	5	6
Speed/m/s	0	5	10	15	20	25	30

The speed increases by 5 m/s every second and the acceleration of 5 m/s² is constant.

An acceleration is positive if the velocity increases, and negative if it decreases. A negative acceleration is also called a **deceleration** or **retardation**.

> ▶ **Test yourself**
>
> 1 What is the average speed of
>   a a car that travels 400 m in 20 s
>   b an athlete who runs 1500 m in 4 minutes?
> 2 A train increases its speed steadily from 10 m/s to 20 m/s in 1 minute.
>   a What is its average speed during this time, in m/s?
>   b How far does it travel while increasing its speed?
>
> 3 a A motorcyclist starts from rest and reaches a speed of 6 m/s after travelling with constant acceleration for 3 s. What is his acceleration?
>   b The motorcyclist then decelerates at a constant rate for 2 s. What is his acceleration?
> 4 An aircraft travelling at 600 km/h accelerates steadily at 10 km/h per second. Taking the speed of sound as 1100 km/h at the aircraft's altitude, how long will it take to reach the 'sound barrier'?

Speed–time graphs

If the speed of an object is plotted against the time, the graph obtained is a **speed–time graph**. It provides a way of solving motion problems.

In Figure 1.2.2, AB is the speed–time graph for an object moving with a **constant speed** of 20 m/s.

Values for the speed of the object at 1 s intervals can be read from the graph and are given in Table 1.2.1. The data shows that the speed is constant over the 5 s time interval.

▼ **Table 1.2.1**

Speed/m/s	20	20	20	20	20	20
Time/s	0	1	2	3	4	5

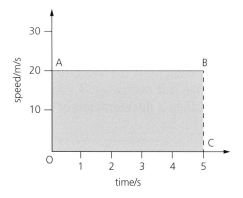

▲ **Figure 1.2.2** Constant speed

The linear shape (PQ) of the speed–time graph shown in Figure 1.2.3a means that the gradient, and hence the acceleration of the body, are constant over the time period OS.

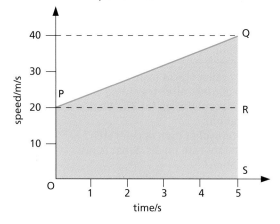

▲ **Figure 1.2.3a** Constant acceleration

Values for the speed of the object at 1 s intervals can be read from the graph and are given in Table 1.2.2. The data shows that the speed increases by the same amount (4 m/s) every second.

▼ **Table 1.2.2**

Speed/m/s	20	24	28	32	36	40
Time/s	0	1	2	3	4	5

You can use the data to plot the speed–time graph. Join up the data points on the graph paper with the best straight line to give the line PQ shown in Figure 1.2.3a. (Details for how to plot a graph are given on pp. 297–8 in the *Mathematics for physics* section.)

Figure 1.2.3b shows the shape of a speed–time graph for an object accelerating from rest over time interval OA, travelling at a constant speed over time interval AB and then decelerating (when the speed is decreasing) over the time interval BC. The steeper gradient in time interval BC than in time interval OA shows that the deceleration is greater than the acceleration. The object remains at rest over the time interval CD when its speed and acceleration are zero.

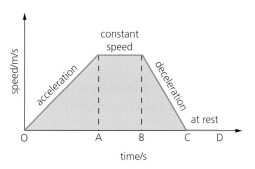

▲ **Figure 1.2.3b** Acceleration, constant speed and deceleration

Figure 1.2.3c shows a speed–time graph for a changing acceleration. The curved shape OX means that the gradient of the graph, and hence the acceleration of the object, change over time period OY – the acceleration is changing.

Values for the speed of the object at 1 s intervals are given in Table 1.2.3. The data shows that the speed is increasing over time interval OY, but by a smaller amount each second so the acceleration is decreasing.

▼ **Table 1.2.3**

Speed/m/s	0	17.5	23.0	26.0	28.5	30.0
Time/s	0	1	2	3	4	5

You can use the data to plot the speed–time graph. Join up the data points on the graph paper with a smooth curve as shown in Figure 1.2.3c.

Note that an object *at rest* will have zero speed and zero acceleration; its speed–time graph is a straight line along the horizontal axis.

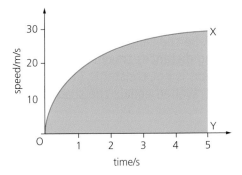

▲ **Figure 1.2.3c** Changing acceleration

Using the gradient of a speed–time graph to calculate acceleration

The gradient of a speed–time graph represents the acceleration of the object.

In Figure 1.2.2, the gradient of AB is zero, as is the acceleration. In Figure 1.2.3a, the gradient of PQ is QR/PR = 20/5 = 4: the acceleration is constant at $4\,m/s^2$. In Figure 1.2.3c, when the gradient along OX changes, so does the acceleration.

An object is accelerating if the speed increases with time and decelerating if the speed decreases with time, as shown in Figure 1.2.3b. In Figure 1.2.3c, the speed is increasing with time and the acceleration of the object is decreasing.

Distance–time graphs

An object travelling with constant speed covers equal distances in equal times. Its **distance–time graph** is a straight line, like OL in Figure 1.2.4a for a constant speed of 10 m/s. The gradient of the graph is

LM/OM = 40 m/4 s = 10 m/s, which is the value of the speed. The following statement is true in general:

The gradient of a distance–time graph represents the speed of the object.

Values for the distance moved by the object recorded at 1 s intervals are given in Table 1.2.4. The data shows it moves 10 m in every second so the speed of the object is constant at 10 m/s.

▼ **Table 1.2.4**

Distance/m	10	20	30	40
Time/s	1	2	3	4

You can use the data to plot the distance–time graph shown in Figure 1.2.4a.

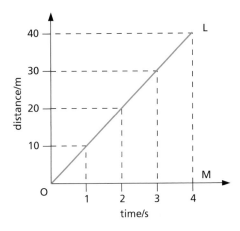

▲ **Figure 1.2.4a** Constant speed

Figure 1.2.4b shows the shape of a distance–time graph for an object that is at rest over time interval OA and then moves at a constant speed in time interval AB. It then stops moving and is at rest over time interval BC before moving at a constant speed in time interval CD.

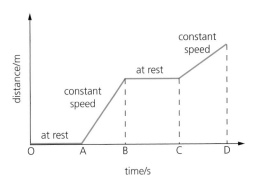

▲ **Figure 1.2.4b** Constant speed

The speed of the object is higher when the gradient of the graph is steeper. The object is travelling faster in time interval AB than it is in time interval CD; it is at rest in time intervals OA and BC when the distance does not change.

When the speed of the object is changing, the gradient of the distance–time graph varies, as in Figure 1.2.5, where the upward curve of increasing gradient of the solid green line shows the object accelerating. The opposite, upward curve of decreasing gradient (indicated by the dashed green line) shows an object decelerating above T.

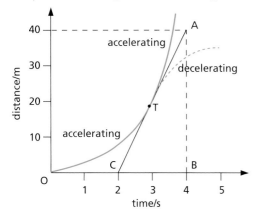

▲ **Figure 1.2.5** Non-constant speed

Speed at any point equals the gradient of the **tangent**. For example, the gradient of the tangent at T is AB/BC = 40 m/2 s = 20 m/s. The speed at the instant corresponding to T is therefore 20 m/s.

Area under a speed–time graph

The area under a speed–time graph measures the distance travelled.

In Figure 1.2.2, AB is the speed–time graph for an object moving with a constant speed of 20 m/s. Since distance = average speed × time, after 5 s it will have moved 20 m/s × 5 s = 100 m. This is the shaded area under the graph, i.e. rectangle OABC.

In Figure 1.2.3a, PQ is the speed–time graph for an object moving with *constant acceleration*.

At the start of the timing the speed is 20 m/s, but it increases steadily to 40 m/s after 5 s. If the distance covered equals the area under PQ, i.e. the shaded area OPQS, then

distance = area of rectangle OPRS + area of triangle PQR

$$= OP \times OS + \frac{1}{2} \times PR \times QR$$

(area of a triangle $= \frac{1}{2}$ base × height)

$$= 20 \, \text{m/s} \times 5 \, \text{s} + \frac{1}{2} \times 5 \, \text{s} \times 20 \, \text{m/s}$$

$$= 100 \, \text{m} + 50 \, \text{m} = 150 \, \text{m}$$

Note that when calculating the area from the graph, the unit of time must be the same on both axes.

The rule for finding distances travelled is true even if the acceleration is not constant. In Figure 1.2.3c, the distance travelled equals the shaded area OXY.

► **Test yourself**

5 The speeds of a bus travelling on a straight road are given below at successive intervals of 1 second.

Time/s	0	1	2	3	4
Speed/m/s	0	4	8	12	16

  a Sketch a speed–time graph using the values.
  b Choose two of the following terms which describe the acceleration of the bus:
     constant   changing   positive   negative

  c Calculate the acceleration of the bus.

  d Calculate the area under your graph.
  e How far does the bus travel in 4 s?
6 The distance of a walker from the start of her walk is given below at successive intervals of 1 second.
  a Sketch a distance–time graph of the following values.

Time/s	0	1	2	3	4	5	6
Distance/m	0	3	6	9	12	15	18

  b How would you describe the speed at which she walks?
     constant   changing   increasing
     accelerating
  c Calculate her average speed.

Equations for constant acceleration

Problems involving bodies moving with constant acceleration in a straight line can often be solved quickly using some *equations of motion*.

First equation

If an object is moving with constant acceleration a in a straight line and its speed increases from u to v in time t, then

$$a = \frac{\text{change of speed}}{\text{time taken}} = \frac{v - u}{t}$$

$$\therefore \quad at = v - u$$

or

$$v = u + at \quad (1)$$

Note that the initial speed u and the final speed v refer to the start and the finish of the *timing* and do not necessarily mean the start and finish of the motion.

Second equation

The speed of an object moving with constant acceleration in a straight line increases steadily. Its average speed therefore equals half the sum of its initial and final speeds, that is,

$$\text{average speed} = \frac{u + v}{2}$$

If s is the distance moved in time t, then since average speed = total distance/total time = s/t,

$$\frac{s}{t} = \frac{u + v}{2}$$

or

$$s = \frac{(u + v)}{2} t \quad (2)$$

Going further

Third equation

From equation (1), $v = u + at$

From equation (2),

$$\frac{s}{t} = \frac{u + v}{2}$$

$$\frac{s}{t} = \frac{u + u + at}{2} = \frac{2u + at}{2}$$

$$= u + \frac{1}{2}at$$

and so

$$s = ut + \frac{1}{2}at^2 \quad (3)$$

Fourth equation

This is obtained by eliminating t from equations (1) and (3). Squaring equation (1) we have

$$v^2 = (u + at)^2$$

$$\therefore \quad v^2 = u^2 + 2uat + a^2t^2$$

$$= u^2 + 2a(ut + \frac{1}{2}at^2)$$

But $\quad s = ut + \frac{1}{2}at^2$

$$\therefore \quad v^2 = u^2 + 2as$$

If we know any *three* of u, v, a, s and t, the others can be found from the equations.

? Worked example

A sprint cyclist starts from rest and accelerates at $1\,\text{m/s}^2$ for 20 seconds. Find her final speed and the distance she travelled.

Since $u = 0 \qquad a = 1\,\text{m/s}^2 \qquad t = 20\,\text{s}$

Using $v = u + at$, we have her maximum speed

$$v = 0 + 1\,\text{m/s}^2 \times 20\,\text{s} = 20\,\text{m/s}$$

and distance travelled

$$s = \frac{(u + v)}{2} t$$

$$= \frac{(0 + 20)\,\text{m/s} \times 20\,\text{s}}{2} = \frac{400}{2} = 200\,\text{m}$$

Now put this into practice

1 An athlete accelerates from rest at a constant rate of $0.8\,\text{m/s}^2$ for $4\,\text{s}$. Calculate the final speed of the athlete.
2 A cyclist increases his speed from $10\,\text{m/s}$ to $20\,\text{m/s}$ in $5\,\text{s}$. Calculate his average speed over this time interval.
3 Calculate the distance moved by a car accelerating from rest at a constant rate of $2\,\text{m/s}^2$ for $5\,\text{s}$.

Falling bodies

In air, a coin falls faster than a small piece of paper. In a vacuum they fall at the same rate, as may be shown with the apparatus of Figure 1.2.6. The difference in air is due to **air resistance** having a greater effect on light bodies than on heavy bodies. The air resistance to a light body is large when compared with the body's weight. With a dense piece of metal, the resistance is negligible at low speeds.

There is a story, untrue we now think, that in the sixteenth century the Italian scientist Galileo Galilei dropped a small iron ball and a large cannonball ten times heavier from the top of the Leaning Tower of Pisa (Figure 1.2.7). And we are told that, to the surprise of onlookers who expected the cannonball to arrive first, they reached the ground almost simultaneously.

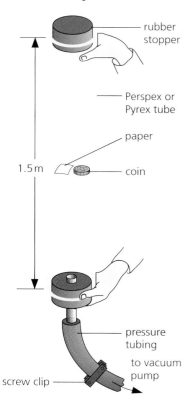

▲ **Figure 1.2.6** A coin and a piece of paper fall at the same rate in a vacuum.

▲ **Figure 1.2.7** The Leaning Tower of Pisa, where Galileo is said to have experimented with falling objects

Practical work

Motion of a falling object

Safety

- Place something soft on the floor to absorb the impact of the masses.
- Take care to keep feet well away from the falling masses.

Arrange your experimental apparatus as shown in Figure 1.2.8 and investigate the motion of a 100 g mass falling from a height of about 2 m.

A tickertape timer has a marker that vibrates at 50 times a second and makes dots at 1/50 s intervals on a paper tape being pulled through it. Ignore the start of the tape where the dots are too close.

Repeat the experiment with a 200 g mass and compare your results with those for the 100 g mass.

1 The spacing between the dots on the tickertape increases as the mass falls. What does this tell you about the speed of the falling mass?
2 The tape has 34 dots on it by the time the mass falls through 2 m. Estimate how long it has taken the mass to fall through 2 m.

3 Why would a stopwatch not be chosen to measure the time of fall in this experiment?
4 How would you expect the times taken for the 100 g and 200 g masses to reach the ground to differ?

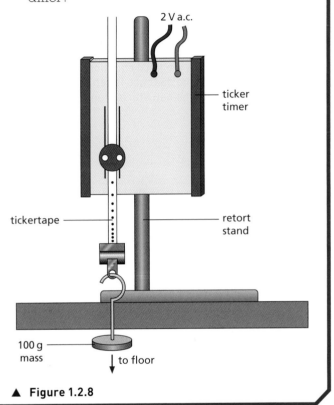

▲ **Figure 1.2.8**

Acceleration of free fall

All bodies falling freely under the force of gravity do so with **uniform acceleration** if air resistance is negligible (i.e. the 'steps' on the tape chart from the practical work should all be equally spaced).

This acceleration, called the **acceleration of free fall**, is denoted by the italic letter g. Its value varies slightly over the Earth but is constant in each place; on average it is about 9.8 m/s², or near enough 10 m/s². The velocity of a free-falling body therefore increases by about 10 m/s every second. A ball shot straight upwards with a velocity of 30 m/s decelerates by about 10 m/s every second and reaches its highest point after 3 s.

> **Key definition**
>
> **Acceleration of free fall** g for an object near to the surface of the Earth, this is approximately constant and is approximately 9.8 m/s²

In calculations using the equations of motion, g replaces a. It is given a positive sign for falling bodies (i.e. $a = g = +9.8$ m/s²) and a negative sign for rising bodies since they are decelerating (i.e. $a = -g = -9.8$ m/s²).

Going further

Measuring g

Using the arrangement in Figure 1.2.9, the time for a steel ball-bearing to fall a known distance is measured by an electronic timer.

When the two-way switch is changed to the 'down' position, the electromagnet releases the ball and simultaneously the clock starts. At the end of its fall the ball opens the 'trap-door' on the impact switch and the clock stops.

The result is found from the third equation of motion $s = ut + \frac{1}{2}at^2$, where s is the distance fallen (in m), t is the time taken (in s), $u = 0$ (the ball starts from rest) and $a = g$ (in m/s^2).

Hence

$$s = \frac{1}{2}gt^2$$

or

$$g = 2s/t^2$$

Air resistance is negligible for a dense object such as a steel ball-bearing falling a short distance.

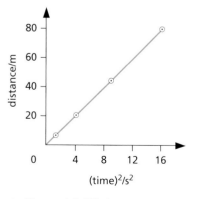

▲ **Figure 1.2.9**

A rough estimate for g can be made by timing the fall of a rubber ball from the top of a building. It will only take a second to reach the ground from a height of 5 m, so you will need fast reactions if you use a stopwatch for the measurement. Watch out that you do not hit anybody below!

Distance–time graphs for a falling object

For an object falling freely from rest in a uniform gravitational field without air resistance, there will be constant acceleration g, so we have

$$s = \frac{1}{2}gt^2$$

A graph of distance s against time t is shown in Figure 1.2.10a. The gradually increasing slope indicates the speed of the object increases steadily. A graph of s against t^2 is shown in Figure 1.2.10b; it is a straight line through the origin since $s \propto t^2$ (g being constant at one place).

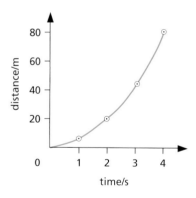

▲ **Figure 1.2.10a** A graph of distance against time for a body falling freely from rest

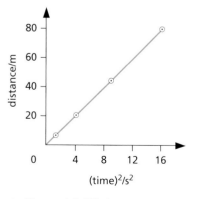

▲ **Figure 1.2.10b** A graph of distance against (time)^2 for a body falling freely from rest

Test yourself

7 An object falls from a hovering helicopter and hits the ground at a speed of 30 m/s. How long does it take the object to reach the ground and how far does it fall? Sketch a speed–time graph for the object (ignore air resistance).

8 A stone falls from rest from the top of a high tower. Ignore air resistance and take $g = 9.8$ m/s^2. Calculate
   a the speed of the stone after 2 seconds
   b how far the stone has fallen after 2 seconds.

9 At a certain instant a ball has a horizontal velocity of 12 m/s and a vertical velocity of 5 m/s. Calculate the resultant velocity of the ball at that instant.

Going further

Projectiles

The photograph in Figure 1.2.11 was taken while a lamp emitted regular flashes of light. One ball was *dropped from rest* and the other, a projectile, was *thrown sideways* at the same time. Their vertical accelerations (due to gravity) are equal, showing that a projectile falls like a body which is dropped from rest. Its horizontal velocity does not affect its vertical motion.

The horizontal and vertical motions of a body are independent and can be treated separately.

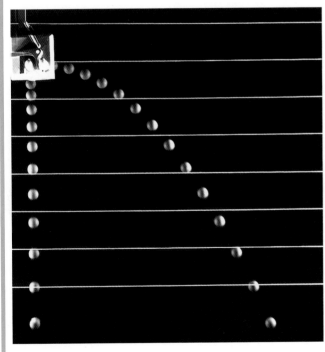

For example, if a ball is thrown horizontally from the top of a cliff and takes 3 s to reach the beach below, we can calculate the height of the cliff by considering the vertical motion only. We have $u = 0$ (since the ball has no vertical velocity initially), $a = g = +9.8$ m/s^2 and $t = 3$ s. The height s of the cliff is given by

$$s = ut + \frac{1}{2}at^2$$
$$= 0 \times 3\,\text{s} + \frac{1}{2}(+9.8\,\text{m/s}^2)3^2\,\text{s}^2$$
$$= 44\,\text{m}$$

Projectiles such as cricket balls and explosive shells are projected from near ground level and at an angle. The horizontal distance they travel, i.e. their range, depends on

- the speed of projection – the greater this is, the greater the range, and
- the angle of projection – it can be shown that, neglecting air resistance, the range is a maximum when the angle is 45° (Figure 1.2.12).

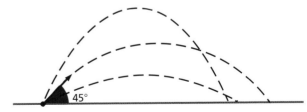

▲ **Figure 1.2.12** The range is greatest for an angle of projection of 45°.

▲ **Figure 1.2.11** Comparing free fall and projectile motion using multiflash photography

Air resistance: terminal velocity

When an object falls in a uniform gravitational field, the air resistance (**fluid** friction) opposing its motion *increases as its speed rises*, so reducing its acceleration. Eventually, air resistance acting upwards equals the weight of the object acting downwards. The resultant force on the object is then zero since the gravitational force balances the frictional force. The object falls at a constant velocity, called its **terminal velocity**, whose value depends on the size, shape and weight of the object.

A small dense object, such as a steel ball-bearing, has a high terminal velocity and falls a considerable distance with a constant acceleration of $9.8\,\text{m/s}^2$ before air resistance equals its weight. A light object, like a raindrop, or an object with a large surface area, such as a parachute, has a low terminal velocity and only accelerates over a comparatively short distance before air resistance equals its weight. A skydiver (Figure 1.2.13) has a terminal velocity of more than $50\,\text{m/s}$ ($180\,\text{km/h}$) before the parachute is opened.

Objects falling in liquids behave similarly to those falling in air.

▲ **Figure 1.2.13** Synchronised skydivers

In the absence of air resistance, an object falls in a uniform gravitational field with a constant acceleration as shown in the distance–time graph of Figure 1.2.10a.

Revision checklist

After studying Topic 1.2 you should know and understand the following:

✓ that a negative acceleration is a deceleration or retardation.

After studying Topic 1.2 you should be able to:
✓ define speed and velocity, and calculate average speed from total distance/total time; sketch, plot, interpret and use speed–time and distance–time graphs to solve problems

✓ define and calculate acceleration and use the fact that deceleration is a negative acceleration in calculations

✓ state that the acceleration of free fall, g, for an object near to the Earth is constant and use the given value of $9.8\,\text{m/s}^2$

✓ describe the motion of objects falling in a uniform gravitational field.

Exam-style questions

1 The speeds of a car travelling on a straight road are given below at successive intervals of 1 second.

Time/s	0	1	2	3	4
Speed/m/s	0	2	4	6	8

Calculate
  a the average speed of the car in m/s [2]
  b the distance the car travels in 4 s [3]
  c the constant acceleration of the car. [2]
  [Total: 7]

2 If a train travelling at 10 m/s starts to accelerate at 1 m/s² for 15 s on a straight track, calculate its final speed in m/s. [Total: 4]

3 The distance–time graph for a girl on a cycle ride is shown in Figure 1.2.14.
  a Calculate
   i how far the girl travelled [1]
   ii how long the ride took [1]
   iii the girl's average speed in km/h [1]
   iv the number of stops the girl made [1]
   v the total time the girl stopped [1]
   vi the average speed of the girl *excluding* stops. [2]
  b Explain how you can tell from the shape of the graph when the girl travelled fastest. Over which stage did this happen? [2]
  [Total: 9]

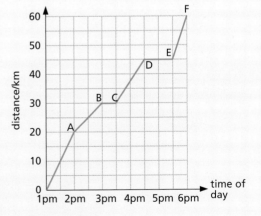
▲ **Figure 1.2.14**

4 The graph in Figure 1.2.15 represents the distance travelled by a car plotted against time.
  a State how far the car has travelled at the end of 5 seconds. [1]
  b Calculate the speed of the car during the first 5 seconds. [1]
  c State what has happened to the car after A. [2]
  d Draw a graph showing the speed of the car plotted against time during the first 5 seconds. [3]
  [Total: 7]

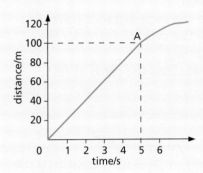
▲ **Figure 1.2.15**

5 Figure 1.2.16 shows an incomplete speed–time graph for a boy running a distance of 100 m.
  a Calculate his acceleration during the first 4 seconds. [2]
  b Calculate how far the boy travels during
   i the first 4 seconds [2]
   ii the next 9 seconds? [2]
  c Copy and complete the graph, showing clearly at what time he has covered the distance of 100 m. Assume his speed remains constant at the value shown by the horizontal portion of the graph. [4]
  [Total: 10]

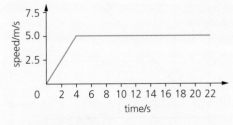
▲ **Figure 1.2.16**

6 The approximate speed–time graph for a car on a 5-hour journey is shown in Figure 1.2.17. (There is a very quick driver change midway to prevent driving fatigue!)
  a State in which of the regions OA, AB, BC, CD, DE the car is
    i accelerating
    ii decelerating
    iii travelling with constant speed. [3]
  b Calculate the value of the acceleration, deceleration or constant speed in each region. [3]
  c Calculate the distance travelled over each region. [3]
  d Calculate the total distance travelled. [1]
  e Calculate the average speed for the whole journey. [1]
  f State what times the car is at rest. [1]
    [Total: 12]

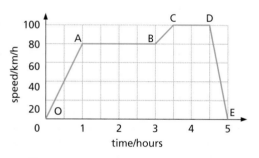

▲ **Figure 1.2.17**

7 The distance–time graph for a motorcyclist riding off from rest is shown in Figure 1.2.18.
  a Describe the motion. [2]
  b Calculate how far the motorbike moves in 30 seconds. [1]
  c Calculate the speed. [2]
    [Total: 5]

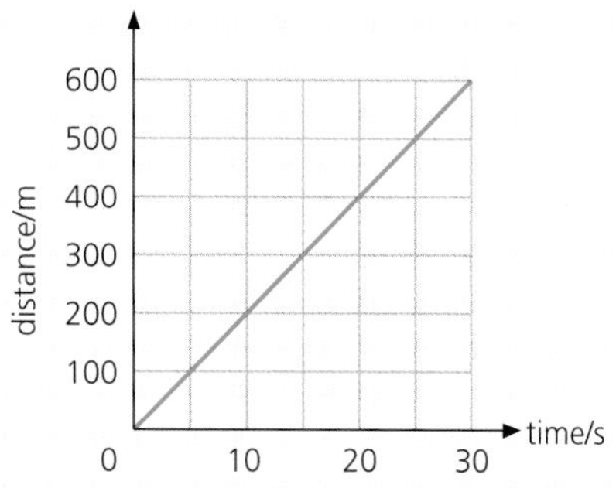

▲ **Figure 1.2.18**

8 A ball is dropped from rest from the top of a high building. Ignore air resistance and take $g = 9.8\,\text{m/s}^2$.
  a Calculate the speed of the ball after
    i 1 s [2]
    ii 3 s. [2]
  b Calculate how far it has fallen after
    i 1 s [2]
    ii 3 s. [2]
    [Total: 8]

9 Figure 1.2.19 shows the forces acting on a raindrop which is falling to the ground.

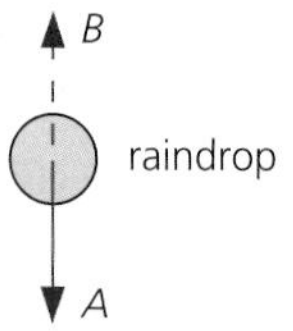

▲ **Figure 1.2.19**

  a i A is the force which causes the raindrop to fall. Give the name of this force. [1]
    ii B is the total force opposing the motion of the drop. State *one* possible cause of this force. [1]
  b What happens to the drop when force A = force B? [2]
    [Total: 4]

1.3 Mass and weight

FOCUS POINTS

★ Define mass and weight and know that weights (and therefore masses) may be compared using a balance or force meter.
★ Define gravitational field strength and know that this is equivalent to the acceleration of free fall.

★ Understand that weight is the effect of a gravitational field on mass.
★ Describe, and use the concept of, weight as the effect of a gravitational field on a mass.

Images of astronauts walking on the surface of the Moon show them walking with bouncing steps. The force of gravity is less on the Moon than it is on the Earth and this accounts for their different movements. In the previous topics you have encountered measurements of space and time, and the rates of change that define speed and acceleration. You will now encounter a further fundamental property, the mass of an object. Mass measures the quantity of matter in a body. In the presence of gravity, mass acquires weight in proportion to its mass and the strength of the gravitational force. Although the mass of an object on the Moon is the same as it is on the Earth, its weight is less on the Moon because the force of gravity there is less.

Mass

The **mass** of an object is the measure of the amount of matter in it. It can be stated that mass is a measure of the quantity of matter in an object at rest relative to an observer.

The standard unit of mass is the **kilogram** (kg) and until 2019 was the mass of a piece of platinum–iridium alloy at the Office of Weights and Measures in Paris. It is now based on a fundamental physical constant which can be measured with great precision. The gram (g) is one-thousandth of a kilogram.

$$1\,g = \frac{1}{1000}\,kg = 10^{-3}\,kg = 0.001\,kg$$

The term **weight** is often used when mass is really meant. In science the two ideas are distinct and have different units. The confusion is not helped by the fact that mass is found on a **balance** by a process we unfortunately call 'weighing'!

Key definitions

Mass a measure of the quantity of matter in an object at rest relative to an observer

Weight a gravitational force on an object that has mass

There are several kinds of balance used to measure mass. In the beam balance the unknown mass in one pan is balanced against known masses in the other pan. In the lever balance a system of levers acts against the mass when it is placed in the pan. A direct reading is obtained from the position on a scale of a pointer joined to the lever system. A digital top-pan balance is shown in Figure 1.3.1.

▲ **Figure 1.3.1** A digital top-pan balance

Weight

We all constantly experience the force of *gravity*, in other words, the pull of the Earth. It causes an unsupported body to fall from rest to the ground. Weight is a gravitational force on an object that has mass.

For an object above or on the Earth's surface, the nearer it is to the centre of the Earth, the more the Earth attracts it. Since the Earth is not a perfect sphere but is flatter at the poles, the weight of a body varies over the Earth's surface. It is greater at the poles than at the equator.

Gravity is a force that can act through space, that is there does not need to be contact between the Earth and the object on which it acts as there does when we push or pull something. Other action-at-a-distance forces which, like gravity, decrease with distance are

(i) **magnetic forces** between magnets and

(ii) **electric forces** between electric charges.

When a mass experiences a gravitational force we say it is in a **gravitational field**. Weight is the result of a gravitational field acting on a mass: weight is a vector quantity and is measured in newtons (N).

The newton

The unit of force is the **newton**. It will be defined later (Topic 1.5); the definition is based on the change of speed a force can produce in a body. Weight is a force and therefore should be measured in newtons.

The weight of an object can be measured by hanging it on a spring balance marked in newtons (Figure 1.3.2) and letting the pull of gravity stretch the spring in the balance. The greater the pull, the more the spring stretches.

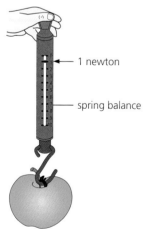

1 newton

spring balance

▲ **Figure 1.3.2** The weight of an average-sized apple is about 1 newton.

On most of the Earth's surface:

The weight of an object of mass 1 kg is 9.8 N.

Often this is taken as 10 N. A mass of 2 kg has a weight of 20 N, and so on. The mass of an object is

the same wherever it is and, unlike weight, does not depend on the presence of the Earth.

> **Test yourself**

1 An object of mass 1 kg has weight of 10 N at a certain place. What is the weight of
   a 100 g
   b 5 kg
   c 50 g?
2 The force of gravity on the Moon is said to be one-sixth of that on the Earth. What would a mass of 12 kg weigh
   a on the Earth
   b on the Moon?

Weight and gravity

The weight W of an object is the force of gravity acting on it which gives it an acceleration g when it is falling freely near the Earth's surface. If the object has mass m, then W can be calculated from $F = ma$ (Newton's second law, see p. 39). We put $F = W$ and $a = g$ to give

$$W = mg$$

Taking $g = 9.8 \text{ m/s}^2$ and $m = 1 \text{ kg}$, this gives $W = 9.8 \text{ N}$, that is an object of mass 1 kg has weight 9.8 N, or near enough 10 N. Similarly, an object of mass 2 kg has weight of about 20 N, and so on.

Gravitational field

The force of gravity acts through space and can cause an object, not in contact with the Earth, to fall to the ground. It is an invisible, action-at-a-distance force. We try to explain its existence by saying that the Earth is surrounded by a gravitational field which exerts a force on any object in the field. Later, magnetic and electric fields will be considered.

The **gravitational field strength** is defined as the force acting per unit mass.

Rearranging the equation $W = mg$ gives $g = \dfrac{W}{m}$.

Key definition
Gravitational field strength force per unit mass

Gravitational field strength is a vector and has both magnitude and direction.

Measurement shows that on the Earth's surface a mass of 1 kg experiences a force of 9.8 N, i.e. its weight is 9.8 N. The strength of the Earth's field is therefore 9.8 N/kg (near enough 10 N/kg). It is denoted by g, the letter also used to denote the acceleration of free fall. Hence

$$g = 9.8\,\text{N/kg} = 9.8\,\text{m/s}^2$$

We now have two ways of regarding g. When considering objects *falling freely*, we can think of it as an acceleration of 9.8 m/s². When an object of known mass is *at rest* and we wish to know the force of gravity (in N) acting on it, we think of g as the Earth's gravitational field strength of 9.8 N/kg. The gravitational field strength is equivalent to the acceleration of free fall.

The weight of an object is directly **proportional** to its mass, which explains why g is the same for all objects. The greater the mass of an object, the greater is the force of gravity on it but it does not accelerate faster when falling because of its greater inertia (i.e. its greater resistance to acceleration).

While the mass of an object is always the same, its weight varies depending on the value of g. On the Moon gravitational field strength is only about 1.6 N/kg, and so a mass of 1 kg has a weight of just 1.6 N there.

▶ Test yourself

3 An astronaut has a mass of 80 kg.
  a Calculate the weight of the astronaut on the Moon where the gravitational field strength is 1.6 N/kg.
  b On the journey back to Earth, the astronaut reaches a point X where the gravitational field strengths due to the Earth and the Moon are equal in magnitude but opposite in direction. State
    i the resultant value of the gravitational field strength at X
    ii the weight of the astronaut at X.

Revision checklist

After studying Topic 1.3 you should know and understand the following:
✓ what is meant by the mass of a body
✓ the difference between mass and weight and that weights (and masses) may be compared using a balance.

After studying Topic 1.3 you should be able to:
✓ state the units of mass and weight and recall that the weight of an object is the force of gravity on it
✓ recall and use the equation $g = \dfrac{W}{m}$

✓ describe and use the concept of weight as the effect of a gravitational field on a mass.

Exam-style questions

1 a i Explain what is meant by the mass of an object.

ii Explain what is meant by the weight of an object.

iii Describe how weights may be compared. [4]

b State which of the following definitions for weight W is correct.

A $W = g/\text{mass}$

B $W = \text{mass}/g$

C $W = \text{mass} \times g$

D $W = \text{force} \times g$ [1]

c Which of the following properties is the same for an object on the Earth and on the Moon?

A weight

B mass

C acceleration of free fall

D gravitational field strength [1]

d State the SI units of

i weight

ii acceleration of free fall

iii gravitational field strength. [3]

[Total: 9]

2 a Define gravitational field strength. [2]

b On the Earth the acceleration of free fall is about $9.8 \, \text{m/s}^2$. On Mars the acceleration of free fall is about $3.7 \, \text{m/s}^2$.

The weight of the Mars Rover Opportunity on the Earth was 1850 N.

i Calculate the mass of the Rover. [2]

ii Calculate the weight of the Rover on Mars. [2]

[Total: 6]

3 a Explain what is meant by a gravitational field. [2]

b State the effect of a gravitational field on a mass. [1]

c Define gravitational field strength. [2]

d The gravitational field strength on Venus is 8.8 N/kg. The mass of a rock is 200 kg. Calculate the weight of the rock on Venus. [2]

[Total: 7]

1.4 Density

FOCUS POINTS
★ Define density and calculate the density of a liquid and both regular- and irregular-shaped solid objects.
★ Use density data to determine whether an object will float or sink.

★ Use density data to determine whether one liquid will float on another liquid.

A pebble thrown into a pond will sink to the bottom of the pond, but a wooden object will float. Objects of the same shape and size but made from different materials have different masses. In this topic you will see how you can quantify such differences with the idea of density. Density specifies the amount of mass in a unit volume. To measure the density of a material you will need to know both its mass and its volume. The mass can be found using a balance, and the volume by measurement. If the density of an object is greater than that of a liquid it will sink, but if the density of the object is less than that of the liquid it will float.

In everyday language, lead is said to be heavier than wood. By this it is meant that a certain volume of lead is heavier than the same volume of wood. In science such comparisons are made by using the term **density**. This is the *mass per unit volume* of a substance and is calculated from

$$\text{density} = \frac{\text{mass}}{\text{volume}}$$

For a mass m of volume V, the density $\rho = m/V$.

> **Key definition**
> **Density** mass per unit volume

The density of lead is 11 grams per cubic **centimetre** ($11\,\text{g/cm}^3$) and this means that a piece of lead of volume $1\,\text{cm}^3$ has mass $11\,\text{g}$. A volume of $5\,\text{cm}^3$ of lead would have mass $55\,\text{g}$. If the density of a substance is known, the mass of *any* volume of it can be calculated. This enables engineers to work out the weight of a structure if they know from the plans the volumes of the materials to be used and their densities. Strong enough foundations can then be made.

The SI unit of density is the kilogram per cubic metre. To convert a density from g/cm^3, normally the most suitable unit for the size of sample we use, to kg/m^3, we multiply by 10^3. For example, the density of water is $1.0\,\text{g/cm}^3$ or $1.0 \times 10^3\,\text{kg/m}^3$.

The approximate densities of some common substances are given in Table 1.4.1.

▼ **Table 1.4.1** Densities of some common substances

Solids	Density/g/cm³	Liquids	Density/g/cm³
aluminium	2.7	paraffin	0.80
copper	8.9	petrol	0.80
iron	7.9	pure water	1.0
gold	19.3	mercury	13.6
glass	2.5	**Gases**	**Density/kg/m³**
wood (teak)	0.80	air	1.3
ice	0.92	hydrogen	0.09
polythene	0.90	carbon dioxide	2.0

Calculations

Using the symbols ρ (rho) for density, m for mass and V for volume, the expression for density is

$$\rho = \frac{m}{V}$$

Rearranging the expression gives

$$m = V \times \rho \text{ and } V = \frac{m}{\rho}$$

These are useful if ρ is known and m or V have to be calculated. If you do not see how they are obtained refer to the *Mathematics for physics* section on p. 295.

The triangle in Figure 1.4.1 is an aid to remembering them. If you cover the quantity you want to know with a finger, such as m, it equals what you can still see, i.e. $\rho \times V$. To find V, cover V and you get $V = m/\rho$.

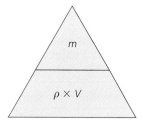

▲ **Figure 1.4.1**

? Worked example

Taking the density of copper as $9\,\text{g/cm}^3$, find **a** the mass of $5\,\text{cm}^3$ and **b** the volume of $63\,\text{g}$.

a $\rho = 9\,\text{g/cm}^3$, $V = 5\,\text{cm}^3$ and m is to be found.
$m = V \times \rho = 5\,\text{cm}^3 \times 9\,\text{g/cm}^3 = 45\,\text{g}$

b $\rho = 9\,\text{g/cm}^3$, $m = 63\,\text{g}$ and V is to be found.

$$\therefore V = \frac{m}{\rho} = \frac{63\,\text{g}}{9\,\text{g/cm}^3} = 7\,\text{cm}^3$$

Now put this into practice

1 A sheet of aluminium has a mass of $200\,\text{g}$ and a volume of $73\,\text{cm}^3$. Calculate the density of aluminium.
2 Taking the density of lead as $11\,\text{g/cm}^3$, find
    a the mass of $4\,\text{cm}^3$
    b the volume of $55\,\text{g}$.

Simple density measurements

If the mass m and volume V of a substance are known, its density can be found from $\rho = m/V$.

Regularly shaped solid

The mass is found on a balance and the volume by measuring its dimensions with a ruler.

Irregularly shaped solid: volume by displacement

Use one of these methods to find the volume of a pebble or glass stopper, for example. The mass of the solid is found on a balance. Its volume is measured by one of the displacement methods shown in Figure 1.4.2. In Figure 1.4.2a the volume is the difference between the first and second readings. In Figure 1.4.2b it is the volume of water collected in the measuring cylinder.

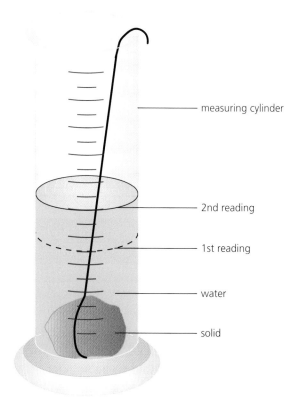

▲ **Figure 1.4.2a** Measuring the volume of an irregular solid: method 1

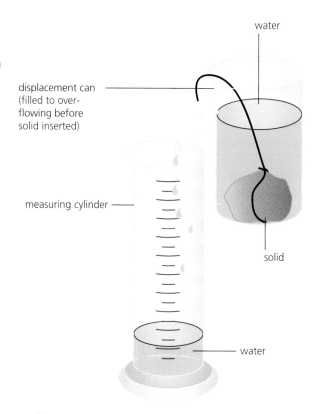

▲ **Figure 1.4.2b** Measuring the volume of an irregular solid: method 2

Liquid

The mass of an empty beaker is found on a balance. A known volume of the liquid is transferred from a burette or a measuring cylinder into the beaker. The mass of the beaker plus liquid is found and the mass of liquid is obtained by subtraction.

Air

Using a balance, the mass of a 500 cm³ round-bottomed flask full of air is found and again after removing the air with a vacuum pump; the difference gives the mass of air in the flask. The volume of air is found by filling the flask with water and pouring it into a measuring cylinder.

Floating and sinking

An object sinks in a liquid of lower density than its own; otherwise it floats, partly or wholly submerged. For example, a piece of glass of density 2.5 g/cm³ sinks in water (density 1.0 g/cm³) but floats in mercury (density 13.6 g/cm³). An iron nail sinks in water but an iron ship floats because its average density is less than that of water, due to the low-density air enclosed in the hull.

> A liquid of low density will float on a liquid of higher density if the two liquids do not mix.

Test yourself

1 a Calculate the density of a substance of
   i   mass 100 g and volume 10 cm³
   ii volume 3 m³ and mass 9 kg.
  b The density of gold is 19 g/cm³. Find the volume of
   i   38 g
   ii 95 g of gold.
2 A rectangular steel bar is 4 cm long, 3 cm wide and 1 cm thick. When weighed it is found to have a mass of 96 g. Calculate its density in
  a g/cm³
  b kg/m³.
3 The water in a measuring cylinder is at the 50 cm³ level. A pebble is dropped into the water and the water level rises to 60 cm³. The pebble is completely covered by water.
  Calculate
  a the volume of the pebble
  b the density of the pebble, if it weighs 60 g.

4 Liquid A has a density of 0.8 g/cm³ and water has a density of 1.0 g/cm³. If the two liquids do not mix, which liquid will float on top of the other?

Revision checklist

After studying Topic 1.4 you should know and understand the following:
✓ how density is defined and how to perform calculations using $\rho = m/V$.

After studying Topic 1.4 you should be able to:
✓ describe methods to measure the density of a liquid and a regularly shaped solid
✓ describe the method of displacement to measure the density of an irregularly shaped solid
✓ predict whether an object will float, based on density data

✓ predict whether one liquid will float on another if they do not mix.

Exam-style questions

1 a Choose which of the following definitions
      for density is correct.
      A mass/volume
      B mass × volume
      C volume/mass
      D weight/area   [1]
   b Calculate
      i   the mass of 5 m³ of cement of density
          3000 kg/m³   [3]
      ii the mass of air in a room measuring
          10 m × 5.0 m × 2.0 m if the density of
          air is 1.3 kg/m³.   [3]
                                        [Total: 7]

2 a Describe how you could determine the
      density of a liquid.   [4]
   b An empty beaker is weighed and found to
      have a mass of 130 g. A measuring cylinder
      contains 50 cm³ of an unknown liquid.
      All the liquid is poured into the beaker
      which is again weighed and found to have
      a mass of 170 g. Calculate the density of
      the liquid.   [4]
   c Explain why ice floats on water.   [1]

   d Explain why oil floats on water.   [1]

                                       [Total: 10]

3 a A block of wood has dimensions of
      10 cm × 8 cm × 20 cm.
      i   Calculate the volume of the block in
          cubic metres.   [2]
      ii The block is placed on a balance and
          found to weigh 1.2 kg. Calculate the
          density of the block in kg/m³.   [3]
   b When a golf ball is lowered into a
      measuring cylinder of water, the water level
      rises by 30 cm³ when the ball is completely
      submerged. If the ball weighs 33 g in air,
      calculate its density in kg/m³.   [3]
                                        [Total: 8]

1.5 Forces

1.5.1 Effects of forces

FOCUS POINTS

★ Understand that the size and shape of objects can be altered by forces.
★ Become familiar with load–extension graphs for an elastic solid and describe an experiment to show how a spring behaves when it is stretched.
★ Understand that when several forces act simultaneously on an object that a resultant can be determined.
★ Know that, unless acted upon by a resultant force, an object will remain at rest or will continue moving with a constant speed in a straight line.
★ Understand that solid friction acts to slow an object and produce heat.
★ Explain the terms 'drag' and 'air resistance' in terms of friction acting on objects.

> ★ Define the spring constant and the limit of proportionality on a load–extension graph.
> ★ Apply the equation $F = ma$ to calculate force and acceleration.
> ★ Describe motion in a circular path and understand the effect on force if speed, radius or mass change.

A gravitational force causes a freely falling object to accelerate and keeps a satellite moving in a circular path. Clearly a force can change the speed or direction of travel of an object. A force can also change the shape or size of an object. If you stand on an empty paper carton it will change its shape and if you pull on a spiral spring it will stretch. Several forces may act on an object at once and it is useful to calculate a resultant force to predict their combined effect; both the size and direction of the forces are needed for this. Friction between a moving object and its surroundings is also important as it acts to reduce the speed of the object and produce heat. You have already learnt how to quantify some of these changes and in this topic you will encounter more ways to do so.

Force

A **force** is a push or a pull. It can cause an object at rest to move, or if the body is already moving it can change its speed or direction of motion.

A force can also change a body's shape or size. For example, a spring (or wire) will stretch when loaded with a weight.

▲ **Figure 1.5.1** A weightlifter in action exerts first a pull and then a push.

Practical work

Stretching a spring

For safe experiments/demonstrations related to this topic, please refer to the *Cambridge IGCSE Physics Practical Skills Workbook* that is also part of this series.

Safety

- Eye protection must be worn (in case the spring snaps).

Arrange a steel spring as in Figure 1.5.2. Read the scale opposite the bottom of the hanger. Add 100 g loads one at a time (thereby increasing the **stretching force** by steps of 1 N) and take readings from the scale after each one. Enter the readings in a table for loads up to 500 g.

Note that at the head of columns (or rows) in data tables it is usual to give the name of the quantity or its symbol followed by / and the unit.

Stretching force/N	Scale reading/mm	Total extension/mm

Sometimes it is easier to discover laws by displaying the results on a graph. Do this on graph paper by plotting total **extension** readings along the *x*-axis (horizontal axis) and **stretching**

force readings along the *y*-axis (vertical axis) in a load–extension graph. Every pair of readings will give a point; mark them by small crosses and draw a smooth line through them.

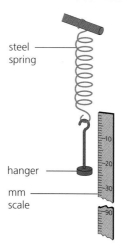

steel spring

hanger

mm scale

▲ **Figure 1.5.2**

1 What is the shape of the graph you plotted?
2 Do the results suggest any rule about how the spring behaves when it is stretched?
3 What precautions could you take to improve the accuracy of the results of this experiment?
4 How could you test if the extension of the spring is proportional to the stretching force?

Extension in springs

Springs were investigated by Robert Hooke just over 350 years ago. He found that the extension was proportional to the stretching force provided the spring was not permanently stretched. This means that doubling the force doubles the extension, trebling the force trebles the extension, and so on. Using the sign for proportionality, ∝, we can write

extension ∝ stretching force

It is true only if the **limit of proportionality** of the spring is not exceeded.

> **Key definition**
>
> **Limit of proportionality** the point at which the load-extension graph becomes non-linear

The graph of Figure 1.5.3 is for a spring stretched beyond its limit of proportionality, E. OE is a

straight line passing through the origin O and is graphical proof that the extension is proportional to the stretching force over this **range**. If the force for point A on the graph is applied to the spring, the proportionality limit is passed and on removing the force some of the extension (OS) remains.

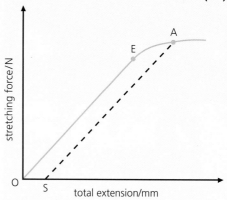

▲ **Figure 1.5.3**

▶ Test yourself

1 In Figure 1.5.3, over which part of the graph does a spring balance work?

Now put this into practice

1 Calculate the spring constant of a spring which is stretched 2 mm by a force of 4 N.
2 A 2 N weight is applied to a spring which has a spring constant of 250 N/m. Calculate the extension of the spring in mm.

Spring constant

The **spring constant**, k, is defined as force per unit extension. It is the force which must be applied to a spring to cause an extension of 1 m.

If a force F produces extension x then

$$k = \frac{F}{x}$$

Rearranging the equation gives

$$F = kx$$

> **Key definition**
>
> **Spring constant** force per unit extension

▶ Test yourself

2 State two effects which a force may have on an object.
3 Make a sketch of a load–extension graph for a spring and indicate the region over which the extension is proportional to the stretching force.

4 Calculate the spring constant of a spring which is stretched 4 cm by a mass of 200 g.
5 Define the limit of proportionality for a stretched spring.

Proportionality also holds when a force is applied to an elastic solid such as a straight metal wire, provided it is not permanently stretched.

Load–extension graphs similar to Figure 1.5.3 are obtained. You should label each axis of your graph with the name of the quantity or its symbol followed by / and the unit, as shown in Figure 1.5.3.

The limit of proportionality can be defined as the point at which the load–extension graph becomes non-linear because the extension is no longer proportional to the stretching force.

❓ Worked example

A spring is stretched 10 mm (0.01 m) by a weight of 2.0 N. Calculate
a the spring constant k
b the weight W of an object that causes an extension of 80 mm (0.08 m).

a $k = \dfrac{F}{x} = \dfrac{2.0\,\text{N}}{0.01\,\text{m}} = 200\,\text{N/m}$

b $W = \text{stretching force } F$
    $= k \times x$
    $= 200\,\text{N/m} \times 0.08\,\text{m}$
    $= 16\,\text{N}$

Forces and resultants

Force has both magnitude (size) and direction. It is represented in diagrams by a straight line with an arrow to show its direction of action.

Usually more than one force acts on an object. As a simple example, an object resting on a table is pulled downwards by its weight W and pushed upwards by a force R due to the table supporting it (Figure 1.5.4). Since the object is at rest, the forces must balance, i.e. $R = W$.

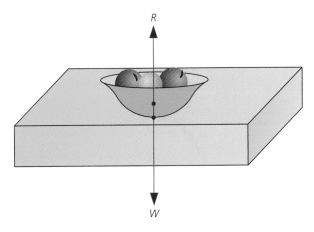

▲ **Figure 1.5.4**

In structures such as a giant oil platform (Figure 1.5.5), two or more forces may act at the same point. It is then often useful for the design

▲ **Figure 1.5.5** The design of an offshore oil platform requires an understanding of the combination of many forces.

engineer to know the value of the single force, i.e. the resultant force, which has exactly the same effect as these forces, If the forces act in the same straight line, the resultant is found by simple addition or subtraction as shown in Figure 1.5.6; if they do not they are added by using the **parallelogram law.**

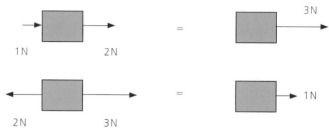

▲ **Figure 1.5.6** The resultant of forces acting in the same straight line is found by addition or subtraction.

Practical work

Parallelogram law

Safety

● Take care when using the mass in case it drops.

Arrange the apparatus as in Figure 1.5.7a with a sheet of paper behind it on a vertical board. We have to find the resultant of forces P and Q.

Read the values of P and Q from the spring balances. Mark on the paper the directions of P, Q and W as shown by the strings. Remove the paper and, using a scale of 1 cm to represent 1 N, draw OA, OB and OD to represent the three forces P, Q and W which act at O, as in Figure 1.5.7b. (W = weight of the 1 kg mass = 9.8 N; therefore OD = 9.8 cm.)

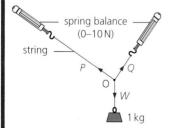

▲ **Figure 1.5.7a**

P and Q together are balanced by W and so their resultant must be a force equal and opposite to W.

Complete the parallelogram OACB. Measure the diagonal OC; if it is equal in size (i.e. 9.8 cm) and opposite in direction to W then it represents the resultant of P and Q.

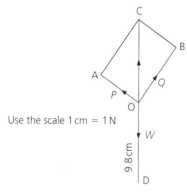

Use the scale 1 cm = 1 N

▲ **Figure 1.5.7b** Finding a resultant by the parallelogram law

The parallelogram law for adding two forces is:

> If two forces acting at a point are represented in size and direction by the sides of a parallelogram drawn from the point, their resultant is represented in size and direction by the diagonal of the parallelogram drawn from the point.

5 List the equipment you would need for this experiment.

6 What quantity would you vary to test the law under different conditions?

Test yourself

6 Jo, Daniel and Helen are pulling a metal ring. Jo pulls with a force of 100 N in one direction and Daniel with a force of 140 N in the opposite direction. If the ring does not move, what force does Helen exert if she pulls in the same direction as Jo?

7 A boy drags a suitcase along the ground with a force of 100 N. If the frictional force opposing the motion of the suitcase is 50 N, what is the resultant forward force on the suitcase?

8 A picture is supported by two vertical strings. If the weight of the picture is 50 N, what is the force exerted by each string?

9 Using a scale of 1 cm to represent 10 N, find the size and direction of the resultant of forces of 30 N and 40 N acting at right angles to each other.

Newton's first law

Friction and air resistance cause a car to come to rest when the engine is switched off. If these forces were absent, we believe that an object, once set in motion, would go on moving forever with a constant speed in a straight line. That is, force is not needed to keep a body moving with **uniform velocity** provided that no opposing forces act on it.

This idea was proposed by Galileo and is summed up in **Isaac Newton's first law of motion**:

An object stays at rest, or continues to move in a straight line at constant speed, unless acted on by a **resultant force**.

It seems that the question we should ask about a moving body is not what keeps it moving but what changes or stops its motion.

The smaller the external forces opposing a moving body, the smaller is the force needed to keep it moving with constant velocity. A hover scooter, which is supported by a cushion of air (Figure 1.5.8), can skim across the ground with little frictional opposition, so that relatively little power is needed to maintain motion.

A resultant force may change the velocity of an object by changing its direction of motion or speed.

> **Key definitions**
> **Resultant force** may change the velocity of an object by changing its direction of motion or its speed

▲ **Figure 1.5.8** Friction is much reduced for a hover scooter.

⮕ **Going further**

Mass and inertia

Newton's first law is another way of saying that all matter has a built-in opposition to being moved if it is at rest or, if it is moving, to having its motion changed. This property of matter is called inertia (from the Latin word for laziness).

Its effect is evident on the occupants of a car that stops suddenly: they lurch forwards in an attempt to continue moving, and this is why seat belts are needed. The reluctance of a stationary object to move can be shown by placing a large coin on a piece of card on your finger (Figure 1.5.9). If the card is flicked *sharply* the coin stays where it is while the card flies off.

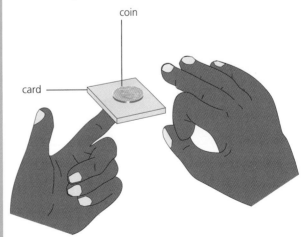

coin

card

▲ **Figure 1.5.9** Flick the card sharply

The larger the mass of a body, the greater is its inertia, i.e. the more difficult it is to move it when at rest and to stop it when in motion. Because of this we consider that *the mass of a body measures its inertia*. This is a better definition of mass than the one given earlier (Topic 1.3) in which it was stated to be the amount of matter in a body.

Practical work

Effect of force and mass on acceleration

For safe experiments/demonstrations related to this topic, please refer to the *Cambridge IGCSE Physics Practical Skills Workbook* that is also part of this series.

Safety

- Take care when rolling the trolley down the ramp. Ensure it is clear at the bottom of the ramp and use a side barrier to prevent the trolley from falling onto the floor.

The apparatus consists of a trolley to which a force is applied by a stretched length of elastic (Figure 1.5.10). The velocity of the trolley is found from a tickertape timer or a motion sensor, datalogger and computer.

First compensate the runway for friction: raise one end until the trolley runs down with constant velocity when given a push. The dots on the tickertape should be equally spaced, or a horizontal trace obtained on a speed–time graph. There is now no resultant force on the trolley and any acceleration produced later will be due only to the force caused by the stretched elastic.

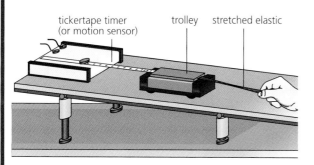

▲ **Figure 1.5.10**

(a) Force and acceleration (mass constant)

Fix one end of a short length of elastic to the rod at the back of the trolley and stretch it until the other end is level with the front of the trolley. Practise pulling the trolley down the runway, keeping the same stretch on the elastic. After a few trials you should be able to produce a steady accelerating force.

Repeat using first two and then three *identical* pieces of elastic, stretched side by side by the same amount, to give two and three units of force.

If you are using tickertape, make a tape chart for each force and use it to find the acceleration produced in cm/ten-tick^2. Ignore the start of the tape (where the dots are too close) and the end (where the force may not be steady). If you use a motion sensor and computer to plot a speed–time graph, the acceleration can be obtained in m/s^2 from the slope of the graph (Topic 1.2).

Put the results in a table.

Force (F)/(no. of pieces of elastic)	1	2	3
Acceleration (a)/cm/ten-tick^2 or m/s^2			

(b) Mass and acceleration (force constant)

Do the experiment as in part (a) using two pieces of elastic (i.e. constant F) to accelerate first one trolley, then two (stacked one above the other) and finally three. Check the friction compensation of the runway each time.

Find the accelerations from the tape charts or computer plots and tabulate the results.

Mass (m)/(no. of trolleys)	1	2	3
Acceleration (a)/cm/ten-tick^2 or m/s^2			

7 For part (a), does a steady force cause a steady acceleration?

8 Do your results in part (a) suggest any relationship between acceleration a and force F?

9 Do your results for part (b) suggest any relationship between a and m?

10 Name the two independent variable quantities in experiments (a) and (b).

11 How could you use the results to verify the equation $F = ma$?

Newton's second law

The previous experiment should show roughly that the acceleration a is

i   directly proportional to the applied force F for a fixed mass, i.e. $a \propto F$, and

ii **inversely proportional** to the mass m for a fixed force, i.e. $a \propto 1/m$.

Combining the results into one equation, we get

$$a \propto \frac{F}{m}$$

or

$$F \propto ma$$

Therefore

$$F = kma$$

where k is the **constant of proportionality.**

One newton is defined as the force which gives a mass of 1 kg an acceleration of $1\,\text{m/s}^2$, i.e. $1\,\text{N} = 1\,\text{kg}\,\text{m/s}^2$, so if $m = 1\,\text{kg}$ and $a = 1\,\text{m/s}^2$, then $F = 1\,\text{N}$.

Substituting in $F = kma$, we get $k = 1$ and so we can write

$$F = ma$$

This is **Newton's second law of motion.** When using it, two points should be noted. First, F is the resultant (or unbalanced) force causing the acceleration a in the same direction as F. Second, F must be in newtons, m in kilograms and a in metres per second squared, otherwise k is not 1. The law shows that a will be largest when F is large and m small.

You should now appreciate that when the forces acting on a body do not balance there is a net (resultant) force which causes a change of motion, i.e. the body accelerates or decelerates. The force and the acceleration are in the same direction. If the forces balance, there is no change in the motion of the body. However, there may be a change of shape, in which case internal forces in the body (i.e. forces between neighbouring atoms) balance the external forces.

? Worked example

A block of mass 2 kg has a constant velocity when it is pushed along a table by a force of 5 N. When the push is increased to 9 N what is

a   the resultant force

b   the acceleration?

When the block moves with constant velocity the forces acting on it are balanced. The force of friction opposing its motion must therefore be 5 N.

a   When the push is increased to 9 N the resultant (unbalanced) force F on the block is $(9 - 5)\,\text{N} = 4\,\text{N}$ (since the frictional force is still 5 N).

b   The acceleration a is obtained from $F = ma$ where $F = 4\,\text{N}$ and $m = 2\,\text{kg}$.

$$a = \frac{F}{m} = \frac{4\,\text{N}}{2\,\text{kg}} = \frac{4\,\text{kg}\,\text{m/s}^2}{2\,\text{kg}} = 2\,\text{m/s}^2$$

Now put this into practice

1   A box of mass 5 kg has a constant velocity when it is pushed along a table by a force of 8 N. When the push is increased to 10 N calculate
    a   the resultant force
    b   the acceleration.

2   A force F produces a constant acceleration in a straight line of $0.5\,\text{m/s}^2$ on a block of mass 7 kg. Calculate the value of F.

▶ Test yourself

10 Which one of the diagrams in Figure 1.5.11 shows the arrangement of forces that gives the block of mass M the greatest acceleration?

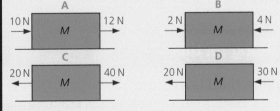

▲ **Figure 1.5.11**

11 In Figure 1.5.12 if P is a force of 20 N and the object moves with constant velocity, what is the value of the opposing force F?

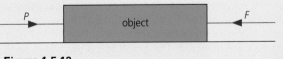

▲ **Figure 1.5.12**

> **12 a** What resultant force produces an acceleration of $5\,\text{m/s}^2$ in a car of mass $1000\,\text{kg}$?
>
> **b** What acceleration is produced in a mass of $2\,\text{kg}$ by a resultant force of $30\,\text{N}$?
>
> **13** A block of mass $500\,\text{g}$ is pulled from rest on a horizontal frictionless bench by a steady force F and reaches a speed of $8\,\text{m/s}$ in $2\,\text{s}$. Calculate
>
> **a** the acceleration
>
> **b** the value of F.

Friction

Friction is the force that opposes one surface moving, or trying to move, over another. It can be a help or a hindrance. We could not walk if there was no friction between the soles of our shoes and the ground. Our feet would slip backwards, as they tend to when we walk on ice. On the other hand, engineers try to reduce friction to a minimum in the moving parts of machinery by using lubricating oils and ball-bearings.

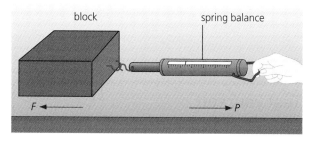

▲ **Figure 1.5.13** Friction opposes motion between surfaces in contact.

When a gradually increasing force P is applied through a spring balance to a block on a table (Figure 1.5.13), the block does not move at first. This is because an equally increasing but opposing frictional force F acts where the block and table touch. At any instant P and F are equal and opposite.

If P is increased further, the block eventually moves; as it does so F has its maximum value, called **starting** or **static friction**. When the block is moving at a steady speed, the balance reading is slightly less than that for starting friction. **Sliding** or **dynamic friction** is therefore less than starting or static friction.

Placing a mass on the block increases the force pressing the surfaces together and increases friction.

When work is done against friction, the temperatures of the bodies in contact rise (as you can test by rubbing your hands together); kinetic energy is transferred to thermal energy by mechanical working (see Topic 1.7).

Solid friction can be described as the force between two surfaces that may impede motion and produce heating.

Friction (drag) acts on an object moving through gas (air resistance), such as a vehicle or falling leaf, which opposes the motion of the object. Similarly, friction (drag) acts on an object moving through a liquid. Drag increases as the speed of the object increases, and acts to reduce acceleration and slow the object down.

 Going further

Newton's third law

If a body A exerts a force on body B, then body B exerts an equal but opposite force on body A.

This is Newton's third law of motion and states that forces never occur singly but always in pairs as a result of the action between two bodies. For example, when you step forwards from rest your foot pushes backwards on the Earth, and the Earth exerts an equal and opposite force forward on you. Two bodies and two forces are involved. The small force you exert on the large mass of the Earth gives no noticeable acceleration to the Earth but the equal force it exerts on your very much smaller mass causes you to accelerate.

Note that the pair of equal and opposite forces *do not act on the same body*; if they did, there could never be any resultant forces and acceleration would be impossible. For a book resting on a table, the book exerts a downward force on the table and the table exerts an equal and opposite upward force on the book; this pair of forces act on different objects and are represented by the red arrows in Figure 1.5.14. The weight of the book (blue arrow) does not form a pair with the upward force on the book (although they are equal numerically) as these two forces act on the same body.

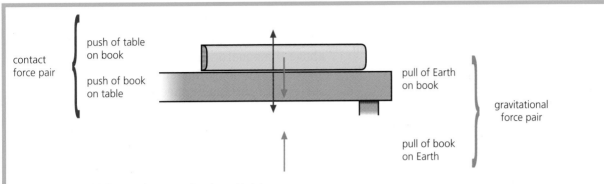

▲ **Figure 1.5.14** Forces between book and table

An appreciation of the third law and the effect of friction is desirable when stepping from a rowing boat (Figure 1.5.15). You push backwards on the boat and, although the boat pushes you forwards with an equal force, it is itself now moving backwards (because friction with the water is slight). This reduces your forwards motion by the same amount – so you may fall in!

▲ **Figure 1.5.15** The boat moves backwards when you step forwards!

Circular motion

There are many examples of bodies moving in circular paths – rides at a funfair, clothes being spun dry in a washing machine, the planets going round the Sun and the Moon circling the Earth. When a car turns a corner, it may follow an arc of a circle. 'Throwing the hammer' is a sport practised at Highland Games in Scotland (Figure 1.5.16), in which the hammer is whirled round and round before it is released.

▲ **Figure 1.5.16** 'Throwing the hammer'

Centripetal force

In Figure 1.5.17 a ball attached to a string is being whirled round in a horizontal circle. Its direction of motion is constantly changing. At A, it is along the tangent at A; shortly afterwards, at B, it is along the tangent at B; and so on. It can be seen that motion in a circular path is due to a force perpendicular to the motion.

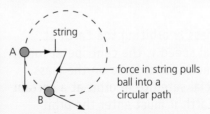

▲ **Figure 1.5.17**

Velocity has both size and direction; speed has only size. Velocity is speed in a stated direction and if the direction of a moving body changes, even if its speed does not, then its velocity has changed. A change of velocity is an acceleration, and so during its whirling motion the ball is accelerating.

It follows from Newton's first law of motion that if we consider a body moving in a circle to be accelerating, then there must be a force acting on it to cause the acceleration. In the case of the whirling ball it is reasonable to say the force is provided by the string pulling inwards on the ball. Like the acceleration, the force acts towards the centre of the circle and keeps the body at a fixed distance from the centre.

A larger force is needed if

- the speed v of the ball is increased, with mass and radius constant
- the radius r of the circle is decreased, with mass and speed constant
- the mass m of the ball is increased, with speed and radius constant.

This force, which acts *towards the centre* and keeps a body moving in a circular path, is called the **centripetal force** (centre-seeking force).

Should the force be greater than the string can bear, the string breaks and the ball flies off with steady speed in a straight line *along the tangent*, i.e. in the direction of travel when the string broke (as Newton's first law of motion predicts). It is not thrown outwards.

Whenever an object moves in a circle (or circular arc) there must be a centripetal force acting on it. In throwing the hammer it is the pull of the athlete's arms acting on the hammer towards the centre of the whirling path. When a car rounds a bend, a frictional force is exerted inwards by the road on the car's tyres.

Satellites

For a satellite of mass m orbiting the Earth at radius r with orbital speed v, the centripetal force, F, is the Earth's gravitational force on the mass.

To put an artificial satellite in orbit at a certain height above the Earth it must enter the orbit at the correct speed. If it does not, the force of gravity, which decreases as height above the Earth increases, will not be equal to the centripetal force needed for the orbit.

Communication satellites

Communication satellites circle the Earth in orbits above the equator. Geostationary satellites have an orbit high above the equator (36 000 km); they travel with the same speed as the Earth rotates, so appear to be stationary at a particular point above the Earth's surface – their orbital period is 24 hours. They are used for transmitting television, intercontinental telephone and data signals. Geostationary satellites need to be well separated so that they do not interfere with each other; there is room for about 400.

Mobile phone networks use many satellites in much lower equatorial orbits; they are slowed by the Earth's atmosphere and their orbit has to be regularly adjusted by firing a rocket engine. Eventually they run out of fuel and burn up in the atmosphere as they fall to Earth.

Monitoring satellites

Monitoring satellites circle the Earth rapidly in low **polar** orbits, i.e. passing over both poles; at a height of 850 km the orbital period is only 100 minutes. The Earth rotates below them so they scan the whole surface at short range in a 24-hour period and can be used to map or monitor regions of the Earth's surface which may be inaccessible by other means. They are widely used in weather forecasting as they continuously transmit infrared pictures of cloud patterns down to Earth (Figure 1.5.18), which are picked up in turn by receiving stations around the world.

▲ **Figure 1.5.18** Satellite image of cloud over Europe

> **Test yourself**
>
> **14 a** Explain the conditions under which friction occurs.
> **b** Name two effects resulting from solid friction.
> **15** A car is moving at a constant speed along a straight road. Describe how the forces acting on the car influence the speed of the car. How is a constant speed achieved?
>
> **16** An apple is whirled round in a horizontal circle on the end of a string which is tied to the stalk. It is whirled faster and faster and at a certain speed the apple is torn from the stalk. Explain why this happens.
> **17** Is the gravitational force on a satellite greater or less when it is in a high orbit than when it is in a low orbit?

1.5.2 Turning effect of forces

FOCUS POINTS

★ Describe and give everyday examples of the turning effect of a force (its moment) and use the appropriate equation to calculate the moment of a force.
★ Apply the principle of moments to different situations.
★ Recall the conditions for an object being in equilibrium.

★ Apply the principle of moments to situations involving more than two forces about a pivot.
★ Be familiar with an experiment showing that an object in equilibrium has no resultant moment.

A seesaw in a children's playground can be balanced if the two children have similar weights or if the lighter child sits further from the pivot than the heavier child. Each child exerts a turning effect on the seesaw, either clockwise or anticlockwise, which depends not only on their weight but also on their distance from the pivot. Forces act in different ways depending on their orientation. In this topic you will discover that the turning effect of a force (its moment) depends on both its magnitude and the perpendicular distance from the pivot point. This means that a small force at a large distance can balance a much larger force applied closer to the pivot. When the combination of all the forces acting on a body is such that there is no net force or turning effect, the body is in equilibrium (the seesaw is level) and will not move unless additional forces are applied.

Moment of a force

The handle on a door is at the outside edge so that it opens and closes easily. A much larger force would be needed if the handle were near the hinge. Similarly, it is easier to loosen a nut with a long spanner than with a short one.

The turning effect of a force is called the **moment of the force**. It depends on both the size of the force and how far it is applied from the pivot. It is measured by multiplying the force by the perpendicular distance of the line of action of the force from the pivot. The unit is the newton metre (N m).

$$\text{moment of a force} = \text{force} \times \text{perpendicular distance from the pivot}$$

> **Key definition**
>
> **Moment of a force** moment = force × perpendicular distance from pivot

In Figure 1.5.19a, a force F acts on a gate at its edge, and in Figure 1.5.19b it acts at the centre.

In Figure 1.5.19a

$$\text{moment of } F \text{ about O} = 5\,\text{N} \times 3\,\text{m} = 15\,\text{N m}$$

In Figure 1.5.19b

$$\text{moment of } F \text{ about O} = 5\,\text{N} \times 1.5\,\text{m} = 7.5\,\text{N m}$$

The turning effect of F is greater in the first case. This agrees with the fact that a gate opens most easily when pushed or pulled at the edge furthest from the hinge.

a

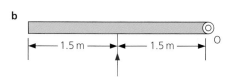

b

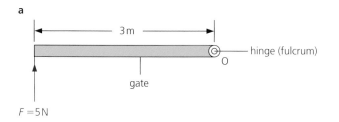

▲ Figure 1.5.19

Balancing a beam

To balance a **beam** about a pivot, like the ruler in Figure 1.5.20, the weights must be moved so that the clockwise turning effect equals the anticlockwise turning effect and the net moment on the beam becomes zero. If the beam tends to swing clockwise, m_1 can be moved further from the pivot to increase its turning effect; alternatively, m_2 can be moved nearer to the pivot to reduce its turning effect. What adjustment would you make to the position of m_2 to balance the beam if it is tending to swing anticlockwise?

Practical work

Law of moments

For safe experiments/demonstrations related to this topic, please refer to the *Cambridge IGCSE Physics Practical Skills Workbook* that is also part of this series.

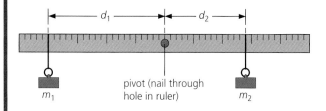

▲ Figure 1.5.20

Balance a half-metre ruler at its centre, adding Plasticine to one side or the other until it is horizontal.

Hang unequal loads m_1 and m_2 from either side of the pivot and alter their distances d_1 and d_2 from the centre until the ruler is again balanced (Figure 1.5.20). Forces F_1 and F_2 are exerted by gravity on m_1 and m_2 and so on the ruler; the force on 100 g is 0.98 N. Record the results in a table and repeat for other loads and distances.

m_1/g	F_1/N	d_1/cm	$F_1 \times d_1$/N cm	m_2/g	F_2/N	d_2/cm	$F_2 \times d_2$/N cm

F_1 is trying to turn the ruler anticlockwise and $F_1 \times d_1$ is its moment. F_2 is trying to cause clockwise turning and its moment is $F_2 \times d_2$. When the ruler is balanced or, as we say, *in equilibrium*, the results should show that the anticlockwise moment $F_1 \times d_1$ equals the clockwise moment $F_2 \times d_2$.

12 Name the variables you will need to measure in this experiment.

13 Calculate the moments of a force of 5 N acting at a perpendicular distance from the pivot of
   a 10 cm
   b 15 cm
   c 30 cm.

Principle of moments

The **law of moments** (also called the law of the lever) is stated as follows:

> When a body is in equilibrium, the sum of the clockwise moments about any point equals the sum of the anticlockwise moments about the same point. There is no resultant moment on an object in equilibrium.

The law of moments is an equivalent statement to the principle of moments. If the clockwise moments are regarded as positive and the anticlockwise moments are regarded as negative, then the sum of the moments is zero when the body is in equilibrium.

? Worked example

The seesaw in Figure 1.5.21 balances when Shani of weight 320 N is at A, Tom of weight 540 N is at B and Harry of weight W is at C. Find W.

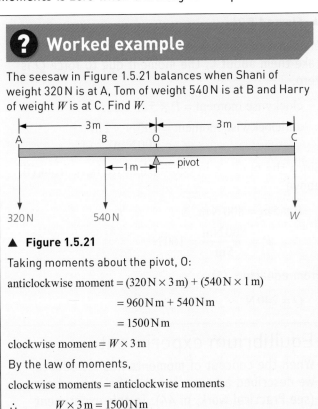

▲ **Figure 1.5.21**

Taking moments about the pivot, O:

anticlockwise moment $= (320\,\text{N} \times 3\,\text{m}) + (540\,\text{N} \times 1\,\text{m})$

$\qquad\qquad\qquad = 960\,\text{N m} + 540\,\text{N m}$

$\qquad\qquad\qquad = 1500\,\text{N m}$

clockwise moment $= W \times 3\,\text{m}$

By the law of moments,

clockwise moments = anticlockwise moments

$\therefore \qquad W \times 3\,\text{m} = 1500\,\text{N m}$

$\therefore \qquad W = \dfrac{1500\,\text{Nm}}{3\,\text{m}} = 500\,\text{N}$

▶ Test yourself

18 A seesaw has a weight of 40 N placed 1 m from the pivot and a weight of 20 N is placed on the opposite side of the pivot at a distance of 2 m from the pivot. Is the seesaw balanced?

19 A half-metre ruler is pivoted at its mid-point and balances when a weight of 20 N is placed at the 10 cm mark and a weight W is placed at the 45 cm mark on the ruler. Calculate the weight W.

Levers

A lever is any device which can turn about a pivot. In a working lever a force called the **effort** is used to overcome a resisting force called the **load**. The pivotal point is called the **fulcrum**.

If we use a crowbar to move a heavy boulder (Figure 1.5.22), our hands apply the effort at one end of the bar and the load is the force exerted by the boulder on the other end. If distances from the fulcrum O are as shown and the load is 1000 N (i.e. the part of the weight of the boulder supported by the crowbar), the effort can be calculated from the law of moments. As the boulder just begins to move, we can say, taking moments about O, that

clockwise moment = anticlockwise moment

effort $\times\ 200\,\text{cm} = 1000\,\text{N} \times 10\,\text{cm}$

effort $= \dfrac{10\,000\,\text{N cm}}{200\,\text{cm}} = 50\,\text{N}$

Examples of other levers are shown in Figure 1.5.23. How does the effort compare with the load for scissors and a spanner in Figures 1.5.23c and d?

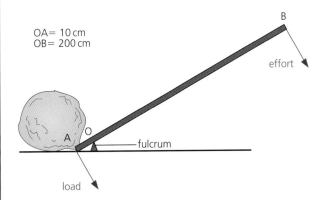

OA = 10 cm
OB = 200 cm

▲ **Figure 1.5.22** Crowbar

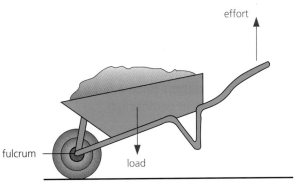

▲ **Figure 1.5.23a** Wheelbarrow

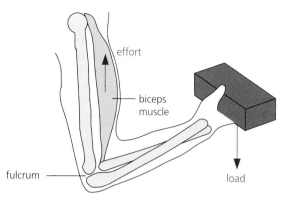

▲ **Figure 1.5.23b** Forearm

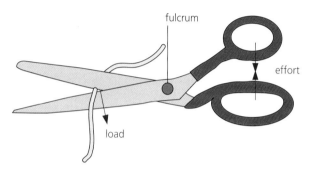

▲ **Figure 1.5.23c** Scissors

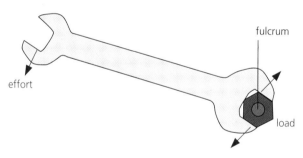

▲ **Figure 1.5.23d** Spanner

Conditions for equilibrium

Sometimes a number of parallel forces act on an object so that it is in **equilibrium**. We can then say:

(i) The sum of the forces in one direction equals the sum of the forces in the opposite direction.

(ii) The law of moments must apply.

When there is no resultant force and no resultant moment, an object is in equilibrium.

> **Key definition**
>
> **Equilibrium** when there is no resultant force and no resultant moment

As an example, consider a heavy plank resting on two trestles, as in Figure 1.5.24. In Topic 1.5.3 we will see that the whole weight of the plank (400 N) may be taken to act vertically downwards at its centre, O. If P and Q are the upward forces exerted by the trestles on the plank (called **reactions**), then we have from (i) above

$$P + Q = 400\,\text{N} \quad (1)$$

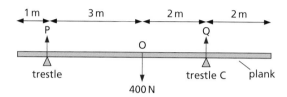

▲ **Figure 1.5.24**

Moments can be taken about any point but if we take them about C, the moment due to force Q is zero.

$$\text{clockwise moment} = P \times 5\,\text{m}$$
$$\text{anticlockwise moment} = 400\,\text{N} \times 2\,\text{m}$$
$$= 800\,\text{N m}$$

Since the plank is in equilibrium we have from (ii) above

$$P \times 5\,\text{m} = 800\,\text{N m}$$
$$\therefore \quad P = \frac{800\,\text{N m}}{5\,\text{m}} = 160\,\text{N}$$

From equation (1)

$$Q = 240\,\text{N}$$

Equilibrium experiment

When the concept of moments was introduced, we described an experiment to balance a beam (see Practical work, p. 44). In this experiment different weights (F) are suspended either side of the central pivot and the distance (d) of each from the pivot is measured when the beam is balanced (in equilibrium). The clockwise and anticlockwise moments ($F \times d$) are then calculated for each weight. It is found that when the beam is in equilibrium, the clockwise and anticlockwise moments are equal in magnitude and there is no resultant moment (i.e. no net turning effect) on the beam.

Test yourself

20 The metre ruler in Figure 1.5.25 is pivoted at its centre. If it balances, which of the following equations gives the mass of M?
  A $M + 50 = 40 + 100$
  B $M \times 40 = 100 \times 50$
  C $M \times 50 = 100 \times 40$
  D $M/50 = 40/100$

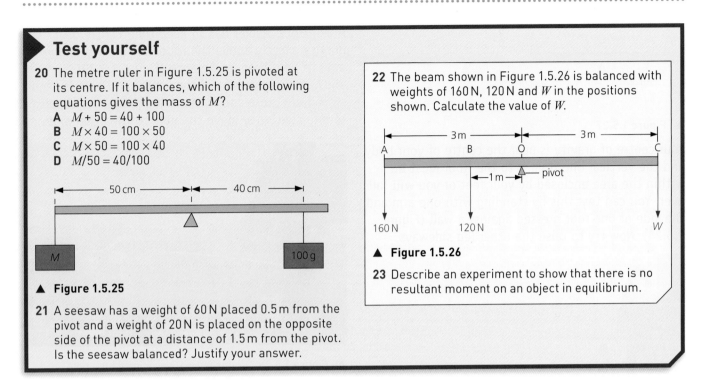

▲ **Figure 1.5.25**

21 A seesaw has a weight of 60 N placed 0.5 m from the pivot and a weight of 20 N is placed on the opposite side of the pivot at a distance of 1.5 m from the pivot. Is the seesaw balanced? Justify your answer.

22 The beam shown in Figure 1.5.26 is balanced with weights of 160 N, 120 N and W in the positions shown. Calculate the value of W.

▲ **Figure 1.5.26**

23 Describe an experiment to show that there is no resultant moment on an object in equilibrium.

1.5.3 Centre of gravity

FOCUS POINTS

★ Define centre of gravity and the effect its position has on the stability of an object.
★ Be familiar with an experiment determining the position of the centre of gravity of an irregularly shaped plane lamina.

Why are tall vehicles more likely to topple over on a slope than less tall ones? The answer lies in the position of the centre of gravity. In the presence of gravity an object behaves as if its entire mass is concentrated at a single point, the centre of gravity. The object's weight appears to act at this point. For a symmetrical object, such as a ball, the centre of gravity will be at its centre. In this topic, you will learn that when an object is suspended so that it can swing freely, it comes to rest with its centre of gravity vertically below the point of suspension. This enables the centre of gravity of unsymmetrical objects to be located. You will discover that it is the position of the centre of gravity that controls stability against toppling. If the centre of gravity remains within the footprint of the base of the object, it remains stable.

Centre of gravity

An object behaves as if its whole mass were concentrated at one point, called its **centre of gravity** even though the Earth attracts every part of it. The object's weight can be considered to act at this point. The centre of gravity of a uniform ruler is at its centre and when supported there it can be balanced, as in Figure 1.5.27a. If it is supported at any other point it topples because the moment of its weight W about the point of support is not zero, as in Figure 1.5.27b. The centre of gravity is sometimes also termed the centre of mass.

> **Key definition**
>
> **Centre of gravity** the point through which all of an object's weight can be considered to act

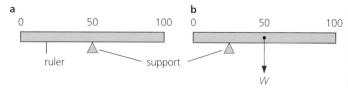

a
0   50   100

b
0   50   100

ruler   support

W

▲ Figure 1.5.27

Your centre of gravity is near the centre of your body and the vertical line from it to the floor must be within the area enclosed by your feet or you will fall over. You can test this by standing with one arm and the side of one foot pressed against a wall (Figure 1.5.28). Now try to raise the other leg sideways.

raise this leg

▲ Figure 1.5.28 Can you do this without falling over?

A tightrope walker has to keep their centre of gravity exactly above the rope. Some carry a long pole to help them to balance (Figure 1.5.29). The combined weight of the walker and pole is then spread out more and if the walker begins to topple to one side, they move the pole to the other side.

▲ Figure 1.5.29 A tightrope walker using a long pole

The centre of gravity of a regularly shaped object that has the same density throughout is at its centre. In other cases, it can be found by experiment.

 Practical work

Centre of gravity of an irregularly shaped lamina

For safe experiments/demonstrations related to this topic, please refer to the *Cambridge IGCSE Physics Practical Skills Workbook* that is also part of this series.

Suppose we have to find the centre of gravity of an irregularly shaped lamina (a thin sheet) of cardboard.

Make a hole A in the lamina and hang it so that it can swing *freely* on a nail clamped in a stand. It will come to rest with its centre of gravity vertically below A. To locate the vertical line through A, tie a plumb line (a thread and a weight) to the nail (Figure 1.5.30), and mark its position AB on the lamina. The centre of gravity lies somewhere on AB.

Hang the lamina from another position, C, and mark the plumb line position CD. The centre

of gravity lies on CD and must be at the point of intersection of AB and CD. Check this by hanging the lamina from a third hole. Also try balancing it at its centre of gravity on the tip of your forefinger.

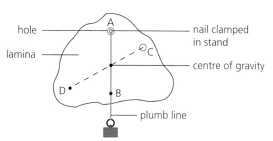

hole — A — nail clamped in stand
lamina — C — centre of gravity
D — B
plumb line

▲ Figure 1.5.30

14 a How could you make a plumb line?
   b Explain the purpose and use of a plumb line.
15 When an object is suspended and allowed to swing freely, where does its centre of gravity lie when it comes to rest?

Toppling

The position of the centre of gravity of an object affects whether or not it topples over easily. This is important in the design of such things as tall vehicles (which tend to overturn when rounding a corner), racing cars, reading lamps and even drinking glasses.

An object topples when the vertical line through its centre of gravity falls outside its base, as in Figure 1.5.31a. Otherwise it remains stable, as in Figure 1.5.31b, where the object will not topple.

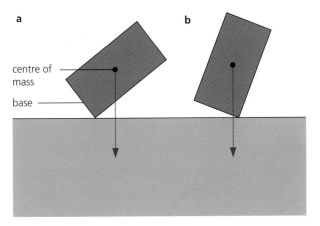

▲ **Figure 1.5.31**

Toppling can be investigated by placing an empty can on a plank (with a rough surface to prevent slipping) which is slowly tilted. The angle of tilt is noted when the can falls over. This is repeated with a mass of 1 kg in the can. How does this affect the position of the centre of gravity? The same procedure is followed with a second can of the same height as the first but of greater width. It will be found that the second can with the mass in it can be tilted through a greater angle.

The stability of a body is therefore increased by
(i) lowering its centre of gravity, and
(ii) increasing the area of its base.
In Figure 1.5.32a the centre of gravity of a fire truck is being found. It is necessary to do this when testing a new design since fire trucks (and other vehicles) may be driven over sloping surfaces, e.g. ditches adjacent to roadways, and any tendency to overturn must be discovered.

The stability of a coach is being tested in Figure 1.5.32b. When the top deck only is fully laden with passengers (represented by sand bags in the test), it must not topple if tilted through an angle of 28°.

Racing cars have a low centre of gravity and a wide wheelbase for maximum stability.

▲ **Figure 1.5.32a** A fire truck under test to find its centre of gravity

▲ **Figure 1.5.32b** A coach being tilted to test its stability

Stability

Three terms are used in connection with stability.

Stable equilibrium

An object is in **stable equilibrium** if when slightly displaced and then released it returns to its previous position. The ball at the bottom of the dish in Figure 1.5.33a is an example. Its centre of gravity rises when it is displaced. It rolls back because its weight has a moment about the point of contact that acts to reduce the displacement.

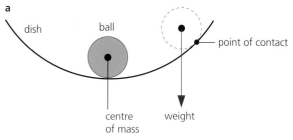

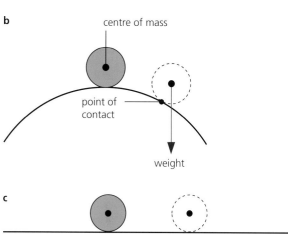

▲ **Figure 1.5.33** States of equilibrium

Unstable equilibrium

An object is in **unstable equilibrium** if it moves further away from its previous position when slightly displaced and released. The ball in Figure 1.5.33b behaves in this way. Its centre of gravity falls when it is displaced slightly because there is a moment which increases the displacement. Similarly, in Figure 1.5.27a, the balanced ruler is in unstable equilibrium.

Neutral equilibrium

An object is in **neutral equilibrium** if it stays in its new position when displaced (Figure 1.5.33c). Its centre of gravity does not rise or fall because there is no moment to increase or decrease the displacement.

Balancing tricks and toys

Some tricks that you can try or toys you can make are shown in Figure 1.5.34. In each case the centre of gravity is vertically below the point of support and equilibrium is stable.

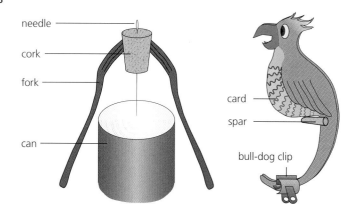

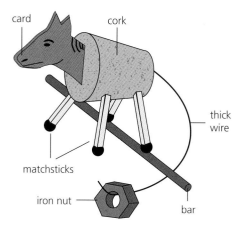

▲ **Figure 1.5.34** Balancing tricks

A self-righting toy (Figure 1.5.35) has a heavy base and, when tilted, the weight acting through the centre of gravity has a moment about the point of contact. This restores it to the upright position.

▲ **Figure 1.5.35** A self-righting toy

Test yourself

24 Where does the centre of gravity lie for
 a a uniform ruler
 b a sphere of uniform density?
25 a When does an object topple?
 b How can the stability of an object be increased?
26 Figure 1.5.36 shows a Bunsen burner in three different positions. State the type of equilibrium when it is in position
 i A
 ii B
 iii C

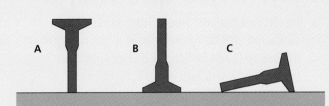

▲ **Figure 1.5.36**

Revision checklist

After studying Topic 1.5 you should know and understand:

✓ the significance of the term limit of proportionality

✓ Newton's first law of motion
✓ friction as the force between two surfaces that impedes motion and results in heating and that friction also acts on an object moving through the air
✓ the conditions for equilibrium
✓ that an object's weight acts through the centre of gravity.

After studying Topic 1.5 you should be able to:
✓ recall that a force can cause a change in the motion, size or shape of a body
✓ describe an experiment to study the relation between force and extension for springs; plot and draw conclusions from load–extension graphs

✓ define the spring constant and use the equation $k = F/x$ to solve problems

✓ combine forces acting along the same straight line to find their resultant

✓ recall the equation $F = ma$ and use it to solve problems
✓ describe qualitatively motion in a circular path due to a perpendicular force and recall that the force required to maintain circular motion changes when the speed, radius of orbit or mass changes

✓ define the moment of a force about a pivot and give everyday examples; recall the law of moments and use it to solve problems, including the balancing of a beam

✓ apply the principle of moments to balance multiple moments (more than two) about a pivot
✓ describe an experiment to verify that there is no resultant moment on an object in equilibrium

✓ recall that an object behaves as if its whole mass acts through its centre of gravity
✓ describe an experiment to find the centre of gravity of an object and connect the stability of an object to the position of its centre of gravity.

Exam-style questions

1 a Describe how you would investigate the variation of the extension of a spring when different loads are applied. Mention two precautions you would take to obtain accurate results. [6]

b The table below shows the results obtained in an experiment to study the stretching of a spring. Copy the table and fill in the missing values. What can you say about the relationship between the extension of the spring and the stretching force? [4]

Mass/g	Stretching force/N	Scale reading/mm	Extension/mm
0		20.0	0
100		20.2	
200		20.4	
300		20.6	
400		20.8	
500		21.0	

[Total: 10]

2 The spring in Figure 1.5.37 stretches from 10 cm to 22 cm when a force of 4 N is applied.

a Calculate the spring constant of the spring. [3]

b If the extension is proportional to the stretching force when a force of 6 N is applied, calculate

i the new extension length of the spring [2]

ii the final length in cm of the spring. [1]

[Total: 6]

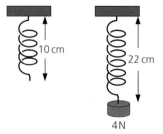

▲ Figure 1.5.37

3 Two forces of 5 N and 12 N act at a point.

a The two forces first act in opposite directions.

i Make a sketch showing the direction of the forces. [2]

ii Calculate the resultant of the forces and mark its direction on your sketch. [2]

b The two forces then act at 90° to each other. Calculate the magnitude and direction of the resultant force by calculation. [6]

[Total: 10]

4 Starting from rest on a level road a girl can reach a speed of 5 m/s in 10 s on her bicycle.

a Find the acceleration. [2]

b Find the average speed during the 10 s. [2]

c Find the distance she travels in 10 s. [2]

d Eventually, even though she still pedals as fast as she can, she stops accelerating and her speed reaches a maximum value. Explain in terms of the forces acting why this happens. [2]

[Total: 8]

5 Explain the following using $F = ma$.

a A racing car has a powerful engine and is made of strong but lightweight material. [3]

b A car with a small engine can still accelerate rapidly. [3]

[Total: 6]

6 A rocket has a mass of 500 kg.

a Write down the weight of the rocket on Earth where $g = 9.8$ N/kg. [1]

b At lift-off the rocket engine exerts an upward force of 25 000 N.

i Calculate the resultant force on the rocket. [2]

ii Calculate the initial acceleration of the rocket. [3]

[Total: 6]

7 A car rounding a bend travels in an arc of a circle.
  a What provides the force to keep the car travelling in a circle? [2]
  b Is a larger or a smaller force required if
   i the car travels faster [1]
   ii the bend is less curved [1]
   iii the car has more passengers? [1]
  c Explain why racing cars are fitted with tyres called 'slicks', which have no tread pattern, for dry tracks and with 'tread' tyres for wet tracks. [2]
  [Total: 7]

8 A satellite close to the Earth (at a height of about 200 km) has an orbital speed of 8 km/s. Take the radius of the orbit to be approximately equal to the Earth's radius of 6400 km.
  a Write down an expression for the circumference of the orbit. [1]
  b Write down an equation for the time for one orbit. [2]
  c Calculate the time it takes for the satellite to complete one orbit. [4]
  [Total: 7]

9 Figure 1.5.38 shows three positions of the pedal on a bicycle which has a crank 0.20 m long. The cyclist exerts the same vertically downward push of 25 N with his foot. Calculate the turning effect in
  a A [2]
  b B [2]
  c C [2]
  [Total: 6]

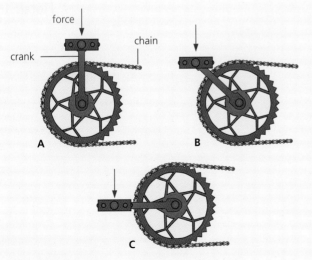

▲ **Figure 1.5.38**

10 The weight of the uniform bar in Figure 1.5.39 is 10 N.
  a Calculate the clockwise moment about the pivot. [3]
  b Calculate the anticlockwise moment about the pivot. [3]
  c Does the beam balance, tip to the right or tip to the left? [2]
  [Total: 8]

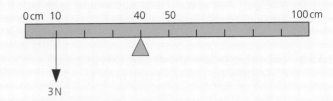

▲ **Figure 1.5.39**

11 a Describe how you could find the centre of gravity of an irregular lamina. [5]
  b A heavy box with a square base and a height twice the length of a side is to be transported by a lorry. Explain how the stability of
   i the lorry [2]
   ii the box [2]
   will be affected if the box lies on its side in the van rather than its base.
  [Total: 9]

Alternative to Practical

12 A physics class is asked to investigate the extension of a stretched spring.

You will be supplied with a spring, a clamp stand, a half-metre ruler, a set square and a hanger with 100 g weights and sticky tape.

a Describe how you would carry out the experiment. [5]

b Mention any precautions you would take to achieve good results. [3]

[Total: 8]

13 a The table below shows the extension of a spring for increasing stretching forces.

Stretching force/N	0	1	2	3	4	5
Extension/mm	0	2	4	6	8.5	12

   i Plot a graph with extension/mm along the x-axis and stretching force/N on the y-axis. [4]

   ii Draw the best line through the points; mark the region over which proportionality holds. [2]

   iii Indicate the limit of proportionality. [1]

  b Calculate the gradient of the graph. [2]

  c Determine the spring constant k. [1]

[Total: 10]

14 In an experiment to investigate the law of moments, a half-metre ruler is balanced at its centre as shown in Figure 1.5.40.

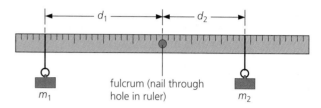

▲ **Figure 1.5.40**

Masses of 50 g, 100 g and 150 g are placed in turn at the positions given in the table below.

a Complete the table, filling in values for

   i the units at the head of each column [1]

   ii force (F) [2]

   iii distance from pivot (d) [2]

   iv moment about pivot ($F \times d$). [2]

b State which combinations of two different masses could be used to balance the beam. [3]

Mass/g	Force/	Ruler reading/cm	d/	$F \times d$/	
50		5			A
50		10			B
50		15			C
50		20			D
100		30			E
100		35			F
100		40			G
150		20			H
150		35			I

[Total: 10]

1.6 Momentum

FOCUS POINTS

★ Define momentum, impulse and resultant force and use the correct equations to calculate them.
★ Solve simple one-dimensional problems using the principle of the conservation of momentum.

When a tennis ball is struck by a racket or a gas molecule rebounds from the side of its container, their behaviour can be understood by introducing the concept of momentum. Momentum is defined as the product of mass and velocity. In a collision, momentum is conserved unless there are external forces acting such as friction. You can demonstrate conservation of momentum with a Newton's cradle (Figure 1.7.10, p. 66); the last ball in the line moves off with the same velocity as the first. Collisions generally occur over a very short interval of time; the shorter the time interval the greater the force on the bodies involved in the collision. Crumple zones at the front and rear of a car help to prolong the collision time and reduce the force of an impact.

Momentum is a useful quantity to consider when bodies are involved in collisions and explosions. It is defined as the mass of the body multiplied by its velocity and is measured in kilogram metre per second (kg m/s) or newton second (N s).

$$\text{momentum} = \text{mass} \times \text{velocity}$$

In symbols, momentum $p = mv$ and the change in momentum

$$\Delta p = \Delta(mv)$$

A 2 kg mass moving at 10 m/s has momentum 20 kg m/s, the same as the momentum of a 5 kg mass moving at 4 m/s.

> **Key definition**
> **Momentum** mass × velocity

Practical work

Collisions and momentum

Safety

● Take care when rolling the trolley down the ramp. Ensure it is clear at the bottom of the ramp and use a side barrier to prevent the trolley from falling onto the floor.

Figure 1.6.1 shows an arrangement which can be used to find the velocity of a trolley before and after a collision. If a trolley of length l takes time t to pass through a photogate, then its velocity = distance/time = l/t.

Two photogates are needed, placed each side of the collision point, to find the velocities before and after the collision. Set them up so that they

will record the time taken for the passage of a trolley.

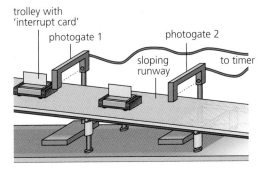

▲ **Figure 1.6.1**

A tickertape timer or motion sensor, placed at the top end of the runway, could be used instead of the photogates if preferred.

Attach a strip of Velcro to each trolley so that they adhere to each other on collision and compensate the runway for friction (see Topic 1.5.1). Place one trolley at rest halfway down the runway and another at the top; give the top trolley a push. It will move forwards with uniform velocity and should hit the second trolley so that they travel on as one. Using the times recorded by the photogate timer, calculate the velocity of the moving trolley before the collision and the common velocity of both trolleys after the collision.

Repeat the experiment with another trolley stacked on top of the one to be pushed so that two are moving before the collision and three after.

Copy and complete the tables of results.

Before collision (m_2 at rest)

Mass m_1 (no. of trolleys)	Velocity v/m/s	Momentum $m_1 v$
1		
2		

After collision (m_1 and m_2 together)

Mass $m_1 + m_2$ (no. of trolleys)	Velocity v_1/m/s	Momentum $(m_1 + m_2)v_1$
2		
3		

1 Do the results suggest any connection between the momentum before the collision and after it in each case?
2 Why is it necessary to tilt the runway slightly before taking measurements?
3 a Calculate the momentum of a 2 kg trolley moving with a velocity of
   i   0.2 m/s
   ii 0.8 m/s
   iii 5 cm/s.
   b Calculate the momentum of a trolley moving at 3 m/s and having a mass of
   i   200 g
   ii 500 g
   iii 1 kg.

Conservation of momentum

When two or more bodies act on one another, as in a collision, the total momentum of the bodies remains constant, provided no external forces act (e.g. friction).

This statement is called the **principle of conservation of momentum**. Experiments like those in the *Practical work* section show that it is true for all types of collisions.

 Worked example

Suppose a truck of mass 60 kg moving with velocity 3 m/s collides and couples with a stationary truck of mass 30 kg (Figure 1.6.2a). The two move off together with the same velocity v which we can find as follows (Figure 1.6.2b).

Total momentum before collision is

$$(60 \text{ kg} \times 3 \text{ m/s}) + (30 \text{ kg} \times 0 \text{ m/s}) = 180 \text{ kg m/s}$$

Total momentum after collision is

$$(60 \text{ kg} + 30 \text{ kg}) \times v = 90 \text{ kg} \times v$$

Since momentum is not lost

$$90 \text{ kg} \times v = 180 \text{ kg m/s} \text{ or } v = 2 \text{ m/s}$$

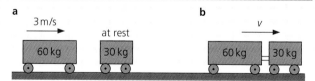

▲ **Figure 1.6.2**

Now put this into practice

1 A trolley of mass 3 kg moving with velocity 5 m/s collides and couples with a stationary trolley of mass 2 kg and the two move off together with the same velocity v. Assuming momentum is not lost in the collision, calculate the value of v.
2 A trolley of mass 5 kg moving with velocity 5 m/s collides with a stationary trolley of mass 2 kg.
   The 5 kg trolley stops and the 2 kg trolley moves off with velocity v. Assuming momentum is not lost in the collision, calculate the value of v.

Explosions

Momentum, like velocity, is a vector since it has both magnitude and direction. Vectors cannot be added by ordinary addition unless they act in the same direction. If they act in exactly opposite directions, such as east and west, the smaller subtracts from the greater, or if they are the same, they cancel out.

Momentum is conserved in an explosion such as occurs when a rifle is fired. Before firing, the total momentum is zero since both rifle and bullet are at rest. During the firing the rifle and bullet receive *equal* but *opposite* amounts of momentum so that the *total* momentum after firing is zero. For example, if a rifle fires a bullet of mass 0.01 kg with a velocity of 300 m/s,

$$\text{forward momentum of bullet} = 0.01 \text{ kg} \times 300 \text{ m/s}$$
$$= 3 \text{ kg m/s}$$

$$\therefore \quad \text{backward momentum of rifle} = 3 \text{ kg m/s}$$

If the rifle has mass m, it recoils (kicks back) with a velocity v such that

$$mv = 3 \text{ kg m/s}$$

Taking $m = 6$ kg gives $v = 3/6$ m/s = 0.5 m/s.

Rockets and jets

If you release an inflated balloon with its neck open, it flies off in the opposite direction to that of the escaping air. In Figure 1.6.3 the air has momentum to the left and the balloon moves to the right with equal momentum.

This is the principle of rockets and jet engines. In both, a high-velocity stream of hot gas is produced by burning fuel and leaves the exhaust with large momentum. The rocket or jet engine itself acquires an equal forward momentum. Space rockets carry their own oxygen supply; jet engines use the surrounding air.

▲ **Figure 1.6.3** A deflating balloon demonstrates the principle of a rocket or a jet engine.

> ## Test yourself
>
> 1 What is the momentum in kg m/s of a 10 kg truck travelling at
>   a 5 m/s
>   b 20 cm/s
>   c 36 km/h?
> 2 A ball X of mass 1 kg travelling at 2 m/s has a head-on collision with an identical ball Y at rest. X stops and Y moves off. What is Y's velocity?
> 3 A boy with mass 50 kg running at 5 m/s jumps on to a 20 kg trolley travelling in the same direction at 1.5 m/s. What is their common velocity?
> 4 A girl of mass 50 kg jumps out of a rowing boat of mass 300 kg on to the bank, with a horizontal velocity of 3 m/s. With what velocity does the boat begin to move backwards?

Force and momentum

If a steady force F acting on an object of mass m increases its velocity from u to v in time Δt, the acceleration a is given by

$$a = (v - u)/\Delta t$$

Substituting for a in $F = ma$,

$$F = \frac{m(v - u)}{\Delta t} = \frac{mv - mu}{\Delta t}$$

We also have

$$\text{impulse} = F\Delta t = mv - mu = \Delta(mv)$$

where mv is the final momentum, mu the initial momentum and $F\Delta t$ is called the **impulse**.

Since $\qquad F\Delta t = mv - mu = \Delta(mv)$

We can write $\qquad F = \dfrac{\Delta(mv)}{\Delta t} = \dfrac{\Delta p}{\Delta t}$

and define the **resultant force** F as the change in momentum per unit time.

This is another version of Newton's second law. For some problems it is more useful than $F = ma$.

> **Key definitions**
>
> **Impulse** force × time for which force acts
>
> **Resultant force** the rate of change in momentum per unit time

Sport: impulse and collision time

The good cricketer or tennis player 'follows through' with the bat or racket when striking the ball (Figure 1.6.4a). The force applied then acts for a longer time, the impulse is greater and so also is the gain of momentum (and velocity) of the ball.

When we want to stop a moving ball such as a cricket ball, however, its momentum has to be reduced to zero. An impulse is then required in the form of an opposing force acting for a certain time. While any number of combinations of force and time will give a particular impulse, the 'sting' can be removed from the catch by drawing back the hands as the ball is caught (Figure 1.6.4b). A smaller average force is then applied for a longer time.

▲ **Figure 1.6.4b** Cricketer drawing back the hands to catch the ball

The use of sand gives a softer landing for long-jumpers (Figure 1.6.5), as a smaller stopping force is applied over a longer time. In a car crash the car's momentum is reduced to zero in a very short time. If the time of impact can be extended by using **crumple zones** (see Figure 1.7.11, p. 66) and extensible seat belts, the average force needed to stop the car is reduced so the injury to passengers should also be less.

▲ **Figure 1.6.5** Sand reduces the athlete's momentum more gently.

▲ **Figure 1.6.4a** This cricketer is 'following through' after hitting the ball.

> ### Test yourself
>
> 5 A force of 5 N is applied to a cricket ball for 0.02 s. Calculate
>   a the impulse on the ball
>   b the change in momentum of the ball.
> 6 In a collision, a car of mass 1000 kg travelling at 24 m/s comes to rest in 1.2 s. Calculate
>   a the change in momentum of the car
>   b the steady stopping force applied to the car.

Revision checklist ✔

After studying Topic 1.6 you should know and understand the following:
✔ the relationship between force and rate of change of momentum and use it to solve problems.

After studying Topic 1.6 you should be able to:
✔ define momentum and apply the principle of conservation of momentum to solve problems
✔ recall that in a collision, impulse = $F\Delta t$ and use the definition to explain how the time of impact affects the force acting in a collision.

Exam-style questions

1 A truck A of mass 500 kg moving at 4 m/s collides with another truck B of mass 1500 kg moving in the same direction at 2 m/s.
  a Write down an expression for momentum.   [1]
  b Calculate the momentum of truck A before the collision.   [2]
  c Calculate the momentum of truck B before the collision.   [2]
  d Determine the common velocity of the trucks after the collision.   [4]
              [Total: 9]

2 The velocity of an object of mass 10 kg increases from 4 m/s to 8 m/s when a force acts on it for 2 s. Write down the
  a initial momentum of the object   [2]
  b final momentum of the object   [2]
  c momentum gained in 2 s   [2]
  d value of the force   [2]
  e impulse of the force.   [2]
              [Total: 10]

3 A rocket of mass 10 000 kg uses 5.0 kg of fuel and oxygen to produce exhaust gases ejected at 5000 m/s.
  a Define momentum.   [1]
  b Calculate the backward momentum of the ejected gas.   [2]
  c Explain what is meant by the principle of conservation of momentum.   [2]
  d Calculate the increase in velocity of the rocket.   [3]
              [Total: 8]

4 A boy hits a stationary billiard ball of mass 30 g head on with a cue. The cue is in contact with the ball for a time of 0.001 s and exerts a force of 50 N on it.
  a Calculate the acceleration of the ball during the time it is in contact with the cue.   [2]
  b Work out the impulse on the ball in the direction of the force.   [2]
  c Calculate the velocity of the ball just after it is struck.   [2]
  d Give *two* ways by which the velocity of the ball could be increased.   [2]
              [Total: 8]

1.7 Energy, work and power

1.7.1 Energy

FOCUS POINTS

★ Identify different energy stores and describe how energy is transferred from one store to another.

★ Use the correct equations for kinetic energy and change in gravitational potential energy.

★ Apply the principle of the conservation of energy to simple examples and use it to interpret flow diagrams.

★ Apply the principle of the conservation of energy to complex examples represented by Sankey diagrams.

Energy is a theme that appears in all branches of science. It links a wide range of phenomena and enables us to explain them. There are different ways in which energy can be stored and, when something happens, it is likely to be due to energy being transferred from one store to another. Energy transfer is needed to enable people, computers, machines and other devices to work and to enable processes and changes to occur. For example, in Figure 1.7.1, the water skier can be pulled along by the boat only if energy is transferred from the burning petrol to its rotating propeller. Although energy can be transferred to different stores, the total energy of a system remains constant. In this topic you will learn in detail about the potential energy associated with the position of an object in a gravitational field and the kinetic energy which is associated with its motion.

▲ **Figure 1.7.1** Energy transfer in action

Energy stores

Energy can be stored in a number of different ways.

> **Key definition**
>
> **Energy** may be stored as kinetic, gravitational potential, chemical, elastic (strain), nuclear, electrostatic and internal (thermal)

Chemical energy

Food and fuels, like oil, gas, coal and wood, are concentrated stores of **chemical energy**. The energy of food is released by chemical reactions in our bodies, and during the transfer to other stores we are able to do useful jobs. Fuels cause **energy transfers** when they are burnt in an engine or a boiler. Batteries are compact sources of chemical energy, which in use is transferred by an electric current.

Gravitational potential energy

This is the energy an object has because of its position. A body above the Earth's surface, like water in a mountain reservoir, has energy stored as **gravitational potential energy.**

Elastic strain energy

This is energy an object has because of its condition. Work has to be done to compress or stretch a spring or elastic material and energy is transferred to *elastic* **strain energy**. If the bow string in Figure 1.7.3c on the next page were released, the strain energy would be transferred to the arrow.

Kinetic energy

Any moving object has **kinetic energy** and the faster it moves, the more kinetic energy it has. As a hammer drives a nail into a piece of wood, there is a transfer of energy from the kinetic energy of the moving hammer to other energy stores.

Electrostatic energy

Energy can be stored by charged objects (see Topic 4.2.1) as **electrostatic energy**. This energy can be transferred by an electric current.

Nuclear energy

The energy stored in the nucleus of an atom is known as **nuclear energy**. It can be transferred to other energy stores in nuclear reactions such as fission and fusion (Topic 5.1.2).

Internal energy

This is also called **thermal energy** and is the final fate of other energy stores. It is transferred by conduction, convection or radiation.

Energy transfers

Demonstration

The apparatus in Figure 1.7.2 can be used to show how energy is transferred between different energy stores. Chemical energy stored in the battery is transferred by an electric current (electrical working) to kinetic energy in the electric motor. The weight is raised when kinetic energy stored in the motor is transferred (by mechanical working) to gravitational potential energy stored in the weight. If the changeover switch is joined to the lamp and the weight allowed to fall, the motor acts as a **generator** of an electric current that transfers (by electrical working) kinetic energy stored in the rotating coil of the generator to internal energy in the lamp. Energy is transferred from the lamp to the environment (by electromagnetic waves and by heating).

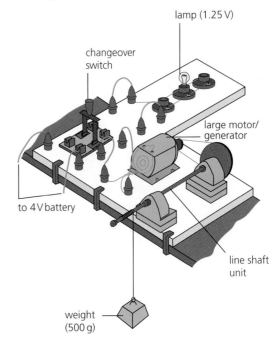

lamp (1.25 V)

changeover switch

large motor/ generator

to 4 V battery

line shaft unit

weight (500 g)

▲ **Figure 1.7.2** Demonstrating energy transfers

Other examples

In addition to electrical working, mechanical working, electromagnetic waves and heating, energy can be transferred between stores by other types of waves, such as sound waves. Sound waves transfer energy from a vibrating source to our eardrums or to a microphone. Heating water in a boiler transfers chemical energy stored in a fuel to internal energy stored in the water.

In summary, energy can be transferred between stores in the following ways:

- mechanical working – by the action of a force (Topic 1.5)
- electrical working – by an electric current (Topic 4.2.2)
- waves – electromagnetic, sound and other waves (Topic 3.3)
- heating – by conduction, convection or radiation (Topic 2.3).

Some energy transfers are shown in Figures 1.7.3a to d:

a Potential energy is transferred to kinetic energy by mechanical working (action of a gravitational force).
b Thermal energy stored in an electric fire element is transferred by electromagnetic waves and by heating to the environment.
c Chemical energy (stored in muscles in the arm) is transferred to elastic energy in the bow by mechanical working.
d Gravitational potential energy stored in the water in the upper reservoir is transferred to the kinetic energy of a turbine by mechanical working.

▲ **Figure 1.7.3** Some energy transfers

Measuring energy transfers

In an energy transfer, work is done. The work done is a measure of the amount of energy transferred. Energy, as well as work, is measured in joules (J).

For example, if you have to exert an upward force of 10 N to raise a stone steadily through a vertical distance of 1.5 m, the mechanical work done is 15 J (see Topic 1.7.2).

work = force × distance moved in the direction of force

This is also the amount of chemical energy transferred from your muscles to the **potential energy** of the stone.

Principle of conservation of energy

The **principle of conservation of energy** is one of the basic laws of physics and is stated as follows:

Energy cannot be created or destroyed; it is always conserved.

However, energy is continually being transferred from one store to another. Some stores, such as those of electrostatic and chemical energy, are easily transferred; for others, such as internal energy, it is hard to arrange a useful transfer.

Ultimately all energy transfers result in the surroundings being heated (as a result of doing work against friction) and the energy is wasted, i.e. spread out and increasingly more difficult to use. For example, when a brick falls, its gravitational potential energy is transferred by mechanical working (gravitational force) to kinetic energy; when the brick hits the ground, kinetic energy is transferred to the surroundings by heating and by sound waves. If it seems in a transfer that some energy has disappeared, the 'lost' energy is often transferred into non-useful thermal energy. This appears to be the fate of all energy in the Universe and is one reason why new sources of useful energy have to be developed.

Representing energy transfers

1 The flow diagram of energy transfers for a hydroelectric scheme like that in Figure 1.7.3d is shown in Figure 1.7.4.

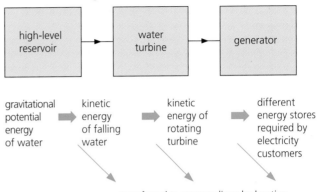

▲ **Figure 1.7.4** Energy transfers in a hydroelectric power station

2 In thermal power stations, thermal energy transferred from burning fossil fuels heats the water in a boiler and turns it into steam. The steam drives turbines which in turn drive the generators that produce electricity as described in Topic 4.5. Figure 1.7.5 shows a **Sankey diagram** for a thermal power station, where the thickness of the bars represents the size of energy transfer at each stage.

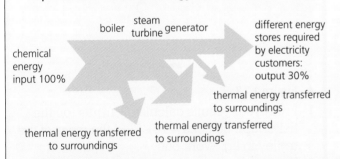

▲ **Figure 1.7.5** Sankey diagram depicting energy transfers in a thermal power station

Kinetic energy (E_k)

Kinetic energy is the energy an object has because of its motion.

For an object of mass m travelling with velocity v,

$$\text{kinetic energy} = E_k = \frac{1}{2}mv^2$$

If m is in kg and v in m/s, then kinetic energy is in J.

Since E_k depends on v^2, a high-speed vehicle travelling at 1000 km/h (Figure 1.7.6), has one hundred times the kinetic energy it has at 100 km/h.

▲ **Figure 1.7.6** Kinetic energy depends on the square of the velocity.

? Worked example

Calculate the kinetic energy of a football of mass 0.4 kg (400 g) moving with a speed of 20 m/s.

$$E_k = \frac{1}{2}mv^2$$
$$= \frac{1}{2} \times 0.4\,\text{kg} \times (20\,\text{m/s})^2$$
$$= 0.2 \times 400\,\text{kg m}^2/\text{s}^2$$
$$= 80\,\text{N m} = 80\,\text{J}$$

Now put this into practice

1 Calculate the kinetic energy of a ball of mass 0.4 kg moving with a speed of 80 m/s.
2 Calculate the kinetic energy of a ball of mass 50 g moving with a speed of 40 m/s.

Potential energy (E_p)

Potential energy is the energy an object has because of its position or condition.

An object above the Earth's surface is considered to have gained an amount of gravitational potential energy equal to the work that has been done against gravity by the force used to raise it. To lift an object of mass m through a vertical height Δh at a place where the Earth's gravitational field strength is g needs a force equal and opposite to the weight mg of the body. Hence

work done by force = force × vertical height
$$= mg \times \Delta h$$

∴ the change in gravitational potential energy
$$= \Delta E_p = mg\Delta h$$

When m is in kg, g in N/kg (or m/s²) and Δh in m, then ΔE_p is in J.

? Worked example

Taking $g = 9.8$ N/kg, calculate the potential energy gained by a 0.1 kg (100 g) mass raised vertically by 1 m.

$$\Delta E_p = mg\Delta h = 0.1\,\text{kg} \times 9.8\,\text{N/kg} \times 1\,\text{m} = 1\,\text{N m} = 1\,\text{J}$$

Now put this into practice

1 Calculate the gravitational potential energy gained by a 0.2 kg mass raised vertically by 2 m.
2 Calculate the gravitational potential energy lost by a 0.4 kg mass which falls vertically by 3 m.

⚡ Practical work

Transfer of gravitational potential energy to kinetic energy

Safety

- Place something soft on the floor to absorb the impact of the masses.
- Take care to keep feet well away from the falling masses.

Friction-compensate a runway by raising the start point slightly so that the trolley maintains a constant speed on the slope when no weight is attached. Arrange the apparatus as in Figure 1.7.7 with the bottom of the 0.1 kg (100 g) mass 0.5 m from the floor.

Start the timer and release the trolley. It will accelerate until the falling mass reaches the floor; after that it moves with *constant* velocity *v*.

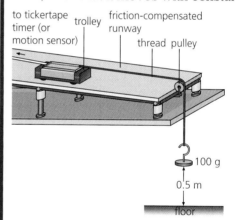

to tickertape timer (or motion sensor) trolley friction-compensated runway thread pulley 100 g 0.5 m floor

▲ Figure 1.7.7

1 From your results calculate *v* in m/s (on the tickertape 50 ticks = 1 s). Find the mass of the trolley in kg. Work out:
   Kinetic energy gained by trolley and 0.1 kg mass = ___ J
   Potential energy lost by 0.1 kg mass = ___ J
   Compare and comment on the results.
2 Explain why
   a the runway should be friction-compensated
   b the trolley in the experiment will move at a constant speed when the mass hits the floor.
3 Calculate the change in gravitational potential energy of a mass of 300 g falling through a distance of 80 cm.
4 Calculate the kinetic energy of a mass of 300 g travelling at a speed of 4.0 m/s.

Conservation of energy

A mass m at height Δh above the ground has gravitational potential energy $= mg\Delta h$ (Figure 1.7.8). When an object falls, its speed increases and it gains kinetic energy at the expense of its gravitational potential energy. If it starts from rest and air resistance is negligible, the kinetic energy it has gained on reaching the ground equals the gravitational potential energy lost by the mass

$$E_k = \Delta E_p$$

or

$$\frac{1}{2}mv^2 = mg\Delta h$$

where v is the speed of the mass when it reaches the ground.

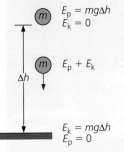

$E_p = mg\Delta h$
$E_k = 0$

$E_p + E_k$

Δh

$E_k = mg\Delta h$
$E_p = 0$

▲ **Figure 1.7.8** Loss of gravitational potential energy = gain of kinetic energy

This is an example of the principle of conservation of energy which was discussed earlier.

In the case of a pendulum (Figure 1.7.9), kinetic and gravitational potential energy are interchanged continually. The energy of the bob is all gravitational potential energy at the end of the swing and all kinetic energy as it passes through its central position. In other positions it has both gravitational potential and kinetic energy. Eventually all the energy is transferred to thermal energy as a result of overcoming air resistance.

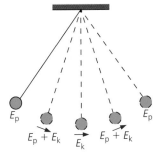

▲ **Figure 1.7.9** Interchange of potential and kinetic energy for a simple pendulum

---

? Worked example

A boulder of mass 4 kg rolls over a cliff and reaches the beach below with a velocity of 20 m/s. Find:
a the kinetic energy of the boulder as it lands
b the potential energy of the boulder when it was at the top of the cliff
c the height of the cliff.

a mass of boulder $= m = 4\,\text{kg}$

velocity of boulder as it lands $= v = 20\,\text{m/s}$

∴ kinetic energy of boulder as it lands $E_k = \frac{1}{2}mv^2$

$$= \frac{1}{2} \times 4\,\text{kg} \times (20)^2\,\text{m}^2/\text{s}^2$$

$$= 800\,\text{kg}\,\text{m}^2/\text{s}^2$$

$$= 800\,\text{N}\,\text{m}$$

$$= 800\,\text{J}$$

b Applying the principle of conservation of energy (and neglecting energy lost in overcoming air resistance)

change in potential energy = kinetic energy of boulder as it lands

∴ $\Delta E_p = E_k = 800\,\text{J}$

c If Δh is the height of the cliff

$$\Delta E_p = mg\Delta h$$

∴ $\Delta h = \dfrac{\Delta E_p}{mg} = \dfrac{800\,\text{J}}{4\,\text{kg} \times 10\,\text{m/s}^2} = \dfrac{800\,\text{N}\,\text{m}}{40\,\text{kg}\,\text{m/s}^2} = 20\,\text{m}$

Now put this into practice

1 A stone of mass 2 kg rolls off the flat roof of a building and reaches the ground with a speed of 10 m/s. Find
   a the kinetic energy of the stone when it reaches the ground
   b the gravitational potential energy of the stone when it was on the roof
   c the height of the roof. Neglect air resistance.
2 A football of mass 0.4 kg rolls off a 30 m high cliff. Calculate the speed of the football when it lands on the beach (neglecting air resistance).

Going further

Elastic and inelastic collisions

In all collisions (where no external force acts) some kinetic energy is usually transferred to thermal energy and, to a small extent, to sound. The greater the proportion of kinetic energy transferred, the less *elastic* is the collision, i.e. the more inelastic it is. In a perfectly elastic collision, kinetic energy is conserved.

▲ **Figure 1.7.10** Newton's cradle is an instructive toy for studying collisions and conservation of energy.

Driving and car safety

Braking distance and speed

For a car moving with speed v, the brakes must be applied over a braking distance s to bring the car to rest. The *braking distance is directly proportional to the square of the speed*, i.e. if v is doubled, s is quadrupled. The thinking distance (i.e. the distance travelled while the driver is reacting before applying the brakes) has to be added to the braking distance to obtain the overall stopping distance, in other words

stopping distance = thinking distance + braking distance

Typical values taken from the Highway Code are given in Table 1.7.1 for different speeds. The greater the speed, the greater the stopping distance for a given braking force. (To stop the car in a given distance, a greater braking force is needed for higher speeds.)

▼ **Table 1.7.1**

Speed/km/h	30	60	90	120
Thinking distance/metres	6	12	18	24
Braking distance/metres	6	24	54	96
Total stopping distance/metres	12	36	72	120

Thinking distance depends on the driver's reaction time – this will vary with factors such as the driver's degree of tiredness, use of alcohol or drugs, eyesight and the visibility of the hazard. Braking distance varies with both the road conditions and the state of the car;

it is longer when the road is wet or icy, when friction between the tyres and the road is low, than when conditions are dry. Efficient brakes and deep tyre tread help to reduce the braking distance.

Car design and safety

When a car stops rapidly in a collision, large forces are produced on the car and its passengers, and their kinetic energy has to be dissipated.

Crumple zones at the front and rear collapse in such a way that the kinetic energy is absorbed gradually (Figure 1.7.11). As we saw in Topic 1.6, this extends the collision time and reduces the decelerating force and hence the potential for injury to the passengers.

▲ **Figure 1.7.11** Cars in an impact test showing the collapse of the front crumple zone

Extensible seat belts exert a backwards force (of 10000N or so) over about 0.5m, which is roughly the distance between the front seat occupants and the windscreen. In a car travelling at 15m/s (54km/h), the effect felt by anyone *not* using a seat belt is the same as that, for example, produced by jumping off a building 12m high.

Air bags in most cars inflate and protect the driver from injury by the steering wheel.

Head restraints ensure that if the car is hit from behind, the head goes forwards with the body and not backwards over the top of the seat. This prevents damage to the top of the spine.

All these are secondary safety devices which aid *survival* in the event of an accident. Primary safety factors help to *prevent* accidents and depend on the car's roadholding, brakes, steering, handling and above all on the driver since most accidents are due to driver error.

The chance of being killed in an accident is about *five times less* if seat belts are worn and head restraints are installed.

> **Test yourself**
>
> 1 Name the way by which energy is transferred in the following processes
>   a a battery is used to light a lamp
>   b a ball is thrown upwards
>   c water is heated in a boiler.
> 2 State how energy is stored in the following
>   a fossil fuels
>   b hot water
>   c a rotating turbine
>   d a stretched spring.
>
> 3 Calculate the kinetic energy of a
>   a 1 kg trolley travelling at 2 m/s
>   b 2 g (0.002 kg) bullet travelling at 400 m/s
>   c 500 kg car travelling at 72 km/h.
> 4 a What is the velocity of an object of mass 1 kg which has 200 J of kinetic energy?
>   b Calculate the potential energy of a 5 kg mass when it is **i** 3 m and **ii** 6 m, above the ground. ($g = 9.8$ N/kg)
> 5 It is estimated that 7×10^6 kg of water pours over the Niagara Falls every second.
>   If the falls are 50 m high, and if all the energy of the falling water could be harnessed, what power would be available? ($g = 9.8$ N/kg)

1.7.2 Work

FOCUS POINTS

★ Understand that when mechanical or electrical work is done, energy is transferred.
★ Use the correct equation to calculate mechanical work.

In science the word work has a different meaning from its everyday use. Here work is associated with the motion of a force. When you lift and move a heavy box upstairs you will have done work in either sense! In the absence of heat being generated, the work done is a measure of the amount of energy transferred. When moving the heavy box, chemical energy from your muscles is transferred to gravitational potential energy. If an electric motor is used to move the box, an equal amount of electrical work will be done. In this topic you will learn how to calculate the mechanical work done in different situations.

Work

In an energy transfer, work is done. *The work done is a measure of the amount of energy transferred.* The same amount of mechanical or electrical work is done in transferring equal amounts of energy.

Mechanical **work** is done when a force moves. No work is done in the scientific sense by someone standing still holding a heavy pile of books: an upward force is exerted, but no motion results.

If a building worker carries ten bricks up to the first floor of a building, they do more work than if they carry only one brick because they have to exert a larger force. Even more work is required if they carry the ten bricks to the second floor.

The amount of work done W depends on the size of the force F applied and the distance d it moves. We therefore measure work by

work = force × distance moved in direction of force

or $W = Fd = \Delta E$

where ΔE is the energy transferred.

The unit of work is the **joule** (J); it is the work done when a force of 1 newton (N) moves through 1 metre (m). For example, if you have to pull with a force of 50 N to move a crate steadily 3 m in the direction of the force (Figure 1.7.12a), the work done is 50 N × 3 m = 150 N m = 150 J. That is

joules = newtons × metres

If you lift a mass of 3 kg vertically through 2 m (Figure 1.7.12b), you have to exert a vertically upward force equal to the weight of the body, i.e. 30 N (approximately), and the work done is 30 N × 2 m = 60 N m = 60 J.

Note that you must always take the distance in the direction in which the force acts.

▲ Figure 1.7.12b

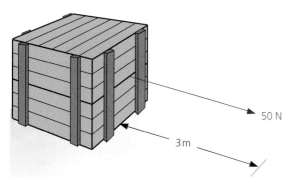

50 N

3 m

▲ Figure 1.7.12a

> ### Test yourself
>
> 6 How much work is done when a mass of 3 kg (g = 9.8 N/kg) is lifted vertically through 6 m?
> 7 A hiker climbs a hill 300 m high. If she has a mass of 51 kg, calculate the work she does in lifting her body to the top of the hill.
> 8 An electric motor does 80 J of work in lifting a box vertically upwards through 5 m. Calculate the weight of the box.

1.7.3 Energy resources

FOCUS POINTS

★ Understand the various ways that useful energy may be obtained or electrical power generated and give advantages and disadvantages of each method.
★ Understand efficiency of energy transfer.

★ Know that energy is released by nuclear fusion in the Sun and that research is being carried out into how energy released from nuclear fusion could produce large-scale electrical energy.
★ Define efficiency and use the correct equations to calculate it.

Energy is needed to heat buildings, to make cars move, to provide artificial light, to make computers work, and so on. The list is endless. This useful energy needs to be produced in controllable energy transfers. For example, in power stations a supply of useful energy is transferred by electric currents to different energy stores required by electricity customers. The raw materials for energy production are energy sources. These may be non-renewable or renewable.

In this topic you will learn that, apart from nuclear, geothermal, hydroelectric or tidal energy, energy released by nuclear fusion in the Sun (Topic 5.1) is the source for all our energy resources. Although energy cannot be destroyed, as you learnt in the previous section, it can be transferred into non-useful stores, such as internal energy. The efficiency of a device measures the useful energy as a percentage of the total energy supplied.

You will be able to recognise many different types of **energy sources**. Such sources may be **renewable** or **non-renewable**; non-renewable sources represent previously stored energy. Much of the energy used in everyday life is ultimately derived from the release of energy in the Sun by nuclear fusion. Sunlight is used in biological processes to store chemical energy and can be harnessed to generate electricity directly in solar cells.

Non-renewable energy sources

Once used up these cannot be replaced.
Two advantages of all non-renewable fuels are
(i) their high **energy density** (i.e. they are concentrated sources) and the relatively small size of the energy transfer device (e.g. a furnace) which releases their energy, and
(ii) their ready **availability** when energy demand increases suddenly or fluctuates seasonally.

Fossil fuels

Fossil fuels include coal, oil and natural gas, formed from the remains of plants and animals which lived millions of years ago and obtained energy originally from the Sun. Their energy is stored as chemical energy and at present they are our main energy source. Predictions vary as to how long they will last since this depends on what reserves are recoverable and on the future demands of a world population expected to increase from about 7700 million in 2019 to about 9700 million by the year 2050. Some estimates say oil and gas will run low early in the present century but coal should last for 200 years or so.

Burning fossil fuels in power stations and in cars pollutes the atmosphere with harmful gases such as carbon dioxide and sulfur dioxide. Carbon dioxide emission aggravates the greenhouse effect (Topic 2.3) and increases global warming. It is not immediately feasible to prevent large amounts of carbon dioxide entering the atmosphere, but less is produced by burning natural gas than by burning oil or coal; burning coal produces most carbon dioxide for each unit of energy produced. When coal and oil are burnt they also produce sulfur dioxide which causes acid rain. The sulfur dioxide can be extracted from the waste gases so it does not enter the atmosphere or the sulfur can be removed from the fuel before combustion, but these are both costly processes which increase the price of electricity produced using these measures.

Nuclear fuels

The energy released in a nuclear reactor (Topic 5.1) from the fission of uranium, found as an ore in the ground, can be used to produce electricity. **Nuclear fuels** do not pollute the atmosphere with carbon dioxide or sulfur dioxide but they do generate **radioactive** waste materials with very long half-lives (Topic 5.2); safe ways of storing this waste for perhaps thousands of years must be found. As long as a reactor is operating normally it does not pose a radiation risk, but if an accident occurs, dangerous radioactive material can leak from the reactor and spread over a large area.

Renewable energy sources

These cannot be exhausted and are generally non-polluting.

Solar energy

The energy falling on the Earth from the Sun is transferred mostly by visible light and infrared radiation and in an hour equals the total energy used by the world in a year. Unfortunately, its low energy density requires large collecting devices and its availability varies. The greatest potential use of **solar energy** is as an energy source for low-temperature water heating. The energy transferred by electromagnetic waves from the Sun is stored as internal energy in **solar panels** and can be transferred by heating to produce domestic hot water at about 70°C and to heat swimming pools.

Solar energy can also be used to produce high-temperature heating, up to 3000°C or so, if a large curved mirror (a solar furnace) focuses the Sun's rays onto a small area. The energy can then be used to turn water to steam for driving the turbine of an electric generator in a power station.

Solar cells, made from semiconducting materials, convert sunlight into electricity directly. A number of cells connected together can be used to supply electricity to homes (Figure 1.7.13) and to the electronic equipment in communication and other satellites. They are also used for small-scale power generation in remote areas where there is no electricity supply. The energy generated by solar cells can be stored in batteries for later use. Recent developments have made large-scale generation more cost effective and large solar power plants are becoming more common. There are many designs for prototype light vehicles run on solar power (Figure 1.7.14).

▲ **Figure 1.7.13** Solar cells on a house provide electricity.

▲ **Figure 1.7.14** Solar-powered car

Wind energy

Infrared radiation from the Sun is also responsible for generating wind energy. Giant windmills called **wind turbines** with two or three blades each up to 30 m long drive electrical generators. Wind farms of 20 to 100 turbines spaced about 400 m apart (Figure 1.7.15) supply about 400 MW (enough electricity for 250 000 homes) in the UK and provide a useful 'top-up' to the National Grid.

Wind turbines can be noisy and are considered unsightly by some people so there is some environmental objection to wind farms, especially as the best sites are often in coastal or upland areas of great natural beauty.

▲ **Figure 1.7.15** Wind farm turbines

Wave energy

The rise and fall of sea waves have to be transferred by some kind of **wave energy** converter into the rotary motion required to drive a generator. It is a difficult problem and the large-scale production of electricity by this means is unlikely in the near future. However, small systems are being developed to supply island communities with power.

Tidal and hydroelectric energy

The flow of water from a higher to a lower level from behind a **tidal barrage** (barrier) or a hydroelectric dam (**tidal energy**) is used to drive a water turbine (water wheel) connected to a generator.

One of the largest working tidal schemes is the La Grande I project in Canada (Figure 1.7.16). Such schemes have significant implications for the environment, as they may destroy wildlife habitats of wading birds for example, and also for shipping routes.

Over 100 years ago, India was one of the first countries to develop hydroelectric power; today such power provides about 14% of the country's electricity supply. China is the world's largest producer of hydroelectricity, generating around 20% of the country's needs. With good management, **hydroelectric energy** is a reliable energy source, but there are risks connected with the construction of dams, and a variety of problems may result from the impact of a dam on the environment. Land previously used for forestry or farming may have to be flooded.

▲ **Figure 1.7.16** Tidal barrage in Canada

▲ **Figure 1.7.17** Filling up with biofuel in Brazil

▲ **Figure 1.7.18** Methane generator in India

Geothermal energy

If cold water is pumped down a shaft into hot rocks below the Earth's surface, it may be forced up another shaft as steam. This can be used to drive a turbine and generate electricity or to heat buildings. The **geothermal energy** that heats the rocks is constantly being released by radioactive elements deep in the Earth as they decay (Topic 5.2).

Geothermal power stations are in operation in the USA, New Zealand and Iceland. A disadvantage is that they can only be built in very specific locations where the underlying rocks are hot enough for the process to be viable.

Biofuels (vegetable fuels)

Biomass includes cultivated crops (e.g. oilseed rape), crop residues (e.g. cereal straw), natural vegetation (e.g. gorse), trees grown for their wood (e.g. spruce), animal dung and sewage. Chemical energy can be stored in **biofuels** such as alcohol (ethanol) and methane gas can be obtained from them by fermentation using enzymes or by decomposition by bacterial action in the absence of air. Liquid biofuels can replace petrol (Figure 1.7.17); although they have up to 50% less energy per litre, they are lead- and sulfur-free and so do not pollute the atmosphere with lead or sulfur dioxide when they are burned. **Biogas** is a mix of methane and carbon dioxide with an energy content about two-thirds that of natural gas. It is produced from animal and human waste in digesters (Figure 1.7.18) and used for heating and cooking. Biogas is cheap to produce on a small scale but not economically viable for large-scale production. It reduces landfills but due to its methane content it is unstable and may explode.

The Sun as an energy source

The Sun is the main source of energy for many of the energy sources described above. The exceptions are geothermal, nuclear and tidal sources. Fossil fuels such as oil, coal and gas are derived from plants which grew millions of years ago in biological processes requiring light from the Sun. Sunlight is also needed by the plants used in biomass energy production today. Energy from the Sun drives the weather systems which enable wind and hydroelectric power to be harnessed. Solar energy is used directly in solar cells for electricity generation.

The source of the Sun's energy is nuclear fusion in the Sun. You will learn more about the fusion process which produces large amounts of energy in Topic 5.1. At present it is not possible to reproduce the fusion process on Earth for the large-scale production of electricity but much research is being directed towards that goal.

Power stations

The processes involved in the production of electricity at power stations depend on the energy source being used.

Non-renewable sources

Fossil fuels and nuclear fuels are used in **thermal power stations** to provide thermal energy that turns water into steam. The steam drives turbines which in turn drive the generators that produce electricity as described in Topic 4.5. If fossil fuels are the energy source (usually coal but natural gas is favoured in new stations), the steam is obtained from a boiler. If nuclear fuel is used, such as uranium or plutonium, the steam is produced in a heat exchanger as explained in Topic 5.1.

The action of a **steam turbine** resembles that of a water wheel but moving steam, not moving water, causes the motion. Steam enters the turbine and is directed by the **stator** or diaphragm (sets of fixed blades) onto the **rotor** (sets of blades on a shaft that can rotate) (Figure 1.7.19). The rotor revolves and drives the electrical generator. The steam expands as it passes through the turbine and the size of the blades increases along the turbine to allow for this.

▲ **Figure 1.7.19** The rotor of a steam turbine

The overall efficiency of thermal power stations is only about 30%. They require cooling towers to condense steam from the turbine to water and this is a waste of energy.

A Sankey diagram (Figure 1.7.5) shows the energy transfers that occur in a thermal power station.

In gas-fired power stations, natural gas is burnt in a **gas turbine** linked directly to an electricity generator. The hot exhaust gases from the turbine are not released into the atmosphere but used to produce steam in a boiler. The steam is then used to generate more electricity from a steam turbine driving another generator. The efficiency is claimed to be over 50% without any extra fuel consumption. Furthermore, the gas turbines have a near 100% combustion efficiency, so very little harmful exhaust gas (i.e. unburnt methane) is produced, and natural gas is almost sulfur-free, so the environmental pollution caused is much less than for coal.

Renewable sources

In most cases the renewable energy source is used to drive turbines directly, as explained earlier in the cases of hydroelectric, wind, wave, tidal and geothermal schemes.

The efficiency of a large installation can be as high as 85–90% since many of the causes of loss in thermal power stations (e.g. water-cooling towers) are absent. In some cases, the generating costs are half those of thermal stations.

A feature of some hydroelectric stations is **pumped storage**. Electricity cannot be stored on a large scale but must be used as it is generated. The demand varies with the time of day and the season (Figure 1.7.20), so in a pumped-storage system electricity generated at off-peak periods is used to pump water back up from a low-level reservoir to a higher-level one. It is easier to do this than to reduce the output of the generator. At peak times the potential energy of the water in the high-level reservoir is converted back into electricity; three-quarters of the electricity that was used to pump the water is generated.

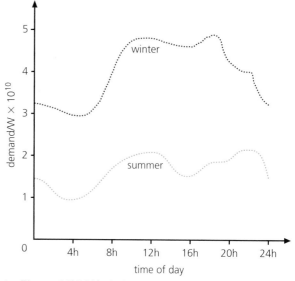

▲ **Figure 1.7.20** Variation in power demand

Economic, environmental and social issues

When considering the large-scale generation of electricity, the economic and environmental costs of using various energy sources have to be weighed against the benefits that electricity brings to society as a clean, convenient and fairly cheap energy supply.

Environmental problems such as polluting emissions that arise with different energy sources were outlined when each was discussed previously. Apart from people using less energy, how far pollution can be reduced by, for example, installing desulfurisation processes in coal-fired power stations, is often a matter of cost.

Although there are no fuel costs associated with electricity generation from renewable energy sources such as wind power, the energy is so dilute that the capital costs of setting up the generating installation are high. Similarly, although fuel costs for nuclear power stations are relatively low, the costs of building the stations and of dismantling them at the end of their useful lives is higher than for gas- or coal-fired stations.

It has been estimated that currently it costs between 9 USc and 22 USc to produce a unit of electricity in a gas- or coal-fired power station in the UK. Wind energy costs vary, depending upon location, but are in the range 7 USc to 16 USc per unit. In the most favourable locations wind competes with coal and gas generation. The cost for a nuclear power station is in excess of 10 USc per unit. After the Tohoku earthquake and tsunami disaster which led to the damage and closure of the Fukushima nuclear reactor in Japan, several countries have reduced their dependence on nuclear energy and Germany plans to phase out nuclear power completely by 2022.

The reliability of a source has also to be considered, as well as how easily production can be started up and shut down as demand for electricity varies. Natural gas power stations have a short start-up time, while coal and then oil power stations take successively longer to start up; nuclear power stations take longest. They are all reliable in that they can produce electricity at any time of day and in any season of the year as long as fuel is available. Hydroelectric power stations are also very reliable and have a very short start-up time, which means they can be switched on when the demand for electricity peaks. The electricity output of a tidal power station, although predictable, is not as reliable because it depends on the height of the tide which varies over daily, monthly and seasonal time scales. The wind and the Sun are even less reliable sources of energy since the output of a wind turbine changes with the strength of the wind and that of a solar cell with the intensity of light falling on it; the output may not be able to match the demand for electricity at a particular time.

Renewable sources are still only being used on a small scale globally. The contribution of the main energy sources to the world's total energy consumption at present is given in Table 1.7.2. (The use of biofuels is not well documented.) The great dependence on fossil fuels worldwide is evident. It is clear the world has an energy problem and new solutions to energy production need to be found.

▼ **Table 1.7.2** World use of energy sources

Oil	Coal	Gas	Nuclear	Hydroelectric
34%	27%	24%	4%	7%

Consumption varies from one country to another; North America and Europe are responsible for about 42% of the world's energy consumption each year. Table 1.7.3 shows approximate values for the annual consumption per head of population for different areas. These figures include the hidden consumption in the manufacturing and transporting of goods. The world average consumption is 76×10^9 J per head per year.

▼ **Table 1.7.3** Energy consumption per head per year/$J \times 10^9$

N. America	UK	Japan	S. America	China	Africa	India	Malaysia
240	121	150	56	97	15	25	130

Efficiency of energy transfers

1   The **efficiency** of a device is the percentage of the energy supplied to it that is usefully transferred.

2   Efficiency is calculated from the expression

$$\text{efficiency} = \frac{\text{useful energy output}}{\text{total energy input}} \times 100\%$$

For example, for the lever shown in Figure 1.5.22 (p. 45)

$$\text{efficiency} = \frac{\text{work done on load}}{\text{work done by effort}} \times 100\%$$

This will be less than 100% if there is friction in the fulcrum.

> **Key definition**
>
> **Efficiency** (%) $\text{efficiency} = \dfrac{(\text{useful energy output})}{(\text{total energy input})} \; (\times 100\%)$

Table 1.7.4 lists the efficiencies of some devices.

▼ **Table 1.7.4**

Device	% efficiency
large electric motor	90
large electric generator	90
domestic gas boiler	75
compact **fluorescent** lamp	50
steam turbine	45
car engine	25
filament lamp	10

A device is efficient if it transfers energy mainly to useful stores and the lost energy is small.

The efficiency of a device can also be defined in terms of power output and input

$$\text{efficiency} = \frac{(\text{useful power output})}{(\text{total power input})} \times 100\%$$

> **Key definition**
>
> **Efficiency** (%) $\text{efficiency} = \dfrac{(\text{useful power output})}{(\text{total power input})} \; (\times 100\%)$

❓ Worked example

a   The energy input to an electric motor is 400 J when raising a load of 200 N through a vertical distance of 1.5 m. Calculate the efficiency of the motor.

$$\text{work done on load} = \text{force} \times \text{distance}$$
$$= 200\,\text{N} \times 1.5\,\text{m}$$
$$= 300\,\text{N m} = 300\,\text{J}$$

useful energy output 300 J

total energy input = 400 J

$$\text{efficiency} = \frac{\text{useful energy output}}{\text{total energy input}} \times 100\%$$
$$= \frac{300\,\text{J}}{400\,\text{J}} \times 100\% = 75\%$$

b   If the energy input to an electric drill is 300 J/s and it transfers 100 J/s of energy to thermal energy when in use, calculate its efficiency.

power supplied to the drill = 300 J/s

$$\text{useful power output} = (300 - 100)\,\text{J/s}$$
$$= 200\,\text{J/s}$$

$$\text{efficiency} = \frac{(\text{useful power output})}{(\text{total power input})} \times 100\%$$
$$= \frac{200}{300} \times 100\% = 67\%$$

Now put this into practice

1   A robot is used to lift a load. If the energy input to the robot is 8000 J in the time it takes to lift a load of 500 N through 12 m, calculate the efficiency of the robot.

2   If the energy input to an electric motor is 560 J/s and 170 J/s of energy is transferred to thermal energy when in use, calculate its efficiency.

▶ Test yourself

9   List six properties which you think the ideal energy source should have for generating electricity in a power station.

10 a   List six social everyday benefits for which electricity generation is responsible.

   b   Draw up two lists of suggestions for saving energy
      i   in the home, and
      ii   globally.

11   Calculate the efficiency of a compact fluorescent light if a power input of 20 J/s gives an output power of 9 J/s.

1.7.4 Power

FOCUS POINT

★ Define power and use the correct equations to calculate power in terms of the rate at which work is done or energy transferred.

To heat up a frozen dinner in a microwave oven you need to know the power of the oven, if over- or under-cooking is to be avoided. Similarly, one needs to check the power rating of a light bulb before inserting it into a socket to ensure over-heating does not occur. Most electrical appliances have their power rating marked on them, usually at the rear or base of the device. The power of a device is the rate at which it does work and so is equal to the rate at which it transfers energy to different stores.

Power

The more powerful a car is, the faster it can accelerate or climb a hill, i.e. the more rapidly it does work. The **power** of a device is the work it does per second, i.e. the rate at which it does work. This is *the same as the rate at which it transfers energy from one store to another.*

$$\text{power} = \frac{\text{work done}}{\text{time taken}} = \frac{\text{energy transferred}}{\text{time taken}}$$

$$\text{power } P = \frac{W}{t}$$

where W is the work done in time t

$$\text{also } P = \frac{\Delta E}{t}$$

where ΔE is the energy transferred in time t.

> **Key definition**
> **Power** the work done per unit time and the energy transferred per unit time

The unit of power is the **watt** (W) and is *a rate of working of 1 joule per second*, i.e. $1\,\text{W} = 1\,\text{J/s}$. Larger units are the **kilowatt** (kW) and the **megawatt** (MW):

$$1\,\text{kW} = 1000\,\text{W} = 10^3\,\text{W}$$

$$1\,\text{MW} = 1\,000\,000\,\text{W} = 10^6\,\text{W}$$

If a machine does 500 J of work in 10 s, its power is 500 J/10 s = 50 J/s = 50 W. A small car develops a maximum power of about 25 kW.

> **Test yourself**
>
> 12 A boy whose weight is 600 N runs up a flight of stairs 10 m high in 12 s. What is his average power?
> 13 Calculate the power of a lamp that transfers 2400 J to thermal energy in 1 minute.
> 14 An escalator carries 60 people of average mass 70 kg to a height of 5 m in one minute. Find the power needed to do this.

> **Practical work**
>
> Measuring your own power
>
> Safety
> ● You should only volunteer for this if you feel able to. No one should pressure you into taking part.
>
> Get someone with a stopwatch to time you running up a flight of stairs; the more steps the better. Find your weight (in newtons). Calculate the total vertical height (in metres) you have climbed by measuring the height of one step and counting the number of steps.
>
> The work you do (in joules) in lifting your weight to the top of the stairs is (your weight) × (vertical height of stairs). Calculate your power (in watts).
>
> 5 Name the stores between which energy is transferred as you run up the stairs.
> 6 How is energy transferred when you run up the stairs?

Revision checklist

After studying Topic 1.7 you should know and understand the following:

✓ work is done when energy is transferred

> ✓ the Sun is the main source of energy for all our energy resources except geothermal, nuclear and tidal
>
> ✓ energy is released by nuclear fusion in the Sun

✓ the different ways of harnessing solar, wind, wave, tidal, hydroelectric, geothermal and biofuel energy

✓ the difference between renewable and non-renewable energy sources

✓ how nuclear fuel and the chemical energy stored in fuel are used to generate electricity in power stations

✓ the meaning of efficiency in energy transfers and that power is the rate of energy transfer.

After studying Topic 1.7 you should be able to:

✓ recall different stores of energy and describe energy transfers in given examples

✓ recall the principle of conservation of energy and apply it to simple systems including the interpretation of flow diagrams

> ✓ define kinetic energy and perform calculations using $E_{\mathrm{k}} = \dfrac{1}{2}\,mv^2$
>
> ✓ define gravitational potential energy and perform calculations using $\Delta E_{\mathrm{p}} = mgh$
>
> ✓ apply the principle of conservation of energy to complex systems and interpret Sankey diagrams

✓ relate work done to the magnitude of a force and the distance moved, and recall the units of work, energy and power

✓ recall and use the equation $W = F \times d = \Delta E$ to calculate energy transfer

✓ compare and contrast the advantages and disadvantages of using different energy sources to generate electricity

> ✓ define efficiency in relation to the transfer of energy and power.

Exam-style questions

1 State how energy is transferred from
  a a toaster   [2]
  b a refrigerator   [2]
  c an audio system.   [2]
                                [Total: 6]

2 A 100 g steel ball falls from a height of 1.8 m onto a metal plate and rebounds to a height of 1.25 m. Calculate the
  a potential energy of the ball before the fall ($g = 9.8\,m/s^2$)   [2]
  b kinetic energy of the ball as it hits the plate   [1]
  c velocity of the ball on hitting the plate   [3]
  d kinetic energy of the ball as it leaves the plate on the rebound   [2]
  e velocity of rebound.   [3]
                        [Total: 11]

3 A ball of mass 0.5 kg is thrown vertically upwards with a kinetic energy of 100 J. Neglecting air resistance calculate
  a the initial speed of the ball   [3]
  b the potential energy of the ball at its highest point   [1]
  c the maximum height to which the ball rises.   [3]
                        [Total: 7]

4 In loading a lorry a man lifts boxes each of weight 100 N through a height of 1.5 m.
  a Calculate the work done in lifting one box.   [2]
  b Calculate how much energy is transferred when one box is lifted.   [1]
  c If he lifts four boxes per minute, at what power is he working?   [3]
                        [Total: 6]

5 The pie chart in Figure 1.7.21 shows the percentages of the main energy sources used by a certain country.

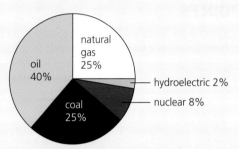

▲ **Figure 1.7.21**

  a Give the percentage supplied by water.   [1]
  b Name any of the sources that are renewable.   [1]
  c Explain what is meant by a renewable source.   [1]
  d Name two other renewable sources.   [2]
  e If energy is always conserved, explain the importance of developing renewable sources.   [2]
                        [Total: 7]

6 a Give
      i two advantages and
      ii two disadvantages
    of using fossil fuels in electricity generating stations.   [4]
  b Give
      i two advantages and
      ii two disadvantages
    of using solar energy in electricity generating stations.   [4]
                        [Total: 8]

7 When the energy input to a gas-fired power station is 1000 MJ, the energy output is 300 MJ.
  a Calculate the efficiency of the power station.   [3]
  b Calculate how much energy is lost and name the energy store to which it is moved.   [2]
  c Describe where the lost energy goes.   [2]
                        [Total: 7]

1.8 Pressure

FOCUS POINTS
★ Define pressure as force per unit area, and illustrate with examples.
★ Describe how pressure varies with depth in a liquid.

★ Calculate the change in pressure beneath the surface of a liquid using the correct equation.

The large flat feet of an Arabian camel prevent it sinking into the soft sand of the desert. This is because the weight of the camel is spread over the area of its four large feet. It appears that the effect of a force depends on the area over which it acts. The effect can be quantified by introducing the concept of pressure. In this topic you will learn that pressure increases as the force increases and the area over which the force acts becomes less. Pressure in a liquid is found to increase with both density and depth. If a deep-sea diver surfaces too quickly, the pressure changes can lead to a condition called the bends. The properties of liquid pressure are utilised in applications ranging from ≈zwater supply systems and dam construction to hydraulic lifts.

Pressure

To make sense of some effects in which a force acts on an object we have to consider not only the force but also the area on which it acts. For example, wearing skis prevents you sinking into soft snow because your weight is spread over a greater area. We say the **pressure** is less.

Pressure is defined as the force per unit area (i.e. $1\,\text{m}^2$) and is calculated from

$$\text{pressure} = \frac{\text{force}}{\text{area}}$$

$$p = \frac{F}{A}$$

> **Key definition**
> **Pressure** the force per unit area

The unit of pressure is the **pascal** (Pa). It equals 1 newton per square metre (N/m^2) and is quite a small pressure. An apple in your hand exerts about 1000 Pa.

The greater the area over which a force acts, the less the pressure. This is why a tractor with wide wheels can move over soft ground. The pressure is large when the area is small and this is why nails are made with sharp points. Walnuts can be broken in the hand by squeezing two together, rather than one alone, because the area of contact is smaller leading to a higher pressure on the shells (Figure 1.8.1).

▲ **Figure 1.8.1** Cracking walnuts by hand

❓ Worked example

Figure 1.8.2 shows the pressure exerted on the floor by the same box standing on end (Figure 1.8.2a) and lying flat (Figure 1.8.2b). If the box has a weight of 24 N, calculate the pressure on the floor when the box is

a   standing on end as in Figure 1.8.2a
b   lying flat as in Figure 1.8.2b.

a   area $= 3\,\text{m} \times 2\,\text{m} = 6\,\text{m}^2$

$$\text{pressure} = \frac{\text{force}}{\text{area}} = \frac{24\,\text{N}}{6\,\text{m}^2} = 4\,\text{Pa}$$

b   area $= 3\,\text{m} \times 4\,\text{m} = 12\,\text{m}^2$

$$\text{pressure} = \frac{\text{force}}{\text{area}} = \frac{24\,\text{N}}{12\,\text{m}^2} = 2\,\text{Pa}$$

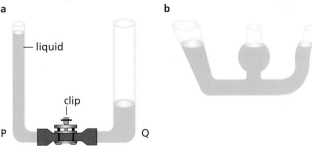

▲ **Figure 1.8.2b**

Now put this into practice

1   A rectangular box has a width of 2 m, a height of 5 m and a depth of 2 m.
   a   Calculate the area of
      i   the base of the box and
      ii   one of the sides of the box.
   b   If the box has a weight of 80 N, calculate the pressure on
      i   the base of the box
      ii   one of the sides of the box.
2   a   Calculate the pressure on a surface when a force of 50 N acts on an area of
      i $2.0\,\text{m}^2$
      ii $100\,\text{m}^2$
      iii $0.50\,\text{m}^2$.
   b   A pressure of 10 Pa acts on an area of $3.0\,\text{m}^2$. What is the force acting on the area?

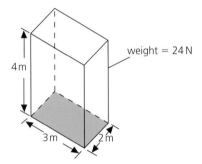

▲ **Figure 1.8.2a**

Liquid pressure

1   *Pressure in a liquid increases with depth*. This is because the further down you go, the greater the weight of liquid above. In Figure 1.8.3a water spurts out fastest and furthest from the lowest hole.
2   *Pressure at one depth acts equally in all directions.* The can of water in Figure 1.8.3b has similar holes all round it at the same level. Water comes out equally fast and spurts equally far from each hole. Hence the pressure exerted by the water at this depth is the same in all directions.

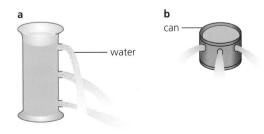

▲   **Figure 1.8.3**

3   *A liquid finds its own level.* In the U-tube of Figure 1.8.4a the liquid pressure at the foot of P is greater than at the foot of Q because the left-hand column is higher than the right-hand one. When the clip is opened, the liquid flows from P to Q until the pressure and the levels are the same, i.e. the liquid 'finds its own level'. Although the weight of liquid in Q is now greater than in P, it acts over a greater area because tube Q is wider.

In Figure 1.8.4b the liquid is at the same level in each tube and confirms that the pressure at the foot of a liquid column depends only on the *vertical* depth of the liquid and not on the tube width or shape.

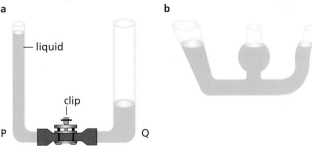

▲ **Figure 1.8.4**

4 *Pressure depends on the density of the liquid.* The denser the liquid, the greater the pressure at any given depth. The densities of some different liquids are listed in Table 1.4.1 in Topic 1.4.

Water supply system

A town's water supply often comes from a reservoir on high ground. Water flows from it through pipes to any tap or storage tank that is below the level of water in the reservoir (Figure 1.8.5). The lower the place supplied, the greater the water pressure. In very tall buildings it may be necessary to pump the water to a large tank in the roof.

Reservoirs for water supply or for hydroelectric power stations are often made in mountainous regions by building a dam at one end of a valley. The dam must be thicker at the bottom than at the top due to the large water pressure at the bottom.

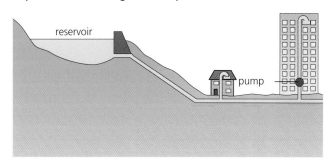

▲ **Figure 1.8.5** Water supply system

Test yourself

1 Why is the pump needed in the high-rise building shown in Figure 1.8.5?
2 Why are dam walls built to be thicker at the bottom than the top?

Hydraulic machines

Liquids are almost incompressible (i.e. their volume cannot be reduced by squeezing) and they 'pass on' any pressure applied to them. Use is made of these facts in hydraulic machines. Figure 1.8.6 shows the principle on which they work.

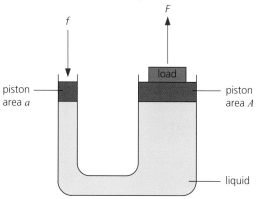

▲ **Figure 1.8.6** The hydraulic principle

Suppose a downward force f acts on a piston of area a. The pressure transmitted through the liquid is

$$\text{pressure} = \frac{\text{force}}{\text{area}} = \frac{f}{a}$$

This pressure acts on a second piston of larger area A, producing an upward force, $F = \text{pressure} \times \text{area}$:

$$F = \frac{f}{a} \times A$$

or

$$F = f \times \frac{A}{a}$$

Since A is larger than a, F must be larger than f and the hydraulic system is a force multiplier; the **multiplying factor** is A/a.

❓ Worked example

A hydraulic jack is used to lift a heavy load. A force of 1 N is applied to a piston of area $0.01\,\text{m}^2$ and pressure is transmitted through the liquid to a second piston of area of $0.5\,\text{m}^2$. Calculate the load which can be lifted.

Taking $f = 1\,\text{N}$, $a = 0.01\,\text{m}^2$ and $A = 0.5\,\text{m}^2$ then

$$F = f \times \frac{A}{a}$$

so

$$F = 1\,\text{N} \times \frac{0.5\,\text{m}^2}{0.01\,\text{m}^2} = 50\,\text{N}$$

A force of 1 N could lift a load of 50 N; the hydraulic system multiplies the force 50 times.

Now put this into practice

1 In a hydraulic jack a force of 20 N is applied to a piston of area $0.1\,\text{m}^2$. Calculate the load which can be lifted by a second piston of area $1.5\,\text{m}^2$.
2 In a hydraulic jack a load of 70 N is required to be lifted on an area of $1.0\,\text{m}^2$. Calculate the force that must be applied to a piston of area $0.1\,\text{m}^2$ to lift the load.
3 Name the property of a liquid on which a hydraulic jack relies.

A **hydraulic jack** (Figure 1.8.7) has a platform on top of piston B and is used in garages to lift cars. Both valves open only to the right and they allow B to be raised a long way when piston A moves up and down repeatedly.

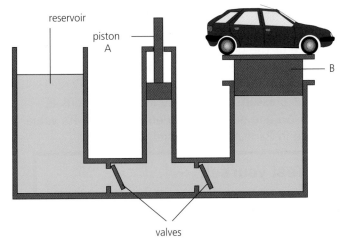

▲ **Figure 1.8.7** A hydraulic jack

Hydraulic fork-lift trucks and similar machines such as loaders (Figure 1.8.8) work in the same way.

▲ **Figure 1.8.8** A hydraulic machine in action

Hydraulic car brakes are shown in Figure 1.8.9. When the brake pedal is pushed, the piston in the master cylinder exerts a force on the brake fluid and the resulting pressure is transmitted equally to eight other pistons (four are shown). These force the brake shoes or pads against the wheels and stop the car.

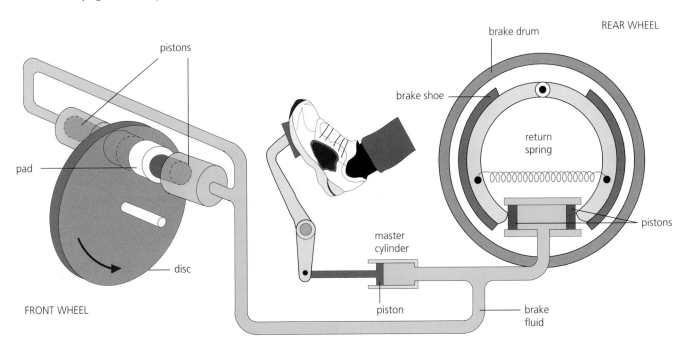

▲ **Figure 1.8.9** Hydraulic car brakes

Expression for liquid pressure

In designing a dam an engineer has to calculate the pressure at various depths below the water surface. The pressure increases with depth and density.

An expression for the change in pressure Δp at a depth Δh below the surface of a liquid of density ρ can be found by considering a horizontal area A (Figure 1.8.10). The force acting vertically downwards on A equals the weight of a liquid column of height Δh and cross-sectional area A above it. Then

volume of liquid column $= \Delta hA$

Since mass = density × volume, then

mass of liquid column $m = \rho \Delta hA$

weight of liquid column $= mg = \rho \Delta hAg$

$\therefore$ force on area $A = \rho \Delta hAg$

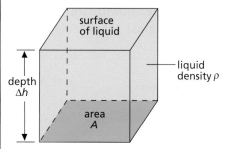

▲ Figure 1.8.10

As

pressure due to liquid column = force/area
$$= g\rho \Delta hA/A$$

we can write

$\Delta p = \rho g \Delta h$

where Δp is the change in pressure beneath the surface of the liquid at depth Δh due to the weight of a liquid of density ρ and g is the gravitational field strength.

This pressure acts equally in all directions at depth Δh and depends only on Δh and ρ. Its value will be in Pa if Δh is in m and ρ in kg/m³.

> ### Test yourself
>
> 3   Calculate the increase in pressure at a depth of 2 m below the surface of water of density 1000 kg/m³.
> 4   Calculate the depth of water of density 1020 kg/m³ where the pressure is 3.0×10^6 Pa.

 Going further

Pressure gauges
These measure the pressure exerted by a fluid, in other words by a liquid or a gas.

Bourdon gauge
This works like the toy shown in Figure 1.8.11, where the harder you blow into the paper tube, the more it uncurls. In a Bourdon gauge (Figure 1.8.12), when a fluid pressure is applied, the curved metal tube tries to straighten out and rotates a pointer over a scale. Car oil-pressure gauges and the gauges on gas cylinders are of this type.

▲ Figure 1.8.11

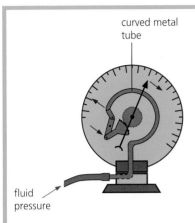

▲ **Figure 1.8.12** A Bourdon gauge

U-tube manometer

In Figure 1.8.13a each surface of the liquid is acted on equally by atmospheric pressure and the levels are the same. If one side is connected to, for example, the gas supply (Figure 1.8.13b), the gas exerts a pressure on surface A and level B rises until

> pressure of gas = atmospheric pressure + pressure due to liquid column BC

The pressure of the liquid column BC therefore equals the amount by which the gas pressure *exceeds* atmospheric pressure. It equals $h\rho g$ (in Pa) where h is the vertical height of BC (in m) and ρ is the density of the liquid (in kg/m³). The height h is called the head of liquid and sometimes, instead of stating a pressure in Pa, we say that it is so many cm of water (or mercury for higher pressures).

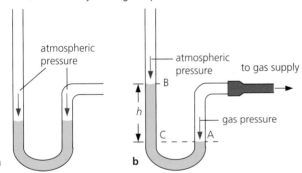

▲ **Figure 1.8.13** A U-tube manometer

Mercury barometer

A barometer is a manometer which measures atmospheric pressure. A simple barometer is shown in Figure 1.8.14. The pressure at X due to the weight of the column of mercury XY equals the atmospheric pressure on the surface of the mercury in the bowl. The height XY measures the atmospheric pressure in mm of mercury (mmHg).

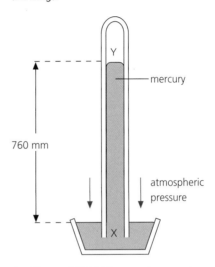

▲ **Figure 1.8.14** Mercury barometer

The *vertical* height of the column is unchanged if the tube is tilted. Would it be different with a wider tube? The space above the mercury in the tube is a vacuum (except for a little mercury vapour).

Revision checklist

After studying Topic 1.8 you should know and understand:

✓ that the pressure beneath a liquid surface increases with depth and density and that pressure is transmitted through a liquid.

After studying Topic 1.8 you should be able to:

✓ define pressure from the equation $p = F/A$ and give everyday examples of its use; recall the units of pressure

✓ calculate the change in pressure below the surface of a liquid.

Exam-style questions

1 The following statements relate to definitions of pressure. In each case write down if the statement is *true* or *false*.

   A Pressure is the force acting on unit area.   [1]
   B Pressure is calculated from force/area.   [1]
   C The SI unit of pressure is the pascal (Pa) which equals 1 newton per square metre ($1\,N/m^2$).   [1]
   D The greater the area over which a force acts, the greater the pressure.   [1]
   E Force = pressure × area.   [1]
   F The SI unit of pressure is the pascal (Pa) which equals 1 newton per metre ($1\,N/m$).   [1]
   [Total: 6]

2 a Calculate the pressure exerted on a wood-block floor by each of the following.
      i   A box weighing 2000 kN standing on an area of $2\,m^2$.   [2]
      ii An elephant weighing 200 kN standing on an area of $0.2\,m^2$.   [2]
      iii A girl of weight 0.5 kN wearing high-heeled shoes standing on an area of $0.0002\,m^2$.   [2]
   b A wood-block floor can withstand a pressure of 2000 kPa ($2000\,kN/m^2$).
      State which of the objects in **a** will damage the floor and explain why.   [2]
      [Total: 8]

3 In a hydraulic press a force of 20 N is applied to a piston of area $0.20\,m^2$. The area of the other piston is $2.0\,m^2$.
   a Calculate the pressure transmitted through the liquid.   [2]
   b Calculate the force on the other piston.   [2]
   c Explain why a liquid and not a gas is used as the 'fluid' in a hydraulic machine.   [1]
   d State another property of a liquid on which hydraulic machines depend.   [1]
      [Total: 6]

4 a The pressure in a liquid varies with depth and density.
      State whether the following statements are *true* or *false*.
      A The pressure in a liquid increases with depth.   [1]
      B The pressure in a liquid increases with density.   [1]
      C The pressure in a liquid is greater vertically than horizontally.   [1]
   b Calculate the increase in pressure at a depth of 100 m below the surface of sea water of density $1150\,kg/m^3$.   [4]
      [Total: 7]

5 a State the equation which relates the change in pressure in a liquid to the depth below the liquid surface.   [2]
   b Name the unit of pressure.   [1]
   c Calculate the depth of water of density $1030\,kg/m^3$ where the pressure is $7.5 \times 10^6\,Pa$.   [3]
      [Total: 6]

Going further

6 Figure 1.8.15 shows a simple barometer.
   a What is the region A?   [1]
   b What keeps the mercury in the tube?   [1]
   c What is the value of the atmospheric pressure being shown by the barometer?   [1]
   d State what would happen to this reading if the barometer were taken up a high mountain? Give a reason.   [2]
      [Total: 5]

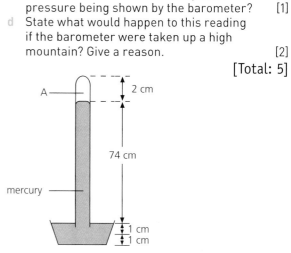

▲ **Figure 1.8.15**

SECTION 2

Thermal physics

Topics

Kinetic particle model of matter

2.1.1 States of matter

FOCUS POINTS

★ Know the properties of solids, liquids and gases.
★ Understand that changes in state can occur and know the terms to describe these changes.

In this topic you will learn about the three states of matter: solids, liquids and gases. The particles in each are ordered differently and this leads to each state having different properties. You will find that solids have a high level of internal order, a liquid has less, and in a gas the particles have no order and move about randomly. The state of a material can be altered by heating or cooling. In a solid the bonds between particles break down on heating and it melts into a liquid; for example, ice melts into water. Boiling a liquid produces a gas with well separated particles; water turns into steam. The three states of matter can be represented in a particle diagram.

Matter is made up of tiny particles (atoms, **molecules**, ions and electrons) which are too small for us to see directly. But they can be 'seen' by scientific 'eyes'. One of these is the electron microscope. Figure 2.1.1 is a photograph taken with such an instrument, showing molecules of a protein. Molecules consist of even smaller particles called **atoms** and these are in continuous motion.

▲ **Figure 2.1.1** Protein molecules

Properties of solids, liquids and gases

Matter can exist in different states and each state has different characteristics.

Solids

Solids have a definite shape and volume and are not easily compressed. The particles in a solid are close together and in fixed positions.

When a force is applied to a solid the atoms move slightly further apart in the direction of the force and stretching occurs (see Topic 1.5.1). When a solid is heated (see Topic 2.2), the distance between atoms increases. If enough heat is supplied to the solid the atoms move even further apart and **melting** into a liquid occurs.

Liquids

Liquids have a definite volume but their shape depends on the container they are kept in. They are more easily compressed or stretched than solids and also expand more when heat is applied. The particles in a liquid are further apart than they are in a solid and have a less ordered structure. They are not fixed in position and can slide over each other when the liquid is poured. The liquid then takes on the shape of the new container.

When a liquid is cooled sufficiently, *solidification* occurs and it returns to the solid state. The density of a material in its solid state is usually higher than it is in its liquid state. When a liquid is heated, particles can escape from its surface by a process called **evaporation**. When sufficient heat is supplied to the liquid **boiling** occurs and the liquid turns into a gas.

Gases

Gases have no definite shape or volume as these depend on the dimensions of the container. The particles in a gas are much further apart than they are in a liquid and the density of a gas is much lower than that of a liquid. The particles have no ordered structure and are able to move about freely in a random manner. Gases are more easily compressed than solids or liquids and expand more when they are heated. When a gas is cooled sufficiently it will return to the liquid state in a process known as **condensation.**

Drops of water are formed when steam condenses on a cold window pane, for example.

> ### Test yourself
> 1 Using what you know about the compressibility (squeezability) of the different states of matter, explain why
>   a air is used to inflate tyres
>   b steel is used to make railway lines.
> 2 Name the processes in which
>   a a solid turns into a liquid
>   b a liquid turns into a gas
>   c a liquid turns into a solid
>   d a gas turns into a liquid.

2.1.2 Particle model

FOCUS POINTS
★ Describe the particle structures of solids, liquids and gases and represent these states using particle diagrams.
★ Understand how temperature affects the movement of particles.

★ Understand the factors that affect the properties of solids, liquids and gases.

★ Understand the relationship between the kinetic energy of particles and temperature, including the concept of absolute zero.
★ Know how a change in pressure in a gas affects the motion and number of collisions of its particles.

★ Describe how a change in pressure of a gas affects the forces exerted by particles colliding with surfaces (force per unit area).

★ Describe Brownian motion and know that it is evidence for the kinetic particle model of matter.

★ Distinguish atoms or molecules from microscopic particles, and understand how these microscopic particles may be moved by collisions with much lighter molecules.

The properties of solids, liquids and gases can be related to the arrangement, separation and motion of the particles in each. In the previous section, you learnt about the properties of solids, liquids and gases. In this topic, you will learn that in a gas, the particles are well separated and in constant random motion, producing pressure on a container by their collisions with its surfaces. In a solid, the particles are closely arranged and firmly bound together, with a regular pattern in crystals. In a liquid the particles are further apart, with only local ordering between particles that have more freedom of movement than those in a solid. Although particles are too small to be seen with the unaided eye, their influence can be detected. When tiny particles in a fluid are observed under a microscope, they can be seen to move slightly in a random manner under the impact of collisions with many much lighter molecules. This effect is known as Brownian motion and provides evidence for the kinetic particle model of matter.

Particle model of matter

As well as being in continuous motion, particles (atoms, molecules, ions and electrons) also exert strong electric forces on one another when they are close together. The forces are both attractive and repulsive. The former hold particles together and the latter cause matter to resist compression.

The *particle model* can explain the existence of the solid, liquid and gaseous states.

Solids

Structure

In solids the particles are close together and the attractive and repulsive forces between neighbouring molecules balance. Also, each particle vibrates about a fixed position. Particles in a solid can be arranged in a regular, repeating pattern like those formed by crystalline substances. Figure 2.1.2a represents the arrangement of particles in a solid.

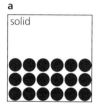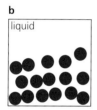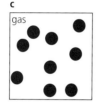

▲ **Figure 2.1.2** Arrangements of particles in a solid, liquid and gas

Properties

We can imagine springs (Figure 2.1.3) representing the electric forces between particles that hold them together and determine the forces and distances between them. These forces enable the solid to keep a definite shape and volume, while still allowing the individual particles to vibrate backwards and forwards. Owing to the strong intermolecular forces, solids resist compression and expand very little when heated.

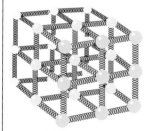

▲ **Figure 2.1.3** The electric forces between particles in a solid can be represented by springs.

Liquids

Structure

In liquids the particles are slightly further apart than in solids but still close enough together to have a definite volume (Figure 2.1.2b). As well as vibrating, they can at the same time move rapidly over short distances, slipping past each other in all directions.

A model to represent the liquid state can be made by covering about a third of a tilted tray with marbles ('particles') (Figure 2.1.4). It is then shaken back and forth and the motion of the marbles observed. The marbles are able to move around but most stay in the lower half of the tray, so the liquid has a fairly definite volume. A few energetic marbles escape from the 'liquid' into the space above. They represent particles that have evaporated from the liquid surface and become gas or vapour particles. The thinning out of the marbles near the liquid surface can also be seen.

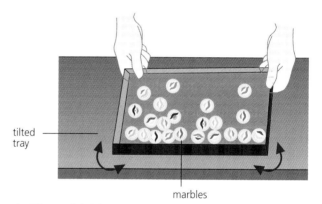

▲ **Figure 2.1.4** A model of particle behaviour in a liquid

Properties

In a liquid the forces between particles are less than in a solid and so the distance between particles is greater. Liquids have a definite volume but individual particles can slide over each other and are never near another particle long enough to get trapped in a regular pattern. This allows liquids to flow and take on the shape of the vessel containing them. The forces between particles are strong enough that liquids are only slightly more easily compressed than solids. When heated, the particles move further apart, enabling liquids to expand more easily than solids. As the temperature increases some particles may have sufficient energy to escape from the surface of the liquid resulting in evaporation of the liquid.

Gases

Structure

The particles in gases are much further apart than in solids or liquids (about ten times; see Figure 2.1.2c) and so gases are much less dense and can be squeezed (compressed) into a smaller space.
The particles dash around at very high speed (about 500 m/s for air molecules at 0°C) in all the space available. It is only during the brief spells when they collide with other particles or with the surfaces of the container that the particle forces act.

A model of a gas is shown in Figure 2.1.5. The faster the vibrator works, the more often the ball-bearings have collisions with the lid, the tube and with each other, representing a gas at a higher temperature. Adding more ball-bearings is like pumping more air into a tyre; it increases the pressure. If a polystyrene ball (1 cm diameter) is dropped into the tube, its irregular motion represents Brownian motion. Brownian motion provides evidence for the kinetic particle model of matter.

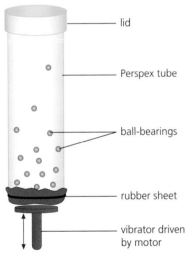

— lid

— Perspex tube

— ball-bearings

— rubber sheet

— vibrator driven by motor

▲ **Figure 2.1.5** A model of particle behaviour in a gas

Properties
Owing to the high speed and the large distance between particles in a gas the interaction between them is small. Gases have no shape and their volume is determined by the volume of the container. They are easily compressed, and expand much more than solids or liquids when heated.

Temperature and kinetic energy

In a solid at room temperature, the particles vibrate about fixed positions. When heat is supplied to the solid and its temperature increases, the particles vibrate more strongly and the average kinetic energy of the particles increases. When the temperature is reduced, the average kinetic energy of the particles reduces, and eventually a temperature is reached where particle motion ceases and the kinetic energy of the particles is zero. We call this temperature **absolute zero** and it occurs at −273°C.

Pressure and kinetic energy

The particle model can explain the behaviour of gases.

Gas pressure

All the particles in a gas are in rapid random motion, with a wide range of speeds, and repeatedly hit and rebound from the surfaces of the container in huge numbers per second.

This causes a pressure on the surfaces of the container. When the temperature of the gas rises, so does the average speed and kinetic energy of the particles. Collisions with the surfaces of the container occur more frequently and so the pressure of the gas increases.

Force and gas pressure
At each collision of a gas particle with a surface of the container, it undergoes a change of momentum which produces a force on the surface (see Topic 1.6). At a constant temperature the average force and hence the pressure exerted on the surface is constant, since pressure is force per unit area. When the temperature rises and the rate at which collisions with the surfaces of the container increases, so does the average force and hence the gas pressure.

Practical work

Brownian motion

For safe experiments/demonstrations related to this topic, please refer to the *Cambridge IGCSE Physics Practical Skills Workbook* that is also part of this series.

a

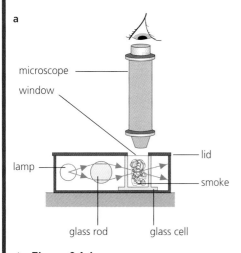

▲ **Figure 2.1.6**

Carefully adjust the microscope until you see bright specks dancing around haphazardly (Figure 2.1.6c). The specks are smoke particles seen by reflected light; their random motion is called **Brownian motion** and is evidence for the kinetic particle model of matter. This motion is due to collisions between the microscopic particles in a suspension and the particles of the gas or liquid.

The apparatus is shown in Figure 2.1.6a. First fill the glass cell with smoke using a match (Figure 2.1.6b). Replace the lid on the apparatus and set it on the microscope platform. Connect the lamp to a 12 V supply; the glass rod acts as a lens and focuses light on the smoke.

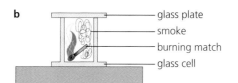

1 What are the specks of light in the glass cell of the Brownian motion experiment?
2 In a glass cell set up to show Brownian motion, describe how the specks of light move.

3 What do you think might cause microscopic particles to move in the way they do in a Brownian motion experiment?

Explanation of Brownian motion

The random motion of the microscopic smoke particles in the cell in Figure 2.1.6 is due to random molecular collisions of fast-moving air molecules in the cell. A smoke particle is massive compared with an air molecule, but if there are more high-speed molecules striking one side of it than the other at a given instant, the particle will move in the direction in which there is a net force. The imbalance, and hence the direction of the net force, changes rapidly in a random manner.

Test yourself

3 Explain the structure of
  a solids
  b liquids
  c gases.
  in terms of the particle model.
4 Explain what is meant by the term absolute zero.

5 Explain how smoke particles can be moved by air molecules in a Brownian motion experiment.

2.1.3 Gases and the absolute scale of temperature

FOCUS POINTS

★ Describe, in terms of particles, the effect of change of temperature or volume on the pressure of a fixed mass of gas.

★ Use the correct equation to calculate pressure and volume of a fixed mass of gas, and be able to represent this relationship graphically.

★ Convert temperatures between the Celsius and Kelvin temperature scales using the correct equation.

Gases show the simplest behaviour of the three states of matter and respond to changes of temperature or volume by a change of pressure. By keeping either volume or temperature constant in an experiment, their relationships with pressure can be determined and explained in terms of the kinetic particle model of matter. The properties of gases can be exploited for use in **thermometers** to measure temperature. You will be familiar with the Celsius scale of temperature for everyday measurements; the freezing temperature of water is set at 0°C and the boiling temperature of water at 100°C. In both the Kelvin and Celsius temperature scales, there are 100 degrees between the freezing temperature and boiling temperature of water, but the Kelvin scale starts from −273°C where the motion of particles ceases.

Pressure of a gas

The air forming the Earth's atmosphere stretches upwards a long way. Air has weight; the air in a normal room weighs about the same as you do, about 500 N. Because of its weight the atmosphere exerts a large pressure at sea level, about $100\,000\,\text{N/m}^2 = 10^5\,\text{Pa}$ (or 100 kPa). This pressure acts equally in all directions.

A gas in a container exerts a pressure on the surfaces of the container. If air is removed from a can by a vacuum pump (Figure 2.1.7), the can collapses because the air pressure outside is greater than that inside. A space from which all the air has been removed is a **vacuum**. Alternatively, the pressure in a container can be increased, for example by pumping more gas into the can; a Bourdon gauge (Topic 1.8) is used for measuring gas pressures.

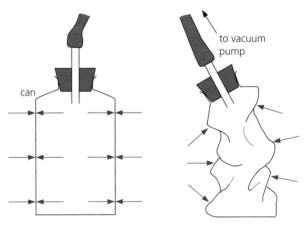

▲ **Figure 2.1.7** Atmospheric pressure collapses the evacuated can.

When a gas is heated, as air is in a jet engine, its pressure as well as its volume may change. To study the effect of temperature on these two quantities we must keep one fixed while the other is changed. When investigating relationships between properties only one **variable** should be changed at a time.

Effect on pressure of a change in temperature (constant volume)

When a gas is heated and its temperature rises, the average speed of its particles increases. If the volume of a fixed mass of gas stays constant, its pressure increases because there are more frequent and more violent collisions of the particles with the surfaces.

Effect on pressure of a change in volume (constant temperature)

If the volume of a fixed mass of gas is halved by halving the volume of the container (Figure 2.1.8),

the number of particles per cm^3 will be doubled. There will be twice as many collisions per second with the surfaces, i.e. the pressure is doubled.

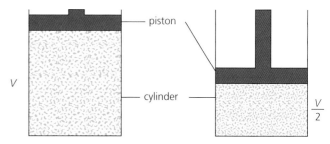

▲ **Figure 2.1.8** Halving the volume doubles the pressure.

Practical work

Effect on pressure of temperature (volume constant)

Safety

- The Bourdon gauge should be firmly clamped to prevent toppling.
- Eye protection must be worn.
- Take care with hot apparatus.

The apparatus is shown in Figure 2.1.9. The rubber tubing from the flask to the pressure gauge should be as short as possible. The flask must be in water almost to the top of its neck and be securely clamped to keep it off the bottom of the can. Heat can be supplied to the water by standing the can on a hot plate or on a tripod over a Bunsen burner.

Record the pressure over a wide range of temperatures, but before taking a reading from the **thermometer**, stop heating, stir and allow time for the gauge reading to become steady; the air in the flask will then be at the temperature of the water. Take about six readings and tabulate the results.

Plot a graph of pressure on the y-axis and temperature on the x-axis.

4 a Name the independent variable in the experiment.
  b Name the dependent variable.
5 Why must the volume be kept constant in the experiment?
6 What precautions should you take to obtain accurate results in the experiment?

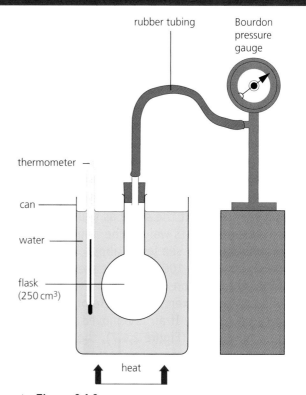

▲ **Figure 2.1.9**

Effect on volume of pressure (temperature constant)

Changes in the volume of a gas due to pressure changes can be studied using the apparatus in Figure 2.1.10. The volume V of air trapped in the glass tube is read off on the scale behind. The pressure is altered by pumping air from a foot pump into the space above the oil reservoir. This forces more oil into the glass tube and increases the pressure p on the air in it; p is measured by the Bourdon gauge. Take about six different measurements. Plot a graph of pressure versus volume as shown in Figure 2.1.11a.

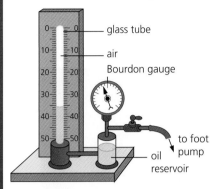

▲ **Figure 2.1.10**

7 a Name the independent variable in the experiment.
 b Name the dependent variable.

8 A graph of pressure against 1/volume for the results of the experiment is shown in Figure 2.1.11b. Name the features of the graph which suggest that pressure is proportional to 1/volume.

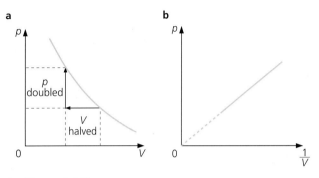

▲ **Figure 2.1.11**

Variations in gas pressure with volume

The variation of the pressure of a fixed mass of gas with changes in volume (at constant temperature) is shown in Figure 2.1.11a. Close examination of the curve shows that if p is doubled, V is halved. That is, p is *inversely proportional to V*. In symbols

$$p \propto \frac{1}{V} \text{ or } p = \text{constant} \times \frac{1}{V}$$

$$\therefore \quad pV = \text{constant}$$

If several pairs of readings, p_1 and V_1, p_2 and V_2, etc. are taken, then it can be confirmed that

$$p_1 V_1 = p_2 V_2 = \text{constant}$$

This is Boyle's law, which is stated as follows:

The pressure of a fixed mass of gas is inversely proportional to its volume if its temperature is kept constant.

Since p is inversely proportional to V, then p is directly proportional to $1/V$. A graph of p against $1/V$ is therefore a straight line through the origin (Figure 2.1.11b).

? Worked example

A certain quantity of gas has a volume of $40\,cm^3$ at a pressure of $1 \times 10^5\,Pa$. Find its volume when the pressure is $2 \times 10^5\,Pa$. Assume the temperature remains constant.

Using the equation $pV = \text{constant}$ we have

$$p_1 V_1 = p_2 V_2$$

Rearranging the equation gives

$$V_2 = p_1 \times V_1 / p_2$$
$$= 1 \times 10^5\,Pa \times 40\,cm^3 / 2 \times 10^5\,Pa$$
$$= 20\,cm^3$$

Now put this into practice

1 A fixed mass of gas has a volume of $9\,cm^3$ at a pressure of $1 \times 10^5\,Pa$. Find its volume when the pressure is $3 \times 10^5\,Pa$.

2 A certain quantity of gas has a volume of $40\,cm^3$ at a pressure of $2 \times 10^5\,Pa$. Find its pressure when the volume is $20\,cm^3$.

Celsius and Kelvin temperature scales

The volume–temperature and pressure–temperature graphs for a gas are straight lines (Figure 2.1.12). They show that gases expand **linearly** with temperature as measured on a mercury thermometer, i.e. equal temperature increases cause equal volume or pressure increases.

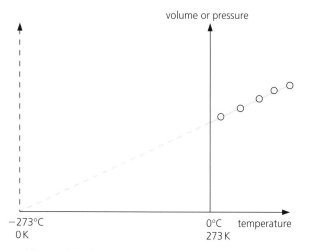

▲ Figure 2.1.12

The graphs do not pass through the Celsius temperature origin (0°C). If graph lines are extrapolated backwards, they cut the temperature axis at about −273°C. This temperature is called absolute zero because we believe it is the lowest temperature possible. It is the zero of the *absolute* or *Kelvin scale of temperature*. At absolute zero molecular motion ceases and a substance has no internal energy.

Degrees on this scale are called **kelvins** and are denoted by K. They are exactly the same size as Celsius degrees. Since −273°C = 0 K, conversions from °C to K are made by adding 273. For example

$$0°C = 273\,K$$

$$15°C = 273 + 15 = 288\,K$$

$$100°C = 273 + 100 = 373\,K$$

Kelvin or absolute temperatures are represented by the letter T, and if θ (Greek letter 'theta') stands for a **degrees Celsius** scale temperature then, in general,

$$T = 273 + \theta$$

Near absolute zero strange things occur. Liquid helium becomes a superfluid. It cannot be kept in an open vessel because it flows up the inside of the vessel, over the edge and down the outside. Some metals and compounds become superconductors of electricity and a **current**, once started in them, flows forever, without a battery. Figure 2.1.13 shows research equipment that is being used to create materials that are superconductors at very much higher temperatures, such as −23°C.

▲ **Figure 2.1.13** This equipment is being used to make films of complex composite materials that are superconducting at temperatures far above absolute zero.

❓ Worked example

a Convert 27°C to K.
   Substitute in the equation $T = 273 + \theta$ to give

   $$T = 273 + 27 = 300\,K$$

b Convert 60 K to °C.
   Rearrange the equation $T = 273 + \theta$ to give

   $$\theta = T - 273 = 60 - 273 = -213°C$$

Now put this into practice

1 Convert 80°C to K.
2 Convert 100 K to °C.

Going further

 Practical work

Effect on volume of temperature (pressure constant): Charles' law

Safety

- Eye protection must be worn.
- Take care as concentrated sulfuric acid is highly corrosive. Do not touch it if any leaks out of the tube.

Arrange the apparatus as in Figure 2.1.14. The index of concentrated sulfuric acid traps the air column to be investigated and also dries it. Adjust the capillary tube so that the bottom of the air column is opposite a convenient mark on the ruler.

Note the length of the air column (between the lower end of the index and the sealed end of the capillary tube) at different temperatures but, before taking a reading, stop heating and stir well to make sure that the air has reached the temperature of the water. Put the results in a table.

Plot a graph of volume (in cm, since the length of the air column is a measure of it) on the *y*-axis and temperature (in °C) on the *x*-axis.

The pressure of (and on) the air column is constant and equals atmospheric pressure plus the pressure of the acid index.

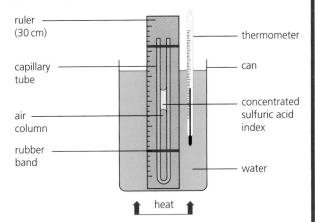

▲ **Figure 2.1.14**

9 a Name the independent variable in the experiment.
 b Name the dependent variable.
10 The results of the experiment are plotted in a graph of volume versus temperature and appear similar to those shown in Figure 2.1.12. What do the results indicate about the relationship between volume and temperature?

The gas laws

Using absolute temperatures, the gas laws can be stated in a convenient form for calculations.

Charles' law

In Figure 2.1.12 the volume–temperature graph passes through the origin if temperatures are measured on the Kelvin scale, that is, if we take 0 K as the origin. We can then say that the volume V is directly proportional to the absolute temperature T, i.e. doubling T doubles V, etc. Therefore

$$V \propto T \text{ or } V = \text{constant} \times T$$

or

$$\frac{V}{T} = \text{constant} \quad (1)$$

Charles' law may be stated as follows:

The volume of a fixed mass of gas is directly proportional to its absolute temperature if the pressure is kept constant.

Pressure law

From Figure 2.1.12 we can say similarly for the pressure p that

$$p \propto T \text{ or } p = \text{constant} \times T$$

or

$$\frac{p}{T} = \text{constant} \quad (2)$$

The pressure law may be stated as follows:

> The pressure of a fixed mass of gas is directly proportional to its absolute temperature if the volume is kept constant.

Variations in gas pressure with volume

For a fixed mass of gas at constant temperature

$$pV = \text{constant} \quad (3)$$

Combining the laws

The three equations can be combined, giving

$$\frac{pV}{T} = \text{constant}$$

For cases in which p, V and T all change from, say, p_1, V_1 and T_1 to p_2, V_2 and T_2, then

$$\frac{p_1 V_1}{T_1} = \frac{p_2 V_2}{T_2} \quad (4)$$

? Worked example

A bicycle pump contains $50 \, \text{cm}^3$ of air at 17°C and at 1.0 atmosphere pressure (atm). Find the pressure when the air is compressed to $10 \, \text{cm}^3$ and its temperature rises to 27°C.

We have

$p_1 = 1.0 \, \text{atm}$ $p_2 = ?$

$V_1 = 50 \, \text{cm}^3$ $V_2 = 10 \, \text{cm}^3$

$T_1 = 273 + 17 = 290 \, \text{K}$ $T_2 = 273 + 27 = 300 \, \text{K}$

From equation (4) we get

$$p_2 = p_1 \times \frac{V_1}{V_2} \times \frac{T_2}{T_1} = 1 \times \frac{50}{10} \times \frac{300}{290} = 5.2 \, \text{atm}$$

Note that: (i) all temperatures must be in K; (ii) any units can be used for p and V provided the same units are used on both sides of the equation; (iii) in some calculations the volume of the gas has to be found at standard temperature and pressure, or 's.t.p.'. This is temperature 0°C and pressure 1 atmosphere (1 atm = 10^5 Pa).

Now put this into practice

1 A fixed mass of gas has a volume of $9 \, \text{cm}^3$ at a pressure of 1×10^5 Pa at 27°C. Find its pressure when the volume is compressed to $5 \, \text{cm}^3$ and its temperature rises to 37°C.

2 A certain quantity of gas has a volume of $40 \, \text{cm}^3$ at a pressure of 2.0×10^5 Pa at 27°C. Find its temperature when the volume is $30 \, \text{cm}^3$ and the pressure is 3.2×10^5 Pa.

Test yourself

6 In terms of particle motion describe the effect on the pressure of a fixed mass of gas when the temperature rises but the volume is kept constant.

7 Describe the effect on the pressure of a fixed mass of gas if the volume is reduced but the temperature of the gas is kept constant.

8 a Why is −273°C chosen as the starting temperature for the Kelvin scale of temperature?

b How do the size of units on the Celsius and Kelvin scales of temperature compare?

Revision checklist

After studying Topic 2.1 you should know and understand:

✔ the different physical properties of solids, liquids and gases

✔ particle diagrams for the different states of matter

✔ the different particle structure of solids, liquids and gases

✔ how the particle model explains the physical properties of solids, liquids and gases.

After studying Topic 2.1 you should be able to:

✔ recall the terms describing changes in state of solids, liquids and gases

✔ explain temperature, absolute zero and change in pressure in terms of molecular motion

✔ describe and explain an experiment to show Brownian motion

✔ describe the effect on the pressure of a fixed mass of gas caused by a change in temperature (at constant volume) and a change of volume (at constant temperature)

✔ convert temperatures between the Celsius and Kelvin scales of temperature

✔ recall and use the equation pV = constant (for a fixed mass of gas at constant temperature).

Exam-style questions

1 Solids, liquids and gases are composed of particles. Which one of the following statements is *not* true?

   A The particles in a solid vibrate about a fixed position.

   B The particles in a liquid are arranged in a regular pattern.

   C The particles in a gas exert negligibly small forces on each other, except during collisions.

   D The densities of most liquids are about 1000 times greater than those of gases because liquid particles are much closer together than gas particles.

   [Total: 1]

2 Sketch particle diagrams for

   a a solid   [2]

   b a liquid   [2]

   c a gas.   [2]

   [Total: 6]

3 **a** Name the state of matter in which the particles are furthest apart.   [1]

   b Use the particle model of matter to explain how a gas exerts pressure on the surfaces of its container.   [2]

   c State and explain how the pressure changes when the temperature of the gas increases.   [4]

   [Total: 7]

4 Smoke particles and air exist in a sealed glass box. The box is illuminated, and the motion of the smoke particles is observed through a microscope.

   a Describe the motion of the smoke particles. [1]

   b Explain the reason the smoke particles move in this way.   [4]

   [Total: 5]

5 **a** The following statements refer to the pressure exerted by a gas in a container. Write down whether each statement is *true* or *false*.

   i Pressure is due to the particles of the gas bombarding the surfaces of the container.   [1]

   ii The pressure decreases if the gas is cooled at constant volume.   [1]

   iii The pressure increases if the volume of the container increases at constant temperature.   [1]

   b i Explain the significance of a temperature of −273°C in terms of particle motion.   [2]

   ii State the value of a temperature of −273°C on the Kelvin temperature scale.   [1]

   iii Calculate the value of a temperature of −200°C on the Kelvin scale of temperature.   [1]

   [Total: 7]

6 The piston in Figure 2.1.15 is pulled out of the cylinder from position X to position Y, without changing the temperature of the air enclosed. If the original pressure in the cylinder was 1.0×10^5 Pa, calculate

   a the air pressure when the piston is at position Y   [3]

   b the air pressure when the piston is moved a further 10 cm to the left of position Y.   [3]

   [Total: 6]

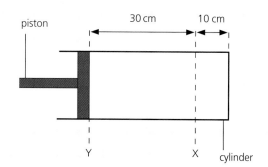

▲   **Figure 2.1.15**

7 A certain quantity of gas has a volume of 30 cm^3 at a pressure of 1×10^5 Pa. Assuming the temperature remains constant, calculate the volume of the gas when the pressure is

   a 2×10^5 Pa   [3]

   b 5×10^5 Pa.   [3]

   [Total: 6]

Alternative to Practical

8 The variation in pressure of a fixed mass of gas is measured for different volumes.
The results obtained are listed in the following table:

Pressure/10^5 Pa	Volume/cm^3	1/volume/cm^3
24	1.0	
12	2.0	
8	3.0	
6	4.0	
4	6.0	

a Plot a graph of pressure against volume. [3]
b Work out values for 1/volume and enter them into the table. [1]
c Plot a graph of pressure against 1/volume. [3]
d Are the results in agreement with the equation pV = constant? [2]

[Total: 9]

2.2 Thermal properties and temperature

2.2.1 Thermal expansion of solids, liquids and gases

FOCUS POINTS

★ Describe thermal expansion in solids, liquids and gases and know some everyday applications of thermal expansion.

★ Use the motion and arrangement of particles in solids, liquids and gases to explain the relative order of magnitudes of their expansion as temperature increases.

As thermal energy is transferred to a material, the particles tend to move further apart. As you saw in Topic 2.1, the effect on heating a gas is large because the particles are free to move and expansion can easily occur. Expansion is much smaller in solids but thermal effects in a solid can still be important in conditions where there are wide temperature variations. Special features to absorb expansion need to be included in railway tracks and engineered structures such as bridges so that they do not distort on very hot days. In this topic you will encounter some everyday applications and consequences of expansion in solids and liquids.

When solids, liquids and gases are heated, the magnitude of the expansion for a given temperature rise is less for a liquid than a gas and even less for a solid where the particles are close together and the force of attraction between them is high.

Thermal expansion

According to the kinetic particle model (Topic 2.1.2) the particles of solids and liquids are in constant vibration. When heated they vibrate faster, so force each other a little further apart and **expansion** results.

The particles in a gas are free to move about rapidly and fill the entire volume of the container. When a gas is heated and its temperature rises, the average speed of its particles increases and there are more frequent collisions with the surfaces of the container. If the pressure of the gas is to remain constant, the volume of the container must increase so that the frequency of collisions does not increase; that means expansion of the gas must occur.

Relative expansion of solids, liquids and gases

The linear (length) expansion of solids is small and for the effect to be noticed, the solid must be long and/or the temperature change must be large. For a 1 m length of steel it is 0.012 mm for a 1°C rise in temperature.

The particles in a liquid are further apart, less ordered and are more mobile than in a solid so the interaction between the particles is less and expansion is easier for liquids than for solids. Liquids typically expand about five times more than solids for a given temperature rise. In gases, the interactions between particles are few because they are far apart and move about very quickly; this means they are able to expand much more easily than liquids. Typically, gases expand about 20 times more than liquids for a given temperature rise. These figures indicate that gases expand much more readily than liquids, and liquids expand more readily than solids.

Uses of expansion

Axles and gear wheels are major components of clocks on the small scale and wheeled vehicles from cars to trains on the large scale.

In Figure 2.2.1 the axles have been shrunk by cooling in liquid nitrogen at −196°C until the gear wheels can be slipped on to them. On regaining normal temperature, the axles expand to give a very tight fit.

▲ **Figure 2.2.1** 'Shrink-fitting' of axles into gear wheels

In the kitchen, a tight metal lid can be removed from a glass jar by immersing the lid in hot water so that it expands and loosens. The expansion of a liquid or a gas can be used in thermometers to measure temperature (see p. 103). An expanding gas drives the pistons in the engine of a motor car.

Bimetallic strip

If equal lengths of two different metals, such as copper and iron, are riveted together so that they cannot move separately, they form a **bimetallic strip** (Figure 2.2.2a). When heated, copper expands more than iron and to allow this the strip bends with copper on the outside (Figure 2.2.2b). If they had expanded equally, the strip would have stayed straight.

Bimetallic strips have many uses.

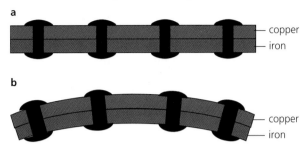

▲ **Figure 2.2.2** A bimetallic strip: **a** before heating; **b** after heating

Fire alarm

Heat from the fire makes the bimetallic strip bend and complete the electrical circuit, so ringing the alarm bell (Figure 2.2.3a).

A bimetallic strip is also used in this way to work the flashing direction indicator lamps in a car, being warmed by an electric heating coil wound round it.

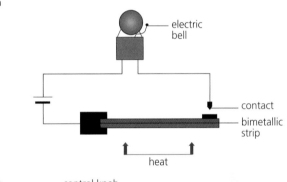

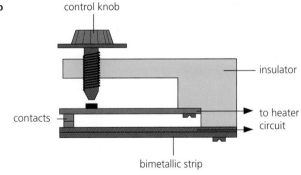

▲ **Figure 2.2.3** Uses of a bimetallic strip: **a** fire alarm; **b** a thermostat in an iron

Thermostat

A **thermostat** keeps the temperature of a room or an appliance constant. The one in Figure 2.2.3b uses a bimetallic strip in the electrical heating circuit of, for example, an iron.

When the iron reaches the required temperature the strip bends down, breaks the circuit at the contacts and switches off the heater. After cooling a little, the strip remakes contact and turns the heater on again. A near-steady temperature results.

If the control knob is screwed down, the strip has to bend more to break the heating circuit and this requires a higher temperature.

Precautions against expansion

In general, when matter is heated it expands and when cooled it contracts. If the changes are resisted large forces are created, which are sometimes useful but at other times are a nuisance.

Gaps used to be left between lengths of railway lines to allow for expansion in summer. They caused a familiar 'clickety-click' sound as the train passed over them. These days rails are welded into lengths of about 1 km and are held by concrete 'sleepers' that can withstand the large forces created without buckling. Also, at the joints the ends are tapered and overlap (Figure 2.2.4a). This gives a smoother journey and allows some expansion near the ends of each length of rail.

For similar reasons slight gaps are left between lengths of aluminium guttering. In central heating pipes 'expansion joints' are used to join lengths of pipe (Figure 2.2.4b); these allow the copper pipes to expand in length inside the joints when carrying very hot water.

▲ **Figure 2.2.4a** Tapered overlap of rails

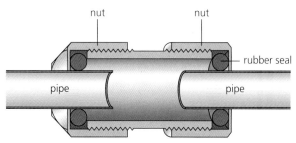

▲ **Figure 2.2.4b** Expansion joint

Test yourself

1 Explain why
   a the metal lid on a glass jam jar can be unscrewed easily if the jar is inverted for a few seconds with the *lid* in very hot water
   b furniture may creak at night after a warm day
   c concrete roads are laid in sections with pitch (a compressible filling) between them.
2 A bimetallic strip is made from aluminium and copper. When heated it bends in the direction shown in Figure 2.2.5.
   a Which metal expands more for the same rise in temperature, aluminium or copper?
   b Draw a diagram to show how the bimetallic strip would appear if it were cooled to below room temperature.

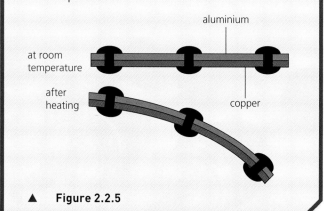

▲ **Figure 2.2.5**

Going further

Linear expansivity

An engineer has to allow for the linear expansion of a bridge when designing it. The expansion can be calculated if all the following are known:
- the length of the bridge,
- the range of temperature it will experience, and
- the linear expansivity of the material to be used.

The linear expansivity of a substance is the increase in length of 1 m for a 1°C rise in temperature.

The linear expansivity of a material is found by experiment. For steel it is 0.000012 per °C. This means that 1 m will become 1.000012 m for a temperature rise of 1°C. A steel bridge 100 m long will expand by 0.000012 × 100 m for each 1°C rise in temperature. If the maximum temperature change expected is 60°C (e.g. from –15°C to +45°C), the expansion will be 0.000012 per °C × 100 m × 60°C = 0.072 m, or 7.2 cm. In general,

$$\text{expansion} = \text{linear expansivity} \times \text{original length} \times \text{temperature rise}$$

Unusual expansion of water

As water is cooled to 4°C it contracts, as we would expect. However, between 4°C and 0°C it expands, surprisingly. Water has a *maximum density* at 4°C (Figure 2.2.6).

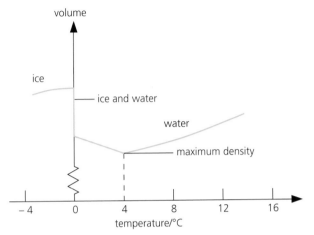

▲ **Figure 2.2.6** Water expands on cooling below 4°C.

At 0°C, when it freezes, a considerable volume expansion occurs and every 100 cm³ of water becomes 109 cm³ of ice. This accounts for the bursting of unlagged water pipes in very cold weather and for the fact that ice is less dense than cold water and so floats. Figure 2.2.7 shows a bottle of frozen milk, the main constituent of which is water.

▲ **Figure 2.2.7** Result of the expansion of water on freezing

The unusual (anomalous) expansion of water between 4°C and 0°C explains why fish survive in a frozen pond. The water at the top of the pond cools first, contracts and being denser sinks to the bottom. Warmer, less dense water rises to the surface to be cooled. When all the water is at 4°C the circulation stops. If the temperature of the surface water falls below 4°C, it becomes less dense and remains at the top, eventually forming a layer of ice at 0°C. Temperatures in the pond are then as in Figure 2.2.8.

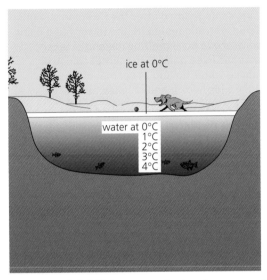

▲ **Figure 2.2.8** Fish can survive in a frozen pond.

The volume expansion of water between 4°C and 0°C is due to the breaking up of the groups that water particles form above 4°C. The new arrangement requires a larger volume and more than cancels out the contraction due to the fall in temperature.

Liquid-in-glass thermometer

The temperature of a body tells us how hot the body is. It is measured using a thermometer usually in degrees Celsius (0°C).

In the liquid-in-glass thermometer the liquid in a glass bulb expands up a capillary tube when the bulb is heated. The liquid must be easily seen and must expand (or contract) rapidly and by a large amount over a wide range of temperature. It must not stick to the inside of the tube or the reading will be too high when the temperature is falling.

Mercury and coloured alcohol are commonly used liquids in this type of thermometer. Mercury freezes at −39°C and boils at 357°C but is a toxic material. A non-toxic metal alloy substitute for mercury, such as Galinstan, is often used nowadays; it melts at −19°C and boils at 1300°C. Alcohol freezes at −115°C and boils at 78°C and is therefore more suitable for low temperatures.

Going further

Scale of temperature

A scale and unit of temperature are obtained by choosing two temperatures, called the fixed points, and dividing the range between them into a number of equal divisions or degrees.

On the Celsius scale (named after the Swedish scientist Anders Celsius who suggested it), the lower fixed point is the temperature of pure melting ice and is taken as 0°C. The upper fixed point is the temperature of the steam above water boiling at normal atmospheric pressure, 10^5 Pa (or N/m^2), and is taken as 100°C.

When the fixed points have been marked on the thermometer, the distance between them is divided into 100 equal degrees (Figure 2.2.9). The thermometer now has a linear scale, in other words it has been *calibrated* or *graduated*.

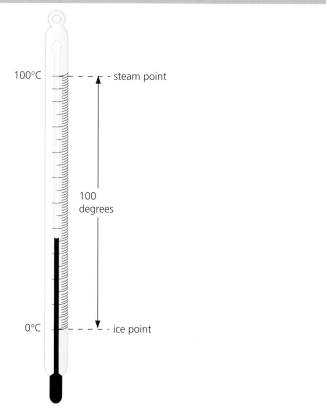

▲ **Figure 2.2.9** A temperature scale in degrees Celsius

Test yourself

3 Explain the relative order of magnitude of the expansion of solids, liquids and gases.

4 a What is meant by the anomalous expansion of water?
  b Name two consequences of the unusual expansion of water.
5 Discuss the action of a liquid-in-glass thermometer.

2.2.2 Specific heat capacity

FOCUS POINTS

★ Know that an object's internal energy is increased when its temperature rises.

★ Explain a change in an object's temperature in terms of the change in kinetic energy of all its particles.
★ Define specific heat capacity, use the correct equation in calculations and describe experiments to measure it.

Some materials require more heat than others to raise their temperature. As discussed in the previous topic, when the temperature of an object rises, its particles move more rapidly. The increase in the kinetic energy associated with this motion raises the internal energy of the object.

The extent of the increase in kinetic energy of the particles in a material when it is heated depends on the nature of the material and its state, and is measured in terms of specific heat capacity. The specific heat capacity of aluminium is higher than that of copper, so copper is a more energy efficient material to use for saucepans. In this topic you will find out how to measure and calculate the specific heat capacity of some solids and liquids.

The **internal energy** of an object is the energy associated with the motion of its particles.

When an object is heated and its temperature rises, there is an increase in the internal energy of the object.

Internal energy

The kinetic particle theory (Topic 2.1.2) regards temperature as a measure of the average kinetic energy (E_k) of the particles of the body. The greater this is, the faster the particles move and the higher the temperature of the body. Increasing the temperature of an object increases its internal energy because the kinetic energy of all the particles increases.

Thermal energy and temperature

It is important not to confuse the temperature of a body with the thermal energy that can be obtained from it. For example, a red-hot spark from a fire is at a higher temperature than the boiling water in a saucepan. In the boiling water the average kinetic energy of the particles is lower than in the spark; but since there are many more water particles, their total energy is greater, and therefore more thermal energy can be supplied by the water than by the spark.

Thermal energy is transferred from a body at a higher temperature to one at a lower temperature.

This is because the average kinetic energy (and speed) of the particles in the hot body falls as a result of the collisions with particles of the cold body whose average kinetic energy, and therefore temperature, increases. When the average kinetic energy of the particles is the same in both bodies, they are at the same temperature. For example, if the red-hot spark landed in the boiling water, thermal energy would be transferred from it to the water even though much more thermal energy could be obtained from the water.

Specific heat capacity

If 1 kg of water and 1 kg of paraffin are heated in turn for the same time by the same heater, the temperature rise of the paraffin is about *twice* that of the water. Since the heater gives equal amounts of thermal energy to each liquid, it seems that different substances require different amounts of energy to cause the same temperature rise in the same mass, say 1°C in 1 kg.

The amount of heat required to raise the temperature of a particular substance by one degree is measured by its **specific heat capacity** (symbol c).

The specific heat capacity of a substance is defined as the energy required per unit mass per unit temperature increase.

In physics, the word 'specific' means that unit mass is being considered.

In general, for a mass m, receiving energy ΔE which results in a temperature rise $\Delta\theta$, this can be written in equation form as

$$c = \frac{\Delta E}{m\Delta\theta}$$

where c is the specific heat capacity of a material of mass m whose temperature rises by $\Delta\theta$ when its internal energy increases by ΔE.

Internal energy is measured in joules (J) and the unit of specific heat capacity is the joule per kilogram per °C, i.e. J/(kg °C).

The equation for specific heat can be rearranged to give the equation:

$$\Delta E = mc\Delta\theta = \text{mass} \times \text{specific heat capacity} \times \text{temperature rise}$$

> **Key definition**
>
> **Specific heat capacity** the energy required per unit mass per unit temperature increase

? Worked example

If 20 000 J is supplied to a mass of 5 kg and its temperature rises from 15°C to 25°C, calculate the specific heat capacity of the mass.

Using $c = \dfrac{\Delta E}{m\Delta\theta}$

$$c = \frac{20\,000\text{ J}}{(5\text{ kg} \times (25-15)°C)} = \frac{20\,000\text{ J}}{50\text{ kg}°C}$$

$$= \frac{400\text{ J}}{\text{kg}°C}$$

Now put this into practice

1 If 25 000 J of energy is supplied to a mass of 2 kg and its temperature rises from 10°C to 35°C, calculate the specific heat capacity of the mass.

2 How much energy must be supplied to a mass of 3 kg of material of specific heat capacity = 500 J/(kg °C) to raise its temperature by 10°C?

Practical work

Finding specific heat capacities

For safe experiments/demonstrations related to this topic, please refer to the *Cambridge IGCSE Physics Practical Skills Workbook* that is also part of this series.

Safety

- Eye protection must be worn.
- Take care as the pan and water and aluminium block may become hot.

You need to know the power of the 12 V electric immersion heater to be used.

Precaution: Do not use one with a cracked seal.

A 40 W heater transfers 40 joules of energy to thermal energy per second. If the power is not marked on the heater, ask about it.

Water

Weigh out 1 kg of water into a container, such as an aluminium saucepan. Note the temperature of the water, insert the heater (Figure 2.2.10), switch on the 12 V supply and start timing. Stir the water and after 5 minutes switch off but continue stirring and note the *highest* temperature reached.

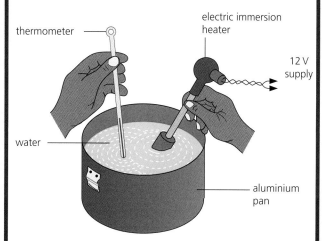

▲ **Figure 2.2.10**

Assuming that the energy supplied by the heater equals the heat received by the water, work out the specific heat capacity of water in J/(kg °C), as shown below:

heat received by water (J) = power of heater (J/s) × time heater on (s)

then the

$$\text{specific heat capacity of water} = \frac{\text{heat received by water (J)}}{\text{mass (kg)} \times \text{temp. rise (°C)}}$$

1 Suggest sources of error in the experiment described above to find the specific heat capacity of water.

2 In an experiment to determine the specific heat capacity of water, the temperature rise of 1 kg of water is found to be 2.5°C when the water is heated by a 40 W heater for 5 minutes. Calculate the specific heat capacity of water.

Aluminium

An aluminium cylinder weighing 1 kg and having two holes drilled in it is used. Place the immersion heater in the central hole and a thermometer in the other hole (Figure 2.2.11).

Note the temperature, connect the heater to a 12 V supply and switch it on for 5 minutes. When the temperature stops rising, record its highest value.

Calculate the specific heat capacity as before.

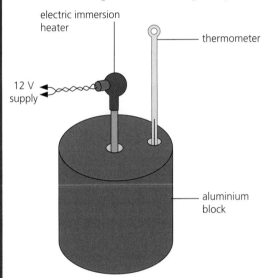

▲ **Figure 2.2.11**

3 Suggest a source of error in the experiment to measure the specific heat capacity of aluminium and suggest how the experiment could be improved.

4 In an experiment to determine the specific heat capacity of aluminium, the temperature rise of an aluminium cylinder weighing 1 kg is found to be 12.5°C when the cylinder is heated by a 40 W heater for 5 minutes. Calculate the specific heat capacity of aluminium.

Importance of the high specific heat capacity of water

The specific heat capacity of water is 4200 J/(kg °C) and that of soil is about 800 J/(kg °C). As a result, the temperature of the sea rises and falls more slowly than that of the land. A certain mass of water needs five times more energy than the same mass of soil for its temperature to rise by 1°C. Water also has to give out more energy to fall 1°C. Since islands are surrounded by water, they experience much smaller changes of temperature from summer to winter than large land masses such as Central Asia.

The high specific heat capacity of water (as well as its cheapness and availability) accounts for its use in cooling engines and in the radiators of central heating systems.

? Worked example

a A tank holding 60 kg of water is heated by a 3 kW electric immersion heater. If the specific heat capacity of water is 4200 J/(kg °C), estimate the time for the temperature to rise from 10°C to 60°C.

Rearranging $c = \dfrac{\Delta E}{m\Delta\theta}$ gives $\Delta E = mc\Delta\theta$

Energy supplied to water = 3000 J/s × *t*, where *t* is the time of heating in seconds.

Assuming energy supplied = energy received by water

$$3000 \text{ J/s} \times t = \Delta E = mc\Delta\theta$$

$$\therefore t = \frac{(60 \times 4200 \times 50)\,\text{J}}{3000 \text{ J/s}} = 4200 \text{ s (70 min)}$$

b A piece of aluminium of mass 0.5 kg is heated to 100°C and then placed in 0.4 kg of water at 10°C. If the resulting temperature of the mixture is 30°C, what is the specific heat capacity of aluminium if that of water is 4200 J/(kg °C)? When two substances at different temperatures are mixed, energy flows from the one at the higher temperature to the one at the lower temperature until both are at the same temperature – the temperature of the mixture. If there is no loss of energy, then in this case:

energy given out by aluminium = energy taken in by water

Using the heat equation $\Delta E = mc\Delta\theta$ and letting c be the specific heat capacity of aluminium in J/(kg °C), we have

energy given out = 0.5 kg × c × (100 − 30)°C

energy taken in = 0.4 kg × 4200 J/(kg °C) × (30 − 10)°C

∴ 0.5 kg × c × 70°C = 0.4 kg × 4200 J/(kg °C) × 20°C

$$c = \frac{(4200 \times 8)\,\text{J}}{35 \text{ kg °C}} = 960 \text{ J/kg °C}$$

Now put this into practice

1 An electric kettle rated at 3 kW containing 1 kg of water is switched on. If the specific heat capacity of water is 4200 J/(kg °C), estimate the time for the water temperature to rise from 30°C to 100°C.

2 A metal sphere of mass 100 g is heated to 100°C and then placed in 200 g of water at 20°C. If the resulting temperature of the mixture is 25°C, what is the specific heat capacity of the metal if that of water is 4200 J/(kg °C)?

> **Test yourself**
>
> 6 Which one of the following statements is *not* true?
>   A Temperature tells us how hot an object is.
>   B When the temperature of an object rises so does its internal energy.
>   C Heat flows naturally from an object at a lower temperature to one at a higher temperature.
>   D The particles of an object move faster when its temperature rises.
>
> 7 How much thermal energy is needed to raise the temperature by 10°C of 5 kg of a substance of specific heat capacity 300 J/(kg °C)?
> 8 How long will it take a 3 kW immersion heater to raise the temperature of 5 kg of water from 30°C to 50°C?

2.2.3 Melting, boiling and evaporation

FOCUS POINTS

★ Describe melting and boiling, including the temperatures for both for water.
★ Describe condensation, solidification and evaporation in terms of particles.

★ Understand the differences between boiling and evaporation.
★ Describe the factors that affect evaporation.

To melt a bar of chocolate you will need to heat it. Melting and boiling require the input of energy to change the state of matter from solid to liquid or from liquid to gas. In the reverse changes, heat is released. During a change of state there is no change in temperature until the process is complete. The kinetic particle model can help us to understand the processes which occur during a change of state. In this section you will also learn how the model explains evaporation and cooling in terms of the escape of energetic particles from the surface of a liquid.

You will learn the differences between the processes of evaporation and boiling and the factors which affect the rate of cooling of an object.

When a solid is heated, it may melt and *change its state* from solid to liquid. If ice is heated it becomes water. The opposite process, freezing, occurs when a liquid solidifies.

A pure substance melts at a definite temperature, called the **melting temperature**; it solidifies at the same temperature – sometimes then called the **freezing temperature**. At standard atmospheric pressure, the melting temperature of water is 0°C.

Practical work

Cooling curve of stearic acid

Safety

● Eye protection must be worn.

Half fill a test tube with stearic acid and place it in a beaker of water (Figure 2.2.12a). Heat the water until all the stearic acid has melted and its temperature reaches about 80°C.

Remove the test tube and arrange it as in Figure 2.2.12b, with a thermometer in the liquid stearic acid. Record the temperature every minute until it has fallen to 60°C.

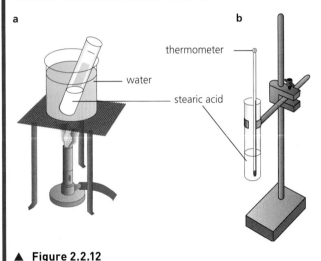

▲ **Figure 2.2.12**

5 Plot a graph of temperature against time (a cooling curve) and identify the freezing temperature of stearic acid.

6 The cooling curve (a plot of temperature against time) for a pure substance is shown in Figure 2.2.13. Why is the cooling curve flat in the region AB?

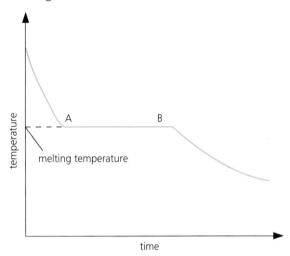

▲ **Figure 2.2.13** Cooling curve

7 What is happening to the liquid over region AB in Figure 2.2.13?

8 Is the rate of cooling faster or slower at higher temperatures in Figure 2.2.13?

Solidifying, melting and boiling

The previous experiment shows that the temperature of liquid stearic acid falls until it starts to solidify (at 69°C) and remains constant until it has all solidified. The cooling curve in Figure 2.2.13 is for a pure substance; the flat part AB occurs at the melting temperature when the substance is solidifying.

During *solidification* a substance transfers thermal energy to its surroundings but its temperature does not fall. Conversely when a solid is *melting*, the energy supplied does not cause a temperature rise; energy is transferred but the substance does not get hotter. For example, the temperature of a well-stirred ice–water mixture remains at 0°C until all the ice is melted. Similarly when energy is supplied to a boiling liquid, the temperature of the liquid does not change. The temperature of pure water boiling at standard atmospheric pressure is 100°C.

Going further

Latent heat of fusion

Energy that is transferred to a solid during melting or given out by a liquid during solidification is called latent heat of fusion. Latent means hidden and fusion means melting. Latent heat does not cause a temperature change; it seems to disappear.

Specific latent heat of fusion

The specific latent heat of fusion (l_f) of a substance is the quantity of heat needed to change *unit mass* from solid to liquid without temperature change.

Specific latent heat is measured in J/kg or J/g. In general, the quantity of heat ΔE needed to change a mass m from solid to liquid is given by

$$\Delta E = m \times l_f$$

 Practical work

Specific latent heat of fusion for ice

Safety

- Eye protection must be worn.

Through measurement of the mass of water m produced when energy ΔE is transferred to melting ice, the specific latent heat of fusion for ice can be calculated.

Insert a 12 V electric immersion heater of known power P into a funnel, and pack crushed ice around it as shown in Figure 2.2.14.

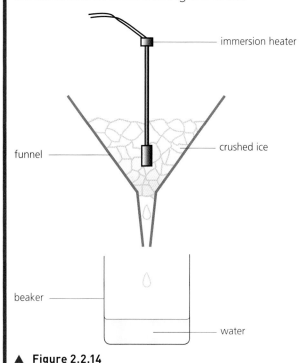

immersion heater

funnel

crushed ice

beaker

water

▲ **Figure 2.2.14**

To correct for heat transferred from the surroundings, collect the melted ice in a beaker for time t (e.g. 4 minutes); weigh the beaker plus the melted ice, m_1. Empty the beaker, switch on the heater, and collect the melted ice for the same time t; re-weigh the beaker plus the melted ice, m_2. The mass of ice melted by the heater is then

$$m = m_2 - m_1$$

The energy supplied by the heater is given by $\Delta E = P \times t$, where P is in J/s and t is in seconds; ΔE will be in joules. Alternatively, a joulemeter can be used to record ΔE directly.

9 Use your data to calculate the specific latent heat of fusion, l_f, for ice from the equation $\Delta E = m \times l_f$

10 What correction is made in the above experiment to measure the specific latent heat of fusion of ice to compensate for heat gained from the surroundings?

11 How could you reduce heat loss to the surroundings in this experiment?

Going further

Latent heat of vaporisation

Latent heat is also needed to change a liquid into a vapour. The reading of a thermometer placed in water that is boiling remains constant at 100°C even though heat, called *latent heat of vaporisation*, is still being transferred to the water from whatever is heating it. When steam condenses to form water, latent heat is given out. A scald from steam is often more serious than one from boiling water (Figure 2.2.15).

▲ **Figure 2.2.15** Steam from boiling water; invisible steam near the spout condenses into visible water droplets higher up.

Specific latent heat of vaporisation

The specific latent heat of vaporisation (l_v) of a substance is the quantity of heat needed to change unit mass from liquid to vapour without change of temperature.

Again, the specific latent heat is measured in J/kg or J/g. In general, the quantity of heat ΔE to change a mass m from liquid to vapour is given by

$$\Delta E = m \times l_v$$

To change 1 kg of water at 100°C to steam at 100°C needs over *five* times as much heat as is needed to raise the temperature of 1 kg of water at 0°C to water at 100°C.

Test yourself

9 1530°C   100°C   55°C   37°C   19°C
   0°C   −12°C   −50°C

From the above list of temperatures choose the most likely value for *each* of the following:
   a the melting temperature of iron
   b the temperature of a room that is comfortably warm
   c the melting temperature of pure ice at normal pressure
   d the boiling temperature of water
   e the normal body temperature of a healthy person.

10 a Why is ice good for cooling drinks?
   b Why do engineers often use superheated steam (steam above 100°C) to transfer heat?

Change of state and the kinetic particle model

Melting and solidification

The kinetic particle model explains the energy absorbed in melting as being the energy that enables the particles of a solid to overcome the intermolecular forces that hold them in place, and when it exceeds a certain value, they break free. Their vibratory motion about fixed positions changes to the slightly greater range of movement they have as liquid particles, and the solid melts. In the reverse process of solidification in which the liquid returns to the solid state, where the range of movement of the particles is less, potential energy is transferred from the particles to thermal energy in the surroundings.

The energy input in melting is used to increase the potential energy (ΔE_p) of the particles, but not their average kinetic energy (E_k) as happens when the energy input causes a temperature rise.

Vaporisation and condensation

If liquid particles are to overcome the forces holding them together and gain the freedom to move around independently as gas particles, they need a large amount of energy. This energy increases the potential energy of the particles but not their kinetic energy. Energy is also required to push back the surrounding atmosphere in the large expansion that occurs when a liquid vaporises. In the reverse process of condensation, in which a vapour returns to the liquid state, where the particles are closer together, potential energy is transferred from the particles to thermal energy in the surroundings.

Boiling and evaporation

At standard atmospheric pressure, the boiling temperature of water is 100°C.

Boiling

For a pure liquid, boiling occurs at a definite temperature called its *boiling temperature* and is accompanied by bubbles that form within the liquid, containing the gaseous or vapour form of the particular substance.

Energy is needed in both evaporation and boiling and is stored in the vapour, from which it is released when the vapour is cooled or compressed and changes to liquid again.

Evaporation

A few energetic particles close to the surface of a liquid may escape and become gas particles. This process of **evaporation** occurs at all temperatures.

Conditions affecting evaporation

Evaporation happens more rapidly when

- the *temperature is higher*, since then more particles in the liquid are moving fast enough to escape from the surface
- the *surface area of the liquid is large*, so giving more particles a chance to escape because more are near the surface
- a *wind* or *draught* is blowing over the surface carrying vapour particles away from the surface, thus stopping them from returning to the liquid and making it easier for more liquid particles to break free. (Evaporation into a vacuum occurs much more rapidly than into a region where there are gas particles.)

Cooling by evaporation

In evaporation, energy is transferred to the liquid from its surroundings, as may be shown by the following demonstration, *done in a fume cupboard*.

Demonstration

Dichloromethane is a volatile liquid, i.e. it has a low boiling temperature and evaporates readily at room temperature, especially when air is blown

through it (Figure 2.2.16). Energy is transferred first from the liquid itself and then from the water below the can. The water soon freezes causing the block and can to stick together.

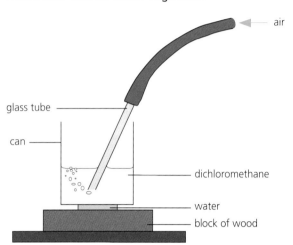

▲ **Figure 2.2.16** Demonstrating cooling by evaporation

Explanation

Evaporation occurs when faster-moving particles escape from the surface of the liquid. The average speed and therefore the average kinetic energy of the particles left behind decreases, i.e. the temperature of the liquid falls.

Cooling by contact

When evaporation occurs from a liquid and the average kinetic energy of the remaining particles decreases, the liquid cools. In Topic 2.3.1 we will see that thermal energy flows from a hotter to a colder object by conduction. If an object is in contact with the liquid during evaporation, thermal energy will flow from the object to the liquid. The object will cool until its temperature equals that of the liquid.

Uses

Water evaporates from the skin when we sweat. This is the body's way of losing unwanted heat and keeping a constant temperature. After vigorous exercise there is a risk of the body being overcooled, especially in a draught; it is then less able to resist infection.

Ether acts as a local anaesthetic by chilling (as well as cleaning) your arm when you are having an injection. Refrigerators, freezers and air-conditioning systems use cooling by evaporation on a large scale.

Volatile liquids are used in perfumes.

Test yourself

11 a When a solid is melting
   i does its temperature increase, decrease or remain constant
   ii is energy absorbed or released or neither?
   b When a liquid is boiling does its temperature increase, decrease or remain constant?
12 a Describe the process of evaporation in particle terms.
   b How does the temperature of a liquid change during evaporation?

13 Some water is stored in a bag of porous material, such as canvas, which is hung where it is exposed to a draught of air. Explain why the temperature of the water is lower than that of the air.

Revision checklist

After studying Topic 2.2 you should know and understand:
✔ that a rise in the temperature of an object increases its internal energy

✔ the relation between an object's temperature and the kinetic energy of the particles

✔ that melting and boiling occur without a change in temperature and recall those temperatures for water.

After studying Topic 2.2 you should be able to:
✔ describe the thermal expansion of solids and liquids
✔ describe precautions taken against expansion and uses of expansion

✔ explain the relative order of magnitude of the expansion of solids, liquids and gases
✔ distinguish between evaporation and boiling
✔ define specific heat capacity, c, and solve problems using the equation $c = \dfrac{\Delta E}{m\Delta\theta}$
✔ describe experiments to measure the specific heat capacity of metals and liquids by electrical heating

✔ describe condensation, solidification and evaporation processes in terms of the kinetic particle model

✔ explain cooling by evaporation
✔ recall the factors which affect evaporation.

Exam-style questions

1

a A gas expands more easily than a liquid. Explain in terms of the motion and arrangement of particles. [3]

b Explain why water pipes may burst in cold weather. [2]
[Total: 5]

2 A bimetallic thermostat for use in an iron is shown in Figure 2.2.17.

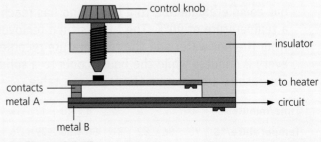

▲ Figure 2.2.17

State if the following statements are *correct* or *incorrect*.

A It operates by the bimetallic strip bending away from the contact. [1]

B Metal A has a greater expansivity than metal B. [1]

C Screwing in the control knob raises the temperature at which the contacts open. [1]

[Total: 3]

3 The same quantity of heat was given to different masses of three substances A, B and C. The temperature rise in each case is shown in the table. Calculate the specific heat capacities of A, B and C.

Material	Mass/kg	Heat given/J	Temp. rise/°C
A	1.0	2000	1.0
B	2.0	2000	5.0
C	0.5	2000	4.0

[3 marks for each of A, B, C]

[Total: 9]

4 a The jam in a hot jam tart always seems hotter than the pastry. Why? [2]

b Calculate the temperature rise of 3 kg of a material of specific heat capacity 500 J/(kg °C) when it is heated with 15 000 J of energy. [3]

[Total: 5]

5 a A certain liquid has a specific heat capacity of 4.0 J/(g °C). How much energy must be supplied to raise the temperature of 10 g of the liquid from 20°C to 50°C? [3]

b Explain why a bottle of milk keeps better when it stands in water in a porous pot in a draught. [3]

[Total: 6]

6 a Define
  i melting temperature
  ii boiling temperature
  iii freezing temperature. [3]

b State
  i the melting temperature of ice
  ii the boiling temperature of water at standard atmospheric pressure. [2]

c State if energy is absorbed or released when
  i a liquid solidifies
  ii a gas condenses. [2]

[Total: 7]

7 A drink is cooled more by ice at 0°C than by the same mass of water at 0°C.
This is because ice
A floats on the drink [1]
B has a smaller specific heat capacity [1]
C gives out heat to the drink as it melts [1]
D absorbs heat from the drink to melt [1]
E is a solid. [1]
State whether each of the above statements is *correct* or *incorrect*.

[Total: 5]

Alternative to Practical

8 In an experiment to investigate the cooling of a liquid to a solid, a test tube containing a pure solid is warmed in a beaker of hot water until it has completely melted to a liquid and has reached a temperature of 90°C. The tube is then removed from the hot water and the temperature recorded every 2 minutes while the liquid cools to a solid. The results are given in the following table.

a Plot a graph of temperature versus time. [4]

b Estimate the melting temperature of the solid and explain your choice. [2]

c Explain what happens to the arrangement of the particles in the liquid during solidification. [2]

[Total: 8]

Time/minutes	0	2	4	6	8	10	12	14	16	18
Temperature/°C	90	86	82	81	80	80	79	76	73	72

2.3 Transfer of thermal energy

2.3.1 Conduction

FOCUS POINTS

★ Know how to investigate whether a material is a good or poor conductor of thermal energy.

★ Use atomic or molecular lattice vibrations and the movement of free (delocalised) electrons in metallic conductors to describe thermal conduction in solids.
★ Describe, in terms of particles, why thermal conduction is poor in gases and most liquids.
★ Understand that thermal conductors conduct thermal energy better than thermal insulators and some solids are better thermal conductors than others.

Heat from a stove is quickly transferred to all parts of a metal saucepan; metals are good conductors of heat. A poor thermal conductor, such as plastic, is often used for the handle of a saucepan to keep it cool. In this topic you will encounter experiments that demonstrate the properties of both good and bad thermal conductors.

Thermal energy can be transferred in various ways. In a solid an increase in temperature produces stronger local vibrations of the particles that are transferred to their neighbours and thermal energy is transferred progressively through the material. This is a slow process but is the main way of transferring energy in poor conductors. In good conductors, the main way of transferring thermal energy is by free electrons in the conductor; these can transfer energy from particle to particle very quickly.

A knowledge of how heat travels is needed to keep a building or a house at a comfortable temperature in winter and in summer, if it is to be done economically and efficiently.

Conduction

The handle of a metal spoon held in a hot drink soon gets warm. Heat passes along the spoon by **conduction**.

Conduction is the flow of thermal energy (heat) through matter from places of higher temperature to places of lower temperature without movement of the matter as a whole.

A simple demonstration of the different conducting powers of various metals is shown in Figure 2.3.1. A match is fixed to one end of each rod using a little melted wax. The other ends of the rods are heated by a burner. When the temperatures of the far ends reach the melting temperature of wax, the matches drop off. The match on copper falls

first, showing it is the best **conductor**, followed by aluminium, brass and then iron.

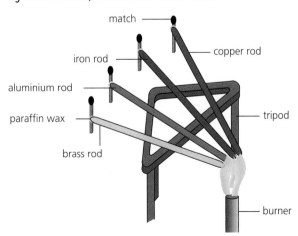

▲ **Figure 2.3.1** Comparing conducting powers

Thermal energy is conducted faster through a rod if it has a large cross-sectional area, is short and has a large temperature difference between its ends.

Most metals are good conductors of thermal energy; materials such as wood, glass, cork, plastics and fabrics are thermal **insulators** (poor conductors). The arrangement in Figure 2.3.2 can be used to show the difference between brass and wood. If the rod is passed through a flame several times, the paper over the wood scorches but not the paper over the brass. The brass conducts the heat away from the paper quickly, preventing the paper from reaching the temperature at which it burns. The wood conducts the thermal energy away only very slowly.

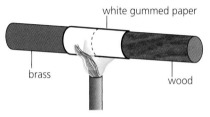

white gummed paper

brass

wood

▲ **Figure 2.3.2** The paper over the brass does not burn.

Metal objects below body temperature *feel* colder than those made of bad conductors – even if all the objects are at exactly the same temperature – because they carry thermal energy away faster from the hand.

Liquids and gases also conduct thermal energy but only very slowly. Water is a very poor thermal conductor, as shown in Figure 2.3.3. The water at the top of the tube can be boiled before the ice at the bottom melts.

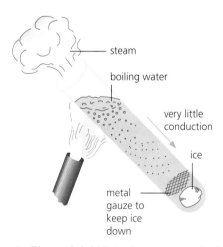

steam

boiling water

very little conduction

ice

metal gauze to keep ice down

▲ **Figure 2.3.3** Water is a poor conductor of thermal energy.

Conduction and the particle model

Two processes occur in metals. Metals have a large number of 'free' (delocalised) electrons (Topic 4.2.2) which move about within the metal. When one part of a metal is heated, the electrons there move faster (their kinetic energy increases) and move further. As a result, they are able to interact with particles in cooler parts, so passing on their energy and raising the temperature of these parts. This process occurs quickly.

The second process is much slower. The atoms or molecules at the hot part make colder neighbouring particles vibrate more vigorously. These atomic or molecular lattice vibrations are less important in metals, but are the only way conduction can occur in non-metals since these do not have free electrons; hence non-metals are poor conductors of heat and are good insulators.

There are many solids which have fewer free electrons available to transfer thermal energy than metals do and so are less good thermal conductors than metals but better thermal conductors than insulators. For example, the semiconductors used in electronic circuits can have a range of thermal conductivities between those of metals and insulators.

Liquids and gases are generally less dense than solids and their particles are further apart. They do not have a regularly ordered particle structure, so it is difficult to set up lattice vibrations, and they do not usually have free electrons. They are therefore less good thermal conductors than solids.

> **Test yourself**
>
> 1   Explain what is meant by thermal conduction.
> 2   Why is thermal conduction poor in gases and most liquids?

2.3.2 Convection

FOCUS POINTS

★ Know that thermal energy transfer in liquids and gases usually occurs by convection.
★ Use density changes to explain convection in liquids and gases.
★ Describe some experiments to show convection.

You may have a convector heater in your home which helps to keep you warm in winter. In convection, heat is transferred by the motion of matter and it is an important method for transferring thermal energy in liquids and gases. When the temperature of a fluid increases, thermal expansion reduces its density and the warmer, less dense parts of the fluid tend to rise, while cooler, denser parts will sink. The combination sets up fluid flows known as convection currents that transfer thermal energy from places of high temperature to those of lower temperature by motion of the fluid itself. In the case of a convector heater, convection currents are set up in the air in the room.

Convection in liquids

Convection is the usual method by which thermal energy (heat) travels through fluids such as liquids and gases. It can be shown in water by dropping a few crystals of potassium permanganate down a tube to the bottom of a beaker or flask of water. When the tube is removed and the beaker heated just below the crystals by a *small* flame (Figure 2.3.4a), purple streaks of water rise upwards and fan outwards.

Streams of warm, moving fluids are called **convection currents**. They arise when a fluid is heated because it expands, becomes less dense and is forced upwards by surrounding cooler, denser fluid which moves under it. We say 'hot water (or hot air) rises'. Warm fluid behaves like a cork released under water: being less dense it bobs up. Lava lamps (Figure 2.3.4b) use this principle.

▲ **Figure 2.3.4a** Convection currents shown by potassium permanganate in water.

▲ **Figure 2.3.4b** Lava lamps make use of convection.

Convection is the flow of thermal energy through a fluid from places of higher temperature to places of lower temperature by movement of the fluid itself.

Convection in air

Black marks often appear on the wall or ceiling above a lamp or a radiator. They are caused by dust being carried upwards in air convection currents produced by the hot lamp or radiator.

A laboratory demonstration of convection currents in air can be given using the apparatus of Figure 2.3.5. The direction of the convection current created by the candle is made visible by the smoke from the touch paper (made by soaking brown paper in strong potassium nitrate solution and drying it).

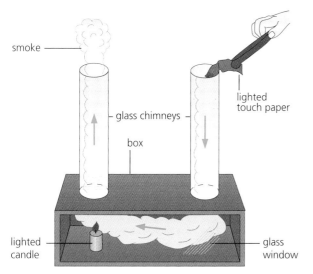

▲ **Figure 2.3.5** Demonstrating convection in air.

Natural convection currents

Coastal breezes

During the day the temperature of the land increases more quickly than that of the sea (because the specific heat capacity of the land is much smaller; see Topic 2.2.2). The hot air above the land rises and is replaced by colder air from the sea. A breeze from the sea results (Figure 2.3.6a).

At night the opposite happens. The sea has more thermal energy to transfer and cools more slowly. The air above the sea is warmer than that over the land and a breeze blows from the land (Figure 2.3.6b).

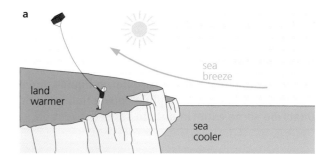

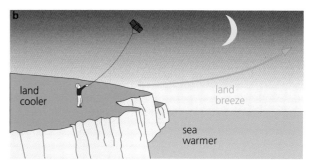

▲ **Figure 2.3.6** Coastal breezes are due to convection: **a** day; **b** night.

Gliding

Gliders, including hang-gliders (Figure 2.3.7), are carried along on hot air currents, called thermals.

▲ **Figure 2.3.7** Once airborne, a hang-glider pilot can stay aloft for several hours by flying from one thermal to another.

> **Test yourself**
>
> 3 Explain the advantage of placing an electric immersion heater in a tank of water
>   a near the top
>   b near the bottom.
> 4 Why does hot air rise?

2.3.3 Radiation

FOCUS POINTS

★ Understand that thermal radiation is infrared radiation that does not require a transmission medium and that this radiation is emitted by all objects.
★ Describe the effects of surface colour and texture on the emission, absorption and reflection of thermal radiation.

★ Understand the factors which affect the amount of radiation emitted by an object.
★ Understand that a balance between rate of energy received and energy transferred must be achieved for an object to maintain a constant temperature.
★ Know that factors controlling the balance between incoming radiation and radiation emitted from the Earth's surface affect the temperature of the Earth.
★ Describe experiments to distinguish between good and bad absorbers and emitters of infrared radiation.
★ Know that surface temperature and surface area of an object affect the rate of emission of radiation.

On a sunny day it is pleasant to feel the warmth of the radiation reaching you from the Sun. Radiation is the third way of transferring thermal energy from one place to another without the need of a transmission medium. On reaching Earth, the Sun's rays are partly reflected, absorbed or transmitted by objects. Shiny white surfaces are good reflectors of radiation but dull black surfaces are good absorbers. The efficiency of emission and absorption of radiation depends on the nature of the surface of the material.

The rate of radiation emission depends on the temperature of the object. The temperature of the Earth is controlled by the balance between radiation absorbed and emitted.

Radiation is a third way in which thermal energy can travel but, whereas conduction and convection both need matter to be present, radiation can occur in a vacuum; particles of matter are not involved. Radiation is the way thermal energy reaches us from the Sun.

Radiation has all the properties of electromagnetic waves (Topic 3.3), and travels with the speed of light. Thermal radiation is **infrared radiation** and all objects emit this radiation. When it falls on an object, it is partly reflected, partly transmitted and partly absorbed: the absorbed radiation raises the temperature of the object.

Buildings in hot countries are often painted white (Figure 2.3.8). This is because white surfaces are good reflectors of radiation and so help to keep the houses cool.

▲ **Figure 2.3.8** White painted buildings.

Good and bad absorbers

Some surfaces absorb radiation better than others, as may be shown using the apparatus in Figure 2.3.9. The inside surface of one lid is shiny and of the other dull black. The coins are stuck on the outside of each lid with candle wax. If the heater is midway between the lids, they each receive the same amount of radiation. After a few minutes the wax on the black lid melts and the coin falls off. The shiny lid stays cool and the wax unmelted.

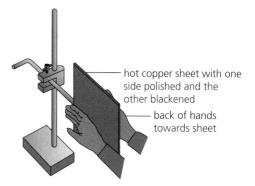

▲ **Figure 2.3.10** Comparing emitters of radiation

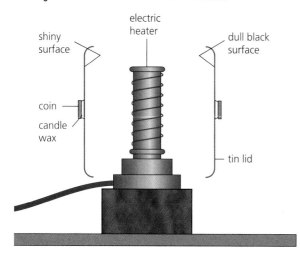

▲ **Figure 2.3.9** Comparing absorbers of radiation

Dull black surfaces are *better* **absorbers** of radiation than *white shiny surfaces*, which are instead good *reflectors* of radiation. Reflectors on electric fires are made of polished metal because of its good reflecting properties.

Good and bad emitters

Some surfaces also emit radiation better than others when they are hot. If you hold the backs of your hands on either side of a hot copper sheet that has one side polished and the other side blackened (Figure 2.3.10), it will be found that your hands feel warmer near the dull black surface. The dull black surface is a *better* **emitter** *of radiation* than the shiny one.

The cooling fins on the heat exchangers at the back of a refrigerator are painted black so are good emitters of radiation and they transfer thermal energy more quickly. By contrast, saucepans that are polished are poor emitters and keep their heat longer.

In general, surfaces that are good absorbers of radiation are good emitters when hot.

Temperature and rate of emission of radiation

Radiation is emitted by all bodies above absolute zero and consists mostly of infrared radiation, but light and ultraviolet are also present if the body is very hot (e.g. the Sun). For an object to maintain a constant temperature, energy must transfer away from the object at the same rate that the object receives energy. If the average energy radiated is less than that absorbed, the temperature of the object will rise. If the average energy radiated is more than that absorbed, the temperature of the object will fall.

The greenhouse effect

The warmth from the Sun is not cut off by a sheet of glass but the warmth from a red-hot fire can be blocked by glass. The radiation from very hot bodies like the Sun is mostly light and short-wavelength infrared. The radiation from less hot objects, such as a fire, is largely long-wavelength infrared which, unlike light and short-wavelength infrared, cannot pass through glass.

Light and short-wavelength infrared from the Sun penetrate the glass of a greenhouse and are absorbed by the soil, plants, etc., raising their temperature. These in turn emit infrared but, because of their relatively low temperature, this has a long wavelength and is not transmitted by the glass. The greenhouse thus acts as a 'heat-trap' and its temperature rises.

Carbon dioxide and other gases such as methane in the Earth's atmosphere act in a similar way to the glass of a greenhouse in trapping heat radiated from the Earth's surface. This has serious implications for the global climate. For the average temperature of the Earth to remain constant, a balance must be achieved between the incoming radiation and the radiation emitted from the Earth's surface.

If there is a build-up of carbon dioxide and methane gases in the atmosphere, the balance between incoming radiation from the Sun and the average power emitted from the Earth will be upset.

Rate of cooling of an object

If the surface temperature of an object is higher than its surroundings, it emits radiation at a faster rate than it absorbs radiation from its surroundings. As a result, it cools until the two rates become equal and a constant temperature is reached. The higher the surface temperature of the object above its surroundings, and the larger its surface area, the greater the quantity of radiation it emits and the greater its rate of cooling.

Practical work

Rate of cooling

For safe experiments/demonstrations related to this topic, please refer to the *Cambridge IGCSE Physics Practical Skills Workbook* that is also part of this series.

Safety

- Eye protection must be worn.
- Take care when handling hot water and its containers.

Place a thermometer in some hot water and wait until the temperature reaches a steady temperature above 80°C. Remove the thermometer from the water, and quickly wipe it dry with a paper towel. Record the temperature on the thermometer every 30 s as it cools away from draughts or any source

of heat. Use your results to plot a graph of temperature against time.

1 In the above experiment a student recorded the following temperatures on the thermometer as it cooled in air.

Time/s	0	30	60	90	120	150	180
Temperature/°C	80	63	51	42	36	31	28

  a Plot a graph of temperature against time using the values given in the table.
  b Calculate the temperature drop
    i between 0 and 90 s
    ii between 90 s and 180 s.
  c State the temperature range over which the thermometer cools most quickly.
  d Does the thermometer emit radiation at a higher rate at the higher or lower temperatures?

Test yourself

5 The door canopy in Figure 2.3.11 shows clearly the difference between white and black surfaces when radiation falls on them. Explain why.

▲ **Figure 2.3.11**

6 What type of radiation is thermal radiation?

7 a The Earth has been warmed by the radiation from the Sun for millions of years yet we think its average temperature has remained fairly steady. Why is this?
  b Why is frost less likely on a cloudy night than a clear one?

2.3.4 Consequences of thermal energy transfer

FOCUS POINTS

★ Explain everyday applications and consequences of thermal energy transfer by conduction, convection and radiation.

★ Explain complex applications and consequences of conduction, convection and radiation involving more than one type of thermal energy transfer.

You have now encountered the three ways in which thermal energy can be transferred from one place to another: conduction, convection and radiation. Such transfers occur in many different situations in everyday living. Transfer of thermal energy by conduction from an external source enables us to heat cooking pots. Convection is often used in water and convector heaters in our homes. Radiation from the Sun can be felt directly and an infrared thermometer allows us to read temperature from a distance. In this topic you will learn more about the uses of both good conductors and poor conductors (insulators).

Applications in which more than one type of thermal energy transfer is involved are explained.

Uses of conductors

Good conductors

These are used whenever heat is required to travel quickly through something. Saucepans, boilers and radiators are made of metals such as aluminium, iron and copper which are all good conductors that transfer thermal energy quickly.

Bad conductors (insulators)

The handles of some saucepans are made of wood or plastic. Cork is used for table mats. These are insulating materials that transfer thermal energy only very slowly.

Air is one of the worst conductors and so one of the best insulators. This is why houses with cavity walls (two layers of bricks separated by an air space) and double-glazed windows keep warmer in winter and cooler in summer.

Because air is such a bad conductor, materials that trap air, such as wool, felt, fur, feathers, polystyrene foam and fibreglass, are also very bad conductors. Some of these materials are used as 'lagging' to insulate water pipes, hot water cylinders, ovens, refrigerators and the walls and roofs of houses (Figures 2.3.12a and 2.3.12b). Others are used to make warm winter clothes like fleece jackets (Figure 2.3.12c on the next page).

▲ **Figure 2.3.12a** Lagging in a cavity wall provides extra insulation.

▲ **Figure 2.3.12b** Laying lagging in a house loft

▲ **Figure 2.3.12c** Fleece jackets help to retain your body warmth.

Wet suits are worn by divers and water skiers to keep them warm. The suit gets wet and a layer of water gathers between the person's body and the suit. The water is warmed by body heat and stays warm because the suit is made of an insulating fabric, such as neoprene (a synthetic rubber).

Energy losses from buildings

The inside of a building can only be kept at a steady temperature above that outside by heating it at a rate which equals the rate at which it is losing energy. The loss occurs mainly by conduction through the walls, roof, floors and windows. For a typical house where no special precautions have been taken, the contribution each of these makes to the total loss is shown in Table 2.3.1a.

As fuels (and electricity) become more expensive and the burning of fuels becomes of greater environmental concern (Topic 1.7.3), more people are considering it worthwhile to reduce heat losses from their homes. The substantial reduction of this loss which can be achieved, especially by wall and roof insulation, is shown in Table 2.3.1b.

▼ **Table 2.3.1** Energy losses from a typical house

a Percentage of total energy loss due to				
walls	roof	floors	windows	draughts
35	25	15	10	15
b Percentage of each loss saved by				
insulating walls	insulating roof	carpets on floors	double glazing	draught excluders
65	80	≈30	50	≈60
Percentage of total loss saved = 60				

 Going further

Ventilation

In addition to supplying heat to compensate for the energy losses from a building, a heating system has also to warm the ventilated cold air, needed for comfort, which comes in to replace stale air.

If the rate of heat loss is, say, 6000 J/s, or 6 kW, and the warming of ventilated air requires 2 kW, then the total power needed to maintain a certain temperature (e.g. 20°C) in the building is 8 kW. Some of this is supplied by each person's 'body heat', estimated to be roughly equal to a 100 W heater.

Uses of convection

Convection currents set up by electric, gas and oil heaters help to warm our homes. Many so-called 'radiators' are really **convector** heaters. Warm air produced by the heater rises because it is less dense than the colder air above. The cold air sinks, is warmed by the heater and itself rises. A convection current is set up which helps to warm the whole room.

Convection currents also form in the water being heated in hot water tanks, kettles and kitchen pans, allowing water to be heated quickly.

Uses of radiation: infrared thermometer

An infrared thermometer detects the thermal radiation emitted by an object and converts it into an electrical signal. The temperature of the object can be determined from the **radiant** power detected and the value is shown on a digital display. It is a non-contact method and allows temperature to be measured at a distance. Infrared thermometers are frequently used to monitor the health of passengers arriving at an airport.

Applications involving more than one type of thermal energy transfer

Car radiator

Both conduction and radiation occur in a car radiator which acts to dissipate the heat generated in the engine. It contains a fluid which circulates between the engine block and the radiator. Thermal energy is transferred to the fluid by conduction as it passes over the engine block. When the fluid enters the radiator, thermal energy is transferred by conduction to the radiator which then radiates energy in the infrared to the surroundings. The metal radiator is black and has a large surface so is a good emitter of radiation. In this way the fluid is cooled before it circulates back to the engine block.

Wood or coal fire

Radiation and convection occur when a room is heated by a wood- or coal-burning fire.

Thermal energy is radiated from the burning wood or coal and heats up objects in the room which absorb it. Air in contact with the hot wood or coal is warmed and rises upwards because it is less dense than the cold air above. Cooler air is drawn down to take its place and a convection current is set up which also serves to transfer heat into the room.

Vacuum flask

A vacuum or Thermos flask keeps hot liquids hot or cold liquids cold. It is very difficult for heat to travel into or out of the flask.

Transfer of thermal energy by conduction and convection is minimised by making the flask a double-walled glass vessel with a vacuum between the walls (Figure 2.3.13). Radiation is reduced by silvering both walls on the vacuum side. Then if, for example, a hot liquid is stored, the small amount of radiation from the hot inside wall is reflected back across the vacuum by the silvering on the outer wall. The slight heat loss that does occur is by conduction up the thin glass walls and through the stopper. If the flask is to be used to store a cold liquid, thermal energy from outside the flask is reflected from the inner wall.

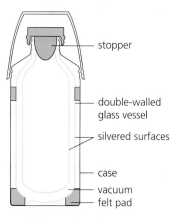

stopper

double-walled glass vessel

silvered surfaces

case

vacuum

felt pad

▲ **Figure 2.3.13** The structure of a vacuum flask

Test yourself

8 Explain why on a cold day the metal handlebars of a bicycle feel colder than the rubber grips.

9 Name the energy transfers which occur
   a when a radiator is used to cool the engine of a car
   b when a room is heated by a coal fire.

Revision checklist

After studying Topic 2.3 you should know and understand:

✔ that thermal energy transferred by radiation does not require a medium and that thermal radiation is infrared radiation emitted by all objects

✔ that for an object at a constant temperature, the rate at which it receives radiation equals the rate that it transfers radiation
✔ the rate of radiation emission increases as the temperature of the object increases

✔ how thermal insulation is used to keep liquids cool and to reduce heat loss from buildings.

After studying Topic 2.3 you should be able to:

✔ describe experiments to show the different conducting powers of various substances and name good and bad conductors

✔ explain conduction using the kinetic particle model

✔ describe experiments to show convection in fluids (liquids and gases) and relate convection in fluids to density changes
✔ describe the effect of surface colour and texture on the emission, absorption and reflection of radiation and recall that good absorbers are also good emitters

✔ describe experiments to study factors affecting the absorption and emission of radiation
✔ explain how a greenhouse acts as a 'heat-trap' and the consequence for the balance between incoming and emitted radiation at the Earth's surface

✔ explain some simple uses and consequences of conduction, convection and radiation

✔ explain applications involving more than one type of thermal energy transfer.

Exam-style questions

1 Describe an experiment to demonstrate the properties of good and bad thermal conductors.

[Total: 4]

2 Explain in terms of particles how thermal energy is transferred by conduction in solids.

[Total: 4]

3 a Explain how thermal energy is transferred by convection. [3]

b Describe an experiment to illustrate convection in a liquid. [3]

[Total: 6]

4 The following statements relate to the absorption and emission of radiation.

State which of the statements are *true* and which are *false*.

A Energy from the Sun reaches the Earth by radiation only. [1]

B A dull black surface is a good absorber of radiation. [1]

C A shiny white surface is a good emitter of radiation. [1]

D The best heat insulation is provided by a vacuum. [1]

[Total: 4]

5 Describe the effect of surface colour and texture on the

a emission of radiation [2]

b reflection of radiation [2]

c absorption of radiation. [2]

[Total: 6]

6 a Describe an experiment to show the properties of good and bad emitters of infrared radiation. [4]

b Describe an experiment to show the properties of good and bad absorbers of infrared radiation. [4]

[Total: 8]

7 Explain why

a newspaper wrapping keeps hot things hot, e.g. fish and chips, and cold things cold, e.g. ice cream [1]

b fur coats would keep their wearers warmer if they were worn inside out [2]

c a string vest helps to keep a person warm even though it is a collection of holes bounded by string. [2]

[Total: 5]

8 Figure 2.3.14 illustrates three ways of reducing heat losses from a house.

a Explain how each of the three methods reduces heat losses. [4]

b Why are fibreglass and plastic foam good substances to use? [2]

c Air is one of the worst conductors of heat. What is the advantage of replacing it by plastic foam as shown in Figure 2.3.14? [1]

d A vacuum is an even better heat insulator than air. Suggest one (scientific) reason why the double glazing should not have a vacuum between the sheets of glass. [1]

[Total: 8]

a

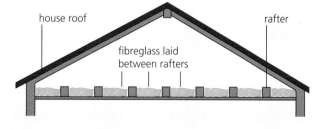

b

c

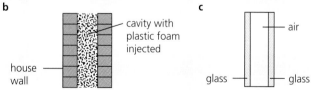

▲ **Figure 2.3.14 a** Roof insulation; **b** cavity wall insulation; **c** double glazing

Alternative to Practical

9 The manufacturers of roof insulation suggest that two layers of fibreglass are more effective than one. Describe how you might set up an experiment in the laboratory to test whether this is true.

[Total: 6]

SECTION 3

Waves

Topics

3.1 General properties of waves

FOCUS POINTS

★ Understand that waves transfer energy without transferring matter and describe and illustrate what is meant by wave motion.

★ Describe the features of waves and use the correct equation to calculate wave speed.

★ Describe transverse and longitudinal waves in terms of their direction of vibration and give examples of each.

★ Describe reflection, refraction and diffraction of waves, including the use of a ripple tank to illustrate these.

★ Describe the factors that affect diffraction through a gap or at an edge.

Ripples spread out from a stone dropped into a pond and you can see that energy is transferred across the water, but a leaf resting on the surface of the pond only bobs up and down when the ripple passes. You will find out in this topic that waves are disturbances that transfer energy from one point to another without transport of matter. The vibrations in a wave can lie along the path, as in sound waves, or be perpendicular, as in water waves. You will learn how to represent waves in terms of general properties that apply to light, sound, seismic and water waves. You will also learn how ripple tank experiments are used to model the reflection of waves when they strike a plane surface or enter a medium in which their speed changes and how waves may spread out when they pass through gaps.

Types of wave motion

Several kinds of wave motion occur in physics. **Mechanical waves** are produced by a disturbance, such as a vibrating object, in a material medium and are transmitted by the particles of the medium vibrating about a fixed position. Such waves can be seen or felt and include waves on a rope or spring, water waves and sound waves in air or in other materials.

A **progressive wave** or travelling wave is a disturbance which carries energy from one place to another without transferring matter. There are two types, **transverse** and **longitudinal waves**.

Transverse waves

In a transverse wave, the direction of the disturbance is at **right angles** to the direction of propagation of the wave, that is the direction in which the wave travels. A transverse wave can be sent along a rope (or a spring) by fixing one end and

moving the other rapidly up and down (Figure 3.1.1). The disturbance generated by the hand is passed on from one part of the rope to the next which performs the same motion but slightly later. The humps and hollows of the wave travel along the rope as each part of the rope vibrates transversely about its undisturbed position.

Water waves can be modelled as transverse waves, as can seismic S-waves (see Topic 3.4). In Topic 3.3 we will find that electromagnetic radiations are also transverse waves.

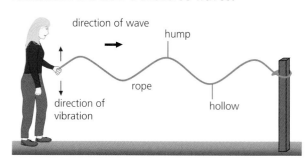

▲ **Figure 3.1.1** A transverse wave

Longitudinal waves

In a progressive longitudinal wave the particles of the transmitting medium vibrate back and forth along the same line as (parallel to) that in which the wave is travelling and not at right angles to it as in a transverse wave. A longitudinal wave can be sent along a spring, stretched out on the bench and fixed at one end, if the free end is repeatedly pushed and pulled sharply, as shown in Figure 3.1.2.

Compressions C (where the coils are closer together) and *rarefactions* R (where the coils are further apart) travel along the spring.

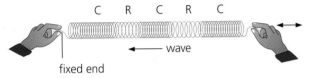

▲ **Figure 3.1.2** A longitudinal wave

We will see in Topic 3.4 that sound waves and seismic P-waves are other examples of longitudinal waves.

Describing waves

Terms used to describe waves can be explained with the aid of a **displacement–distance graph** (Figure 3.1.3). It shows, at a certain instant of time, the distance moved (sideways from their undisturbed positions) by the parts of the medium vibrating at different distances from the cause of the wave.

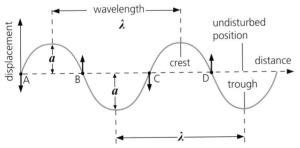

▲ **Figure 3.1.3** Displacement–distance graph for a wave at a particular instant

Wavelength

The **wavelength** of a wave, represented by the Greek letter λ ('lambda'), is the distance between successive **crests** (peaks) (see Figure 3.1.3).

Frequency

The *frequency* f is the number of complete waves generated per second. If the end of a rope is moved up and down twice in a second, two waves are produced in this time. The frequency of the wave is two vibrations per second or 2 hertz (2 Hz; the **hertz** is the unit of frequency), which is the same as the frequency of the movement of the end of the rope. That is, the frequencies of the wave and its source are equal.

The frequency of a wave is also the number of crests passing a chosen point per second.

Wave speed

The *wave speed* v of the wave is the distance moved in the direction of travel of the wave by a crest or any point on the wave in 1 second.

Amplitude

The **amplitude** a is the height of a crest or the depth of a trough measured from the undisturbed position of what is carrying the wave, such as a rope (see Figure 3.1.3).

Phase

The short arrows at A, B, C, D on Figure 3.1.3 show the directions of vibration of the parts of the rope at these points. The parts at A and C have the same speed in the same direction and are in **phase**. At B and D the parts are also in phase with each other but they are out of phase with those at A and C because their directions of vibration are opposite.

The wave equation

The faster the end of a rope is vibrated, the shorter the wavelength of the wave produced. That is, the higher the frequency of a wave, the smaller its wavelength. There is a useful connection between f, λ and v, which is true for all types of wave.

Suppose waves of wavelength $\lambda = 20$ cm travel on a long rope and three crests pass a certain point every second. The frequency $f = 3$ Hz. If Figure 3.1.4 represents this wave motion then, if crest A is at P at a particular time, 1 second later it will be at Q, a distance from P of three wavelengths, i.e. $3 \times 20 = 60$ cm.

The speed of the wave is $v = 60$ cm per second (60 cm/s), obtained by multiplying f by λ. Hence the **wave equation** is

speed of wave = frequency × wavelength

or

$$v = f\lambda$$

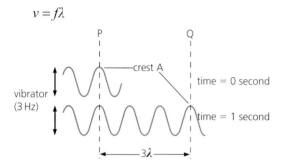

▲ Figure 3.1.4

? Worked example

The speed of a wave is 4 cm/s. If its frequency is 2 Hz, calculate its wavelength.

We can rearrange the equation $v = f\lambda$ to give $\lambda = v/f$.

Now $v = 4$ cm/s and $f = 2$ Hz so

$$\lambda = v/f = 4\,\text{cm/s} / 2\,\text{Hz} = 2\,\text{cm}$$

Now put this into practice

1 A wave travelling at 5 cm/s has a frequency of 50 Hz. Calculate its wavelength.
2 A wave has a speed of 10 m/s and a wavelength of 1 m. Calculate its frequency.
3 A wave travelling at 10 m/s has a wavelength of 25 cm. Calculate its frequency.

⚡ Practical work

The ripple tank

For safe experiments/demonstrations related to this topic, please refer to the *Cambridge IGCSE Physics Practical Skills Workbook* that is also part of this series.

Safety

● Eye protection must be worn.

The behaviour of water waves can be studied in a ripple tank. It consists of a transparent tray containing water, having a light source above and a white screen below to receive the wave images (Figure 3.1.5).

Pulses (i.e. short bursts) of ripples are obtained by dipping a finger in the water for circular ripples and a ruler for straight ripples. **Continuous ripples** are generated using an electric motor and a bar. The bar gives straight ripples if it just touches the water or circular ripples if it is raised and has a small ball fitted to it.

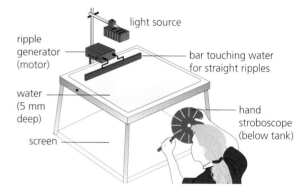

▲ Figure 3.1.5 A ripple tank

Continuous ripples are studied more easily if they are *apparently* stopped ('frozen') by viewing the screen through a disc with equally spaced slits, which can be spun by hand, i.e. a *stroboscope*. If the disc speed is such that the waves have advanced one wavelength each time a slit passes your eye, they appear at rest.

1 Sketch the appearance of the continuous ripples produced by the straight bar.
2 What is the wavelength of the ripples?

Test yourself

1 The lines in Figure 3.1.6 are crests of straight ripples.
   a What is the wavelength of the ripples?
   b If 5 seconds ago ripple A occupied the position now occupied by ripple F, what is the frequency of the ripples?
   c What is the speed of the ripples?

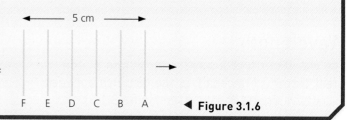

◀ Figure 3.1.6

Wavefronts and rays

In two dimensions, a **wavefront** is a line on which the disturbance has the same phase at all points; the crests of waves in a ripple tank can be thought of as wavefronts. A vibrating source produces a succession of wavefronts, all of the same shape. In a ripple tank, straight wavefronts are produced by a vibrating bar (a line source) and circular wavefronts are produced by a vibrating ball (a point source). A line drawn at right angles to a wavefront, which shows its direction of travel, is called a **ray**. Straight wavefronts and the corresponding rays are shown in Figure 3.1.7; circular wavefronts can be seen in Figure 3.1.16 (p. 134).

The properties of waves (reflection at a plane surface, refraction and diffraction) can be illustrated by the behaviour of water waves in a ripple tank.

Reflection at a plane surface

In Figure 3.1.7 *straight* water waves are falling on a metal strip placed in a ripple tank at an angle of 60°, i.e. the angle i between the direction of travel of the waves and the normal to the strip is 60°, as is the angle between the wavefront and the strip. (The perpendicular to the strip at the point where the incident ray strikes is called the **normal.**) The wavefronts are represented by straight lines and can be thought of as the crests of the waves. They are at right angles to the direction of travel, i.e. to the rays. The angle of *reflection r* is 60°. Incidence at other angles shows that the *angle of incidence and angle of reflection are always equal*.

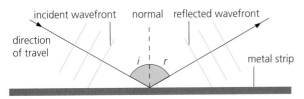

▲ **Figure 3.1.7** Reflection of waves

Refraction

If a glass plate is placed in a ripple tank so that the water over the glass plate is about 1 mm deep but is 5 mm deep elsewhere, continuous straight waves in the shallow region are found to have a shorter wavelength than those in the deeper parts, i.e. the wavefronts are closer together (Figure 3.1.8). Both sets of waves have the frequency of the vibrating bar and, since $v = f\lambda$, if λ has decreased so has v, since f is fixed. Hence *waves travel more slowly in shallow water*.

▲ **Figure 3.1.8** Waves in shallower water have a shorter wavelength.

When the plate is at an angle to the waves (Figure 3.1.9a), their direction of travel in the shallow region is bent towards the normal (Figure 3.1.9b). The change in the direction of travel of the waves, which occurs when their speed and hence wavelength changes, is termed **refraction**.

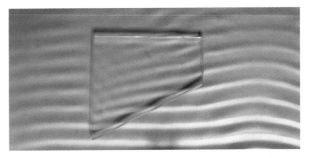

▲ **Figure 3.1.9a** Waves are refracted at the boundary between deep and shallow regions.

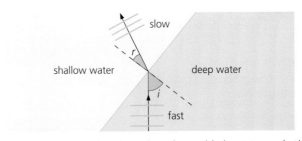

▲ **Figure 3.1.9b** The direction of travel is bent towards the normal in the shallow region.

The speed of waves also changes, and refraction occurs, when waves move from one medium to another. For example, light waves are refracted when they move from air to glass.

Diffraction

Diffraction through a narrow gap

In Figures 3.1.10a and 3.1.10b, straight water waves in a ripple tank are meeting gaps formed by obstacles. In Figure 3.1.10a the gap is narrow and the wavefronts curve around the edges of the gap producing a circular wavefront.

Diffraction due to an edge or wide gap

In Figure 3.1.10b the gap is wide (10 cm) compared with the wavelength and the wavefront remains straight, except at the edges of the gap where some curvature around the edges occurs. The spreading of waves at the edges of obstacles is called **diffraction**; when designing harbours, engineers use models like that in Figure 3.1.11 to study it.

▲ **Figure 3.1.10a** Spreading of waves after passing through a narrow gap

▲ **Figure 3.1.10b** Spreading of waves after passing through a wide gap

▲ **Figure 3.1.11** Model of a harbour used to study wave behaviour

Effect of wavelength and gap size on diffraction

In Figure 3.1.10a the gap width is about the same as the wavelength of the waves (1 cm); the wavefronts that pass through become circular and spread out in all directions. In Figure 3.1.10b the gap is wide (10 cm) compared with the wavelength and the waves continue straight on; some spreading occurs but it is less obvious.

Diffraction at an edge

For a single edge, diffraction will occur at the edge. The spreading and curvature of the wavefront around the edge will be more noticeable for longer wavelengths. Diffraction of radio waves around an obstacle is shown in Figure 3.3.4a.

▶ Test yourself

2 One side of a ripple tank ABCD is raised slightly (Figure 3.1.12), and a ripple is started at P by a finger. After 1 second the shape of the ripple is as shown.
  a Why is it not circular?
  b Which side of the tank has been raised?

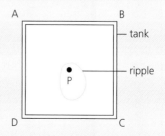

▲ **Figure 3.1.12**

3 The angle of incidence of a ray at a plane reflecting surface is 35°. What will be the angle of reflection?

4 During the refraction of a wave, which one of the following properties does not change?
  A The wave speed
  B The frequency
  C The wavelength
  D Direction of travel

5 When a wave passes through a narrow gap, which of the following terms best describes what occurs?
  A Reflection
  B Refraction
  C Diffraction
  D Inversion

 Going further

Wave theory

If the position of a wavefront is known at one instant, its position at a later time can be found using Huygens' construction. Each point on the wavefront is considered to be a source of *secondary* spherical wavelets (Figure 3.1.13) which spread out at the wave speed; the new wavefront is the surface that touches all the wavelets (in the forward direction). In Figure 3.1.13 the straight wavefront AB is travelling from left to right with speed *v*. At a time *t* later, the spherical wavelets from AB will be a distance *vt* from the secondary sources and the new surface which touches all the wavelets is the straight wavefront CD.

Wave theory can be used to explain reflection, refraction and diffraction effects.

Reflection and wave theory

Figure 3.1.14 shows a straight wavefront AB incident at an angle *i* on a reflecting surface; the wavefront has just reached the surface at A. The position of the wavefront a little later, when B reaches the reflecting surface, can be found using Huygens' construction. A circle of radius BB' is drawn about A; the reflected wavefront is then A'B', the tangent to the wavelet from B'. Measurements of the angle of incidence *i* and the angle of reflection *r* show that they are equal.

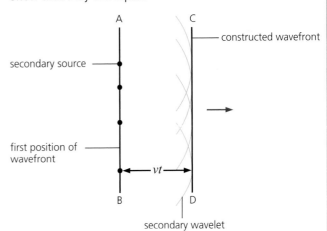

▲ **Figure 3.1.13** Huygens' construction for a straight wavefront

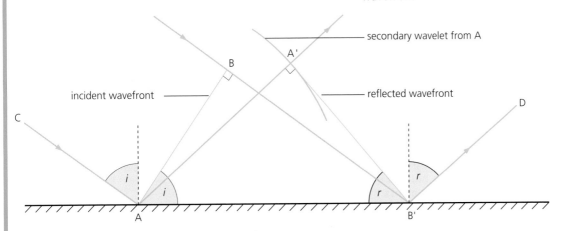

▲ **Figure 3.1.14** Reflection of a straight wavefront

Refraction and wave theory

Huygens' construction can also be used to find the position of a wavefront when it enters a second medium in which the speed of travel of the wave is different from in the first (see Figure 3.1.9b).

In Figure 3.1.15, point A, on the straight wavefront AB, has just reached the boundary between two media. When B reaches the boundary the secondary wavelet from A will have moved on to A'. If the wave travels more slowly in the second medium the distance AA' is shorter than BB'. The new wavefront is then A'B', the tangent to the wavelet from B'; it is clear that the direction of travel of the wave has changed – refraction has occurred.

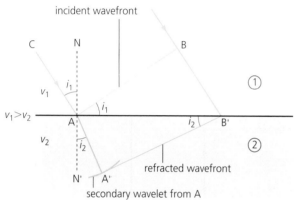

▲ **Figure 3.1.15** Refraction of a straight wavefront

Diffraction and wave theory

Diffraction effects, such as those shown in Figure 3.1.11, can be explained by describing what happens to secondary wavelets arising from point sources on the unrestricted part of the wavefronts in the gaps.

Interference

When two sets of continuous circular waves cross in a ripple tank, a pattern like that in Figure 3.1.16 is obtained.

At points where a crest from one source, S_1, arrives at the same time as a crest from the other source, S_2, a bigger crest is formed, and the waves are said to be *in phase*. At points where a crest and a trough arrive together, they cancel out (if their amplitudes are equal); the waves are exactly *out of phase* (because they have travelled different distances from S_1 and S_2) and the water is undisturbed; in Figure 3.1.16 the blurred lines radiating from between S_1 and S_2 join such points.

▲ **Figure 3.1.16** Interference of circular waves

Interference or superposition is the combination of waves to give a larger or a smaller wave (Figure 3.1.17).

▲ **Figure 3.1.17**

Study this effect with two ball dippers about 3 cm apart on the bar of a ripple tank. Also observe the effect of changing (i) the frequency and (ii) the separation of the dippers; use a stroboscope when necessary. You will find that if the frequency is increased, i.e. the wavelength decreased, the blurred lines are closer together. Increasing the separation has the same effect.

Polarisation

The polarisation effect occurs only with transverse waves. It can be shown by fixing a rope at one end, D in Figure 3.1.18, and passing it through two slits, B and C. If end A is vibrated in all directions (as shown by the short arrowed lines), vibrations of the rope occur in every plane and transverse waves travel towards B.

At B only waves due to vibrations in a vertical plane can emerge from the vertical slit. The wave between B and C is said to be plane polarised (in the vertical plane containing the slit at B). By contrast the waves between A and B are unpolarised. If the slit at C is vertical, the wave can travel on, but if the slit at C is horizontal (as shown), the wave is stopped. The slits are said to be *crossed*.

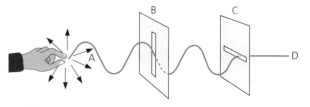

▲ **Figure 3.1.18** Polarising waves on a rope

Revision checklist

After studying Topic 3.1 you should know and understand the following:

✔ the direction of vibration for transverse and longitudinal wave motion and that waves transfer energy without transferring matter

✔ the meaning of wavelength, frequency, wave speed, amplitude, crest (peak), trough and wavefront.

After studying Topic 3.1 you should be able to:

✔ describe the production and give examples of transverse and longitudinal waves and use the wave equation $v = f\lambda$ to solve problems

✔ describe experiments and draw diagrams to show reflection, refraction and diffraction of water waves

✔ predict the effect of changing the wavelength or the size of the gap on diffraction of waves at a single slit.

Exam-style questions

1 Figure 3.1.19 gives a full-scale representation of the water in a ripple tank 1 second after the vibrator was started. The coloured lines represent crests.

▲ **Figure 3.1.19**

a What is represented at A at this instant? [1]

b Estimate
  i the wavelength [2]
  ii the wave speed [3]
  iii the frequency of the vibrator. [2]

c Describe a suitable attachment which could have been vibrated up and down to produce this wave pattern. [2]

[Total: 10]

2 a In the transverse wave shown in Figure 3.1.20 distances are in centimetres. Which pair of entries A to D is correct? [1]

	A	B	C	D
Amplitude	2	4	4	8
Wavelength	4	4	8	8

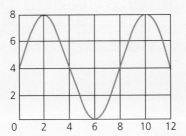

▲ **Figure 3.1.20**

b When a water wave goes from deep to shallow water, the changes (if any) in its wave speed, wavelength and frequency are described by which of the following options: [1]

	Wave speed	Wavelength	Frequency
A	greater	greater	the same
B	less	less	the same
C	the same	less	greater
D	less	the same	less

[Total: 2]

3 Copy Figure 3.1.21 and show on it what happens to the waves as they pass through the gap if the water is much shallower on the right side of the gap than on the left.

▲ **Figure 3.1.21**

[Total: 6]

3.2 Light

3.2.1 Reflection of light

FOCUS POINTS

★ Define and understand the terms normal, angle of incidence and angle of reflection.
★ Describe how an optical image is formed by a plane mirror and give its characteristics.
★ For reflection, know the relationship between angle of incidence and angle of reflection.

★ Construct, measure and calculate reflection by plane mirrors.

In the last topic you learnt about some of the general properties of waves by studying the behaviour of water waves. Water waves are transverse waves but so too are light waves and you can expect them both to have similar properties. In this topic you will explore how light is reflected by a plane surface. Reflection is made use of in mirrors and instruments such as a periscope. You may have noticed that images in a plane mirror are laterally inverted. This happens because reflection produces an apparent image of an object behind the surface of the mirror; this virtual image is the same size as the object but has left and right switched.

Sources of light

You can see an object only if light from it enters your eyes. Some objects such as the Sun, electric lamps and candles make their own light. We call these **luminous** sources.

Most things you see do not make their own light but reflect it from a luminous source. They are **non-luminous** objects. This page, you and the Moon are examples. Figure 3.2.1 shows some others.

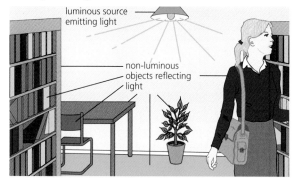

▲ **Figure 3.2.1** Luminous and non-luminous objects

Luminous sources radiate light when their atoms become 'excited' as a result of receiving energy. In a light bulb, for example, the energy comes from electricity. The 'excited' atoms give off their light haphazardly in most luminous sources.

A light source that works differently is the *laser*, invented in 1960. In laser light sources the excited atoms act together and emit a narrow, very bright **beam** of light. The laser has a host of applications. It is used in industry to cut through plate metal, in scanners to read bar codes at shop and library check-outs, in CD players, in optical fibre telecommunication systems, in delicate medical operations on the eye or inner ear (for example Figure 3.2.2), in printing and in surveying and range-finding.

▲ **Figure 3.2.2** Laser surgery in the inner ear

Rays and beams

Sunbeams streaming through trees (Figure 3.2.3) and light from a cinema projector on its way to the screen both suggest that light travels in straight lines. The beams are visible because dust particles in the air reflect light into our eyes.

The direction of the path in which light is travelling is called a *ray* and is represented in diagrams by a straight line with an arrow on it. A beam is a stream of light and is shown by a number of rays, as in Figure 3.2.4. A beam may be **parallel**, **diverging** (spreading out) or **converging** (getting narrower).

▲ **Figure 3.2.3** Light travels in straight lines.

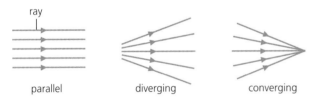

▲ **Figure 3.2.4** Beams of light

Practical work

The pinhole camera

Safety

- Take care when using the needle to make the pinhole.

A simple pinhole camera is shown in Figure 3.2.5a. Make a small pinhole in the centre of the black paper. Half darken the room. Hold the box at arm's length so that the pinhole end is nearer to and about 1 metre from a luminous object, such as a carbon **filament** lamp or a candle. Look at the **image** on the screen

(an image is a likeness of an object and need not be an exact copy). Make several small pinholes round the large hole (Figure 3.2.5b), and view the image again.

The formation of an image is shown in Figure 3.2.6.

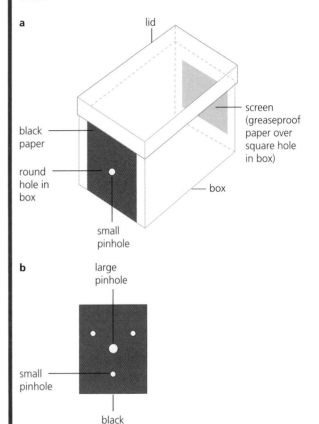

▲ **Figure 3.2.5**

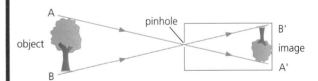

▲ **Figure 3.2.6** Forming an image in a pinhole camera

1 Can you see *three* ways in which the image differs from the object?
2 What is the effect of moving the camera closer to the object?
3 Make the pinhole larger. What happens to the
   a brightness
   b sharpness
   c size of the image?

➡ **Going further**

Shadows

Shadows are formed for two reasons. First, because some objects, which are said to be opaque, do not allow light to pass through them. Secondly, light travels in straight lines.

The sharpness of the shadow depends on the size of the light source. A very small source of light, called a point source, gives a sharp shadow which is equally dark all over. This may be shown as in Figure 3.2.7a where the small hole in the card acts as a point source.

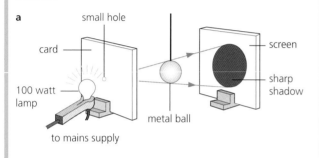

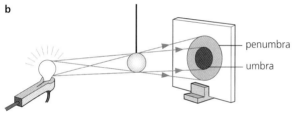

▲ **Figure 3.2.7** Forming a shadow

If the card is removed the lamp acts as a large or extended source (Figure 3.2.7b). The shadow is then larger and has a central dark region, the umbra, surrounded by a ring of partial shadow, the penumbra. You can see by the rays that some light reaches the penumbra, but none reaches the umbra.

Speed of light

Proof that light travels very much faster than sound is provided by a thunderstorm. The flash of lightning is seen before the thunder is heard. The length of the time lapse is greater the further the observer is from the storm.

The speed of light has a definite value; light does not travel instantaneously from one point to another but takes a certain, very small time. Its speed is about 1 million times greater than that of sound.

Reflection of light

If we know how light behaves when it is reflected, we can use a mirror to change the direction in which the light is travelling. This happens when a mirror is placed at the entrance of a concealed drive to give warning of approaching traffic, for example.

An ordinary mirror is made by depositing a thin layer of silver on one side of a piece of glass and protecting it with paint. The silver – at the *back* of the glass – acts as the reflecting surface. A plane mirror is produced when the reflecting surface is flat.

Law of reflection

Terms used in connection with reflection are shown in Figure 3.2.8. The perpendicular to the mirror at the point where the incident ray strikes is called the **normal**. Note that the **angle of incidence** i is the angle between the incident ray and the normal; similarly, the **angle of reflection** r is the angle between the reflected ray and the normal.

Key definitions

Normal line which is perpendicular to a surface

Angle of incidence angle between incident ray and the normal to a surface

Angle of reflection angle between reflected ray and the normal to a surface

Law of reflection the angle of incidence is equal to the angle of reflection

---

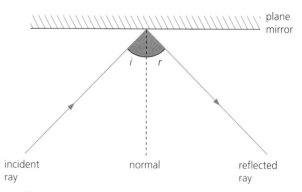

▲ **Figure 3.2.8** Reflection of light by a plane mirror

The **law of reflection** states:

> The angle of incidence is equal to the angle of reflection.

The incident ray, the reflected ray and the normal all lie in the same plane. (This means that they could all be drawn on a flat sheet of paper.)

Practical work

Reflection by a plane mirror

For safe experiments/demonstrations related to this topic, please refer to the *Cambridge IGCSE Physics Practical Skills Workbook* that is also part of this series.

Safety

● Take care as the filament lamp and shield will get hot when in use.

Draw a line AOB on a sheet of paper and using a protractor mark angles on it. Measure them from the perpendicular ON, which is at right angles to AOB. Set up a plane (flat) mirror with its reflecting surface on AOB.

Shine a narrow ray of light along, say, the 30° line onto the mirror (Figure 3.2.9).

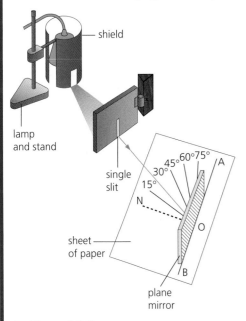

▲ **Figure 3.2.9**

Mark the position of the reflected ray, remove the mirror and measure the angle between the reflected ray and ON. Repeat for rays at other angles.

4 In the experiment what can you conclude about the incident and reflected rays?
5 The silver surface on a mirror is usually on the back surface of the glass. How might this affect the accuracy of your measurements?

Periscope

A simple *periscope* consists of a tube containing two plane mirrors, fixed parallel to and facing each other. Each makes an angle of 45° with the line joining them (Figure 3.2.10). Light from the object is turned through 90° at each reflection and an observer is able to see over a crowd, for example (Figure 3.2.11), or over the top of an obstacle.

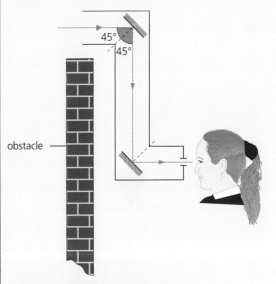

▲ **Figure 3.2.10** Action of a simple periscope

▲ **Figure 3.2.11** Periscopes being used by people in a crowd.

In more elaborate periscopes like those used in submarines, prisms replace mirrors (see p. 148).

Make your own periscope from a long, narrow cardboard box measuring about 40 cm × 5 cm × 5 cm (such as one in which aluminium cooking foil or clingfilm is sold), two plane mirrors (7.5 cm × 5 cm) and sticky tape. When you have got it to work, make modifications that turn it into a 'see-back-o-scope', which lets you see what is behind you.

Regular and diffuse reflection

If a parallel beam of light falls on a plane mirror it is reflected as a parallel beam (Figure 3.2.12a) and **regular reflection** occurs. Most surfaces, however, reflect light irregularly and the rays in an incident parallel beam are reflected in many directions (Figure 3.2.12b): this is known as **diffuse reflection**.

a

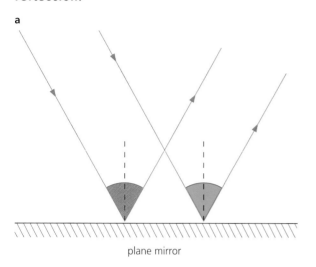

plane mirror

b

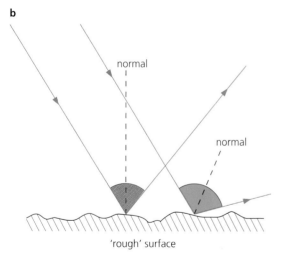

'rough' surface

▲ **Figure 3.2.12**

Plane mirrors

When you look into a plane mirror on the wall of a room you see an image of the room behind the mirror; it is as if there were another room. Restaurants sometimes have a large mirror on one wall just to make them look larger. You may be able to say how much larger after the next experiment.

The position of the image formed by a mirror depends on the position of the object.

 Practical work

Position of the image

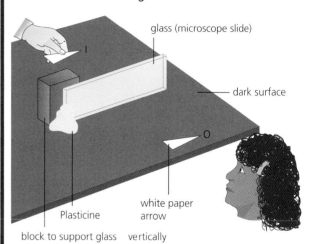

glass (microscope slide)

dark surface

O

white paper arrow

Plasticine

block to support glass   vertically

▲ **Figure 3.2.13**

Support a piece of thin glass on the bench, as in Figure 3.2.13. It must be *vertical* (at 90° to the bench). Place a small paper arrow, O, about 10 cm from the glass. The glass acts as a poor mirror and an image of O will be seen in it; the darker the bench top, the brighter the image will be.

Lay another identical arrow, I, on the bench behind the glass; move it until it coincides with the image of O.

6 How do the sizes of O and its image compare?
7 Imagine a line joining them. What can you say about it?
8 Measure the distances of the points of O and I from the glass along the line joining them. How do they compare?
9 Try placing O at other distances and orientations. How does the orientation of the image relative to the object change if the arrow is turned through 45°?

Real and virtual images

A **real image** is one which can be produced on a screen (as in a pinhole camera) and is formed by rays that actually pass through the screen.

A **virtual image** cannot be formed on a screen. The virtual image is produced by rays which seem to come from it but do not pass through it. The image in a plane mirror is virtual. Rays from a point on an object are reflected at the mirror and appear to our eyes to come from a point behind the mirror where the rays would intersect when extrapolated backwards (Figure 3.2.14). IA and IB are construction lines and are shown as broken lines.

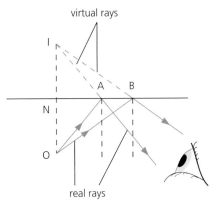

▲ **Figure 3.2.14** A plane mirror forms a virtual image.

➡ Going further

Lateral inversion

If you close your left eye, your image in a plane mirror seems to close the right eye. In a mirror image, left and right are interchanged and the image appears to be laterally inverted. The effect occurs whenever an image is formed by one reflection and is very evident if print is viewed in a mirror (Figure 3.2.15). What happens if two reflections occur, as in a periscope?

▲ **Figure 3.2.15** The image in a plane mirror is laterally inverted.

Properties of the image

The image in a plane mirror is
(i) as far behind the mirror as the object is in front, with the line joining the same points on object and image being perpendicular to the mirror
(ii) the same size as the object
(iii) virtual.

From (i) it follows that the *orientation* of an image depends on the orientation of the object.

When the object is aligned parallel to the surface of the mirror, the image has the same orientation, but when the object is aligned parallel to the normal to the mirror, the orientation of the image is at 180° relative to that of the object. For intermediate positions the relative orientation of object and image is twice the angle between the direction of orientation of the object and the surface of the mirror.

Kaleidoscope

▲ **Figure 3.2.16** A kaleidoscope produces patterns using the images formed by two plane mirrors – the patterns change as the objects move.

Check that the images in Figure 3.2.16 have the expected properties of an image in a plane mirror.

To see how a kaleidoscope works, draw on a sheet of paper two lines at right angles to one another. Using different coloured pens or pencils, draw a design between them (Figure 3.2.17a). Place a small mirror along each line and look into the mirrors (Figure 3.2.17b). You will see three reflections which join up to give a circular pattern. If you make the angle between the mirrors smaller, more reflections appear but you always get a complete design.

In a kaleidoscope the two mirrors are usually kept at the same angle (about 60°) and different designs are made by hundreds of tiny coloured beads which can be moved around between the mirrors.

Now make a kaleidoscope using a cardboard tube (from half a kitchen roll), some thin card, greaseproof paper, clear sticky tape, small pieces of different coloured cellophane and two mirrors (10 cm × 3 cm) or a single plastic mirror (10 cm × 6 cm) bent to form two mirrors at 60° to each other, as shown in Figure 3.2.17c.

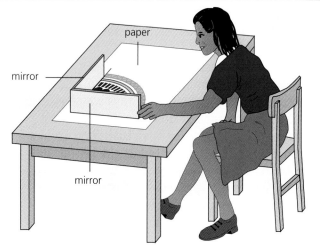

▲ Figure 3.2.17b

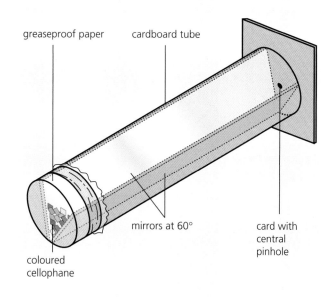

▲ Figure 3.2.17c

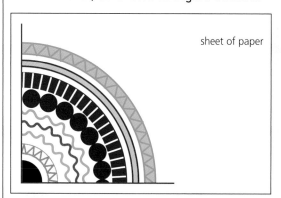

▲ Figure 3.2.17a

Test yourself

1 How would the size and brightness of the image formed by a pinhole camera change if the camera were made longer?
2 What changes would occur in the image if the single pinhole in a camera were replaced by
   a four pinholes close together
   b a hole 1 cm wide?
3 In Figure 3.2.18 the completely dark region on the screen is
   A PQ   **B** PR   **C** QR   **D** QS
4 When watching a distant firework display do you see the cascade of lights before or after hearing the associated bang? Explain your answer.

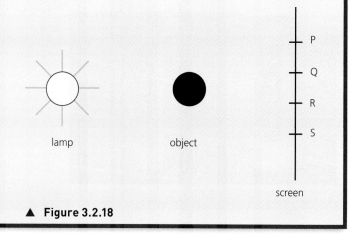

▲ Figure 3.2.18

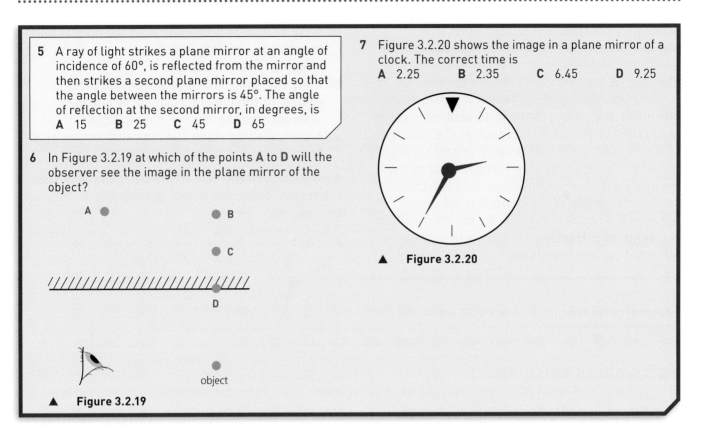

5 A ray of light strikes a plane mirror at an angle of incidence of 60°, is reflected from the mirror and then strikes a second plane mirror placed so that the angle between the mirrors is 45°. The angle of reflection at the second mirror, in degrees, is

A 15   **B** 25   **C** 45   **D** 65

6 In Figure 3.2.19 at which of the points **A** to **D** will the observer see the image in the plane mirror of the object?

▲ **Figure 3.2.19**

7 Figure 3.2.20 shows the image in a plane mirror of a clock. The correct time is

A 2.25   **B** 2.35   **C** 6.45   **D** 9.25

▲ **Figure 3.2.20**

3.2.2 Refraction of light

FOCUS POINTS

★ Understand the terms normal, angle of incidence, angle of refraction and critical angle.
★ Know how to demonstrate the refraction of light by transparent blocks of different shapes and describe the passage of light through a transparent material.

★ Understand the term refractive index and use the correct equations to calculate refractive index and critical angle.

★ Use experimental and everyday examples to describe internal reflection and total internal reflection.

★ Understand the uses of optical fibres.

If you place a coin in an empty dish and move back until you just cannot see it, the result is surprising if someone gently pours water into the dish. Try it! In this topic, you will see that, as with water waves, when a parallel beam of light travels from one medium to another it changes direction if the speed of light differs in the second medium. For passage from air to a transparent medium, the angles to the normal to the surface for the incident and refracted beams are related through the refractive index. At the critical angle of incidence from a transparent medium to air, the refracted beam lies along the surface. At greater angles of incidence, the beam is totally internally reflected. Total internal reflection is used in optical fibres to transmit images and signals in applications from medicine to telecommunications.

Although light travels in straight lines in a transparent material, such as air, if it passes into a different material, such as water, it changes direction at the boundary between the two, i.e. it is bent. The *bending of light* when it passes from one material (called a medium) to another is called **refraction**. It causes effects such as the coin trick.

Terms used in connection with refraction are shown in Figure 3.2.21. The perpendicular to the boundary between two mediums is called the normal. The **angle of incidence** i is the angle between the incident ray and the normal; similarly, the **angle of refraction** r is the angle between the refracted ray and the normal.

> **Key definitions**
>
> **Angle of refraction** angle between refracted ray and the normal to a surface

Facts about refraction

(i) A ray of light is bent *towards* the normal when it enters an optically denser medium at an angle, for example from air to glass as in Figure 3.2.21. So the angle of refraction r is smaller than the angle of incidence i.

(ii) A ray of light is bent *away from* the normal when it enters an optically less dense medium, for example from glass to air.

(iii) A ray emerging from a parallel-sided block is *parallel* to the ray entering, but is *displaced sideways*, like the ray in Figure 3.2.21a.

(iv) A ray travelling along the normal direction at a boundary is *not refracted* (Figure 3.2.21b).

Note that 'optically denser' means having a greater refraction effect; the actual density may or may not be greater.

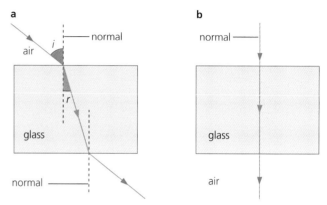

▲ **Figure 3.2.21** Refraction of light in glass

Practical work

Refraction in glass

For safe experiments/demonstrations related to this topic, please refer to the *Cambridge IGCSE Physics Practical Skills Workbook* that is also part of this series.

Safety

● Take care as the filament lamp and shield will get hot when in use.

Shine a ray of light at an angle onto a glass block (which has its lower face painted white or is frosted), as in Figure 3.2.22. Draw the outline ABCD of the block on the sheet of paper under it. Mark the positions of the various rays in air and in glass.

Remove the block and draw the normals on the paper at the points where the ray enters side AB (see Figure 3.2.22) and where it leaves side CD.

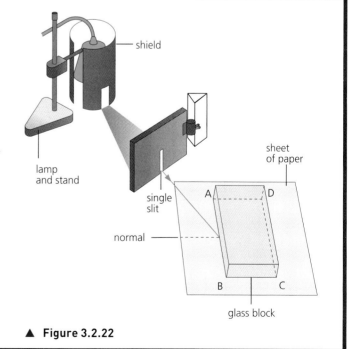

▲ **Figure 3.2.22**

10 What *two* things happen to the light falling on AB?

11 When the ray enters the glass at AB, is it bent towards or away from the part of the normal in the block?

12 How is it bent at CD?

13 What can you say about the direction of the ray falling on AB and the direction of the ray leaving CD?

14 What happens if the ray hits AB at right angles?

Real and apparent depth

Rays of light from a point O on the bottom of a pool are refracted away from the normal at the water surface because they are passing into an optically less dense medium, i.e. air (Figure 3.2.23). On entering the eye, they appear to come from a point I that is *above* O; I is the virtual image of O formed by refraction. The apparent depth of the pool is less than its real depth. Similarly, rays from the submerged part of the pencil in Figure 3.2.24 are refracted at the water surface.

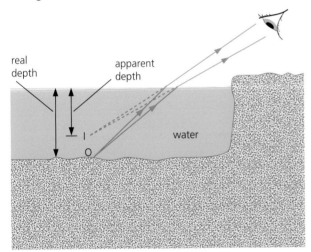

▲ **Figure 3.2.23** A pool of water appears shallower than it is.

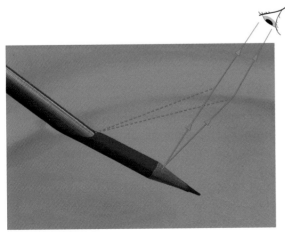

▲ **Figure 3.2.24** A pencil placed in water seems to bend at the water surface.

Refractive index

Light is refracted because its speed changes when it enters another medium. An analogy helps to explain why.

Suppose three people A, B, C are marching in line, with hands linked, on a good road surface. If they approach marshy ground at an angle (see Figure 3.2.25a on the next page), person A is slowed down first, followed by B and then C. This causes the whole line to swing round and change its direction of motion.

In air (and a vacuum) light travels at 300 000 km/s (3×10^8 m/s); in glass its speed falls to 200 000 km/s (2×10^8 m/s) (Figure 3.2.25b).

We define the **refractive index**, n, as the ratio of the speeds of a wave in two different regions.

In Figure 3.2.25b for the light passing from air to glass,

$$\text{refractive index, } n = \frac{\text{speed of light in air (or a vacuum)}}{\text{speed of light in medium}}$$

$$\text{for glass, } n = \frac{300\,000\,\text{km/s}}{200\,000\,\text{km/s}} = \frac{3}{2}$$

> **Key definition**
>
> **Refractive index *n*** the ratio of the speeds of a wave in two different regions

Experiments also show that

$$n = \frac{\text{sine of angle between ray in air and normal}}{\text{sine of angle between ray in glass and normal}}$$

$$= \frac{\sin i}{\sin r}$$

The more light is slowed down when it enters a medium from air, the greater is the refractive index of the medium and the more the light is bent.

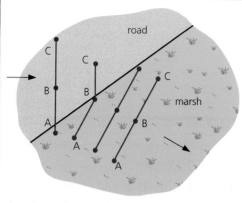

▲ **Figure 3.2.25a**

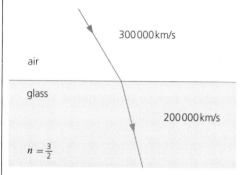

▲ **Figure 3.2.25b**

We saw earlier (Topic 3.1) that water waves are refracted when their speed changes. The change in the direction of travel of a light ray when its speed changes on entering another medium suggests that light may also be a type of wave motion.

> **?** **Worked example**

The refractive index for a certain glass is 1.6.
a Calculate the angle of refraction for an angle of incidence of 24°.

$$n = \frac{\sin i}{\sin r}$$

so $\sin r = \sin i / n$

$$= \sin 24° / 1.6$$

$$= 0.41 / 1.6 = 0.25$$

and $r = 15°$

b Calculate the speed of light in the glass.

$$\text{refractive index, } n = \frac{\text{speed of light in air (or a vacuum)}}{\text{speed of light in medium}}$$

so speed of light in glass = speed of light in air/n

$$= 3.0 \times 10^8 \, \text{m/s} / 1.6$$

$$= 1.9 \times 10^8 \, \text{m/s}$$

Now put this into practice

1 The refractive index for a certain glass is 1.5.
  Calculate the angle of refraction for an angle of incidence of 30°.
2 The refractive index of water is 1.3.
  Calculate the speed of light in water.

Critical angle

When light passes at small angles of incidence from an optically dense to a less dense medium, such as from glass to air, there is a strong refracted ray and a weak ray reflected back into the denser medium (Figure 3.2.26a). As well as refraction, some *internal reflection* occurs. Increasing the angle of incidence increases the angle of refraction.

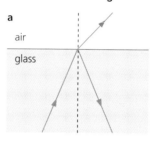

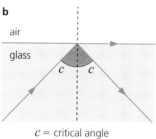

c = critical angle

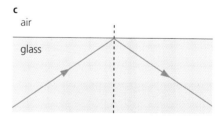

▲ **Figure 3.2.26**

At a certain angle of incidence, called the **critical angle**, c, the angle of refraction is 90° (Figure 3.2.26b) and the refracted ray passes along the boundary between the two media.

For angles of incidence greater than c, the refracted ray disappears and all the incident light is reflected inside the denser medium (Figure 3.2.26c). The light does not cross the boundary and is said to undergo **total internal reflection**.

On a hot day the road ahead may appear to shimmer with water. The layers of air close to the surface of the road are hotter and less dense than those above and refraction of sunlight occurs. When the critical angle of incidence is reached, the light undergoes total internal reflection, resulting in a mirage which disappears as you move towards it.

Practical work

Critical angle of glass

Place a semicircular glass block on a sheet of paper (Figure 3.2.27), and draw the outline LOMN where O is the centre and ON the normal at O to LOM. Direct a narrow ray (at an angle of about 30° to the normal ON) along a radius towards O. The ray is not refracted at the curved surface. Note the refracted ray emerging from LOM into the air and also the weak internally reflected ray in the glass.

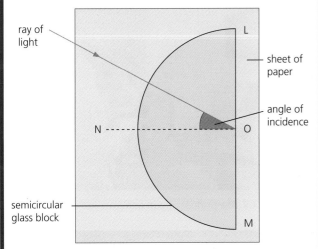

▲ **Figure 3.2.27**

Slowly rotate the paper so that the angle of incidence on LOM increases until total internal reflection *just* occurs. Mark the incident ray. Measure the angle of incidence; this is the critical angle.

15 Why is the ray not refracted at the curved surface?

16 What is the value you obtain for the critical angle?

17 Name the rays at O when the angle of incidence is less than the critical angle.

Key definition

Critical angle c angle of incidence which produces an angle of refraction of 90°

Refractive index and critical angle

From Figure 3.2.26b and the definition of refractive index:

$$n = \frac{\text{sine of angle between ray in air and normal}}{\text{sine of angle between ray in glass and normal}}$$

$$= \frac{\sin 90°}{\sin c}$$

$$= \frac{1}{\sin c} \qquad \text{(because } \sin 90° = 1\text{)}$$

So, if $n = 3/2$, then $\sin c = 2/3$ and c must be 42°.

? Worked example

If the critical angle for diamond is 24°, calculate its refractive index.

$$\text{critical angle, } c = 24°$$
$$\sin 24° = 0.4$$
$$n = \frac{\sin 90°}{\sin c} = \frac{1}{\sin 24°}$$
$$= \frac{1}{0.4} = 2.5$$

Now put this into practice

1 The critical angle for a transparent material is 32°. Calculate its refractive index.
2 The refractive index of a transparent material is 1.7. Work out its critical angle.

Multiple images in a mirror

An ordinary mirror silvered at the back forms several images of one object, because of multiple reflections inside the glass (Figure 3.2.28). These blur the main image I (which is formed by one reflection at the silvering), especially if the glass is thick. The problem is absent in front-silvered mirrors but such mirrors are easily damaged.

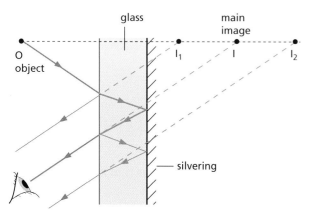

▲ **Figure 3.2.28a** Multiple reflections in a mirror

▲ **Figure 3.2.28b** The multiple images in a mirror cause blurring.

Totally reflecting prisms

The defects of mirrors are overcome if 45° right-angled glass prisms are used. The critical angle of ordinary glass is about 42° and a ray falling normally on face PQ of such a prism (Figure 3.2.29a) hits face PR at 45°. Total internal reflection occurs and the ray is turned through 90°. Totally reflecting prisms replace mirrors in good periscopes.

Light can also be reflected through 180° by a prism (Figure 3.2.29b); this happens in binoculars.

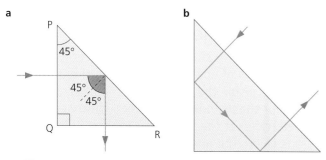

▲ **Figure 3.2.29** Reflection of light by a prism

Light pipes and optical fibres

Light can be trapped by total internal reflection inside a bent glass rod and piped along a curved path (Figure 3.2.30). A single, very thin glass fibre, an *optical fibre*, behaves in the same way.

▲ **Figure 3.2.30** Light travels through a curved glass rod or optical fibre by total internal reflection.

If several thousand such fibres are taped together, a flexible light pipe is obtained that can be used, for example, by doctors as an endoscope (Figure 3.2.31a), to obtain an image from inside the body (Figure 3.2.31b), or by engineers to light up some awkward spot for inspection.

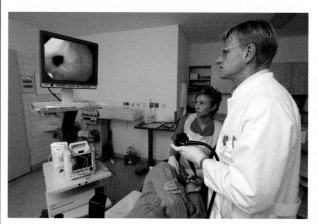

▲ **Figure 3.2.31a** Endoscope in use

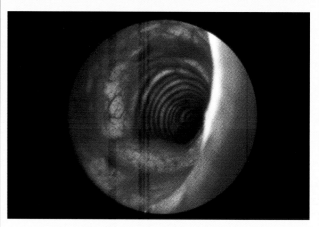

▲ **Figure 3.2.31b** Trachea (windpipe) viewed by an endoscope

Increasingly, optical fibres are being used to carry telephone, high-speed broadband internet and cable TV signals as pulses of visible or infrared light. The advantages over copper cables for telecommunication purposes are that the use of light allows information to be transmitted at a higher rate and the data is more secure because the cables are unaffected by electronic interference. They can be used over longer distances (since they have lower power loss), are made of cheaper material and, as they are lighter and thinner, are easier to handle and install. However, they are not as robust as copper cables and can break if bent too much.

▶ Test yourself

8 Figure 3.2.32 shows a ray of light entering a rectangular block of glass.
   a Copy the diagram and draw the normal at the point of entry.
   b Sketch the approximate path of the ray through the block and out of the other side.

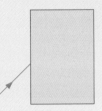

▲ **Figure 3.2.32**

9 Draw two rays from a point on a fish in a stream to show where someone on the bank will see the fish. Where must the person aim to spear the fish?

10 What is the speed of light in a medium of refractive index 6/5 if its speed in air is 300 000 km/s?

11 Which diagram in Figure 3.2.33 shows the ray of light refracted correctly?

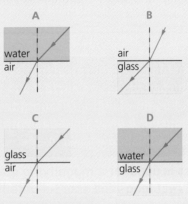

▲ **Figure 3.2.33**

12 Copy Figures 3.2.34a and 3.2.34b and complete the paths of the rays through the glass prisms.

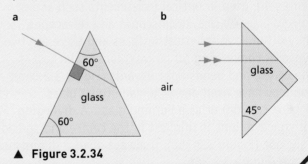

▲ **Figure 3.2.34**

3.2.3 Thin lenses

FOCUS POINTS

★ Understand how thin converging and diverging lenses act on a beam of light.
★ Understand the terms focal length, principal axis and principal focus (focal point).
★ Draw and interpret ray diagrams.

★ Describe how a single lens can be used as a magnifying glass.

★ Describe the characteristics of an image in terms of size, orientation and whether it is real or virtual.
★ Know that when diverging rays are extrapolated backwards a virtual image is formed, which does not form a visible projection on a screen.

★ Describe how long- and short-sightedness can be corrected through the use of converging and diverging lenses.

In the last topic you learnt about refraction at plane surfaces. Lenses make use of the refraction of light at carefully shaped curved surfaces to form images. For a thin converging lens with convex surfaces, a parallel beam of light is brought to a focus; the principal focus is real and a range of images can be produced. A diverging lens with concave surfaces has an apparent principal focus and always produces a diminished virtual image.

A converging lens is used for a magnifying glass. Both converging and diverging lenses are used in spectacles to correct long- and short-sightedness.

Converging and diverging lenses

Lenses are used in optical instruments such as cameras, spectacles, microscopes and telescopes; they often have spherical surfaces and there are two types. A **converging** (or convex) lens is thickest in the centre and bends light inwards (Figure 3.2.35a). You may have used one as a magnifying glass (Figure 3.2.36a) or as a burning glass. A **diverging** (or concave) lens is thinnest in the centre and spreads light out (Figure 3.2.35b); it always gives a diminished image (Figure 3.2.36b).

The centre of a lens is its **optical centre**, C; the line through C at right angles to the lens is the **principal axis**.

The action of a lens can be understood by treating it as a number of prisms (most with the tip removed), each of which bends the ray towards its base, as in Figure 3.2.35c and Figure 3.2.35d. The centre acts as a parallel-sided block.

Key definition
Principal axis line through the optical centre of a lens at right angles to the lens

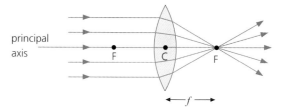

▲ Figure 3.2.35a

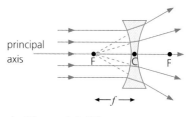

▲ Figure 3.2.35b

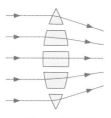

▲ Figure 3.2.35c

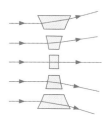

▲ Figure 3.2.35d

▲ **Figure 3.2.36a** A converging lens forms a magnified image of a close object

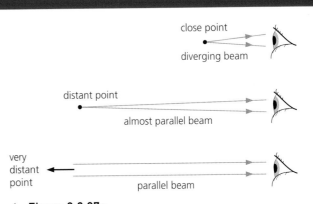

▲ **Figure 3.2.36b** A diverging lens always forms a diminished image

Principal focus

When a beam of light parallel to the principal axis passes through a converging lens it is refracted so as to converge to a point on the axis called the **principal focus (focal point)**, F. It is a real **focus**. A diverging lens has a virtual principal focus behind the lens, from which the refracted beam seems to diverge.

Since light can fall on both faces of a lens it has two principal foci, one on each side, equidistant from C. The distance CF is the **focal length** f of the lens (see Figure 3.2.35a); it is an important property of a lens. The more curved the lens faces are, the smaller is f and the more powerful is the lens.

> **Key definitions**
>
> **Principal focus (focal point)** point on the principal axis of a lens to which light rays parallel to the principal axis converge, or appear to diverge from
>
> **Focal length** distance between the optical centre and the principal focus of a lens

---

 Practical work

Focal length, f, of a converging lens

For safe experiments/demonstrations related to this topic, please refer to the *Cambridge IGCSE Physics Practical Skills Workbook* that is also part of this series.

We use the fact that rays from a point on a very distant object, i.e. at infinity, are nearly parallel (Figure 3.2.37a).

close point

diverging beam

distant point

almost parallel beam

very distant point

parallel beam

▲ **Figure 3.2.37a**

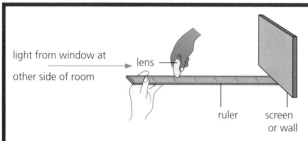

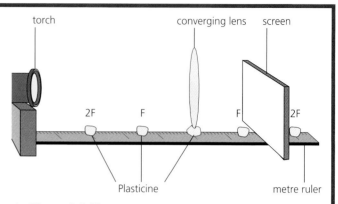

▲ Figure 3.2.37b

Move the lens, arranged as in Figure 3.2.37b, until a *sharp* image of a window at the other side of the room is obtained on the screen. The distance between the lens and the screen is then f, roughly.

18 Why is the distance between the lens and the screen f, roughly?

19 Why could the focal length of a diverging lens not be found by the method suggested in the practical?

Images formed by a converging lens

For safe experiments/demonstrations related to this topic, please refer to the *Cambridge IGCSE Physics Practical Skills Workbook* that is also part of this series.

In the formation of images by lenses, two important points on the principal axis are F and 2F; 2F is at a distance of twice the focal length from C.

First find the focal length of the lens by the distant object method just described, then fix the lens upright with Plasticine at the centre of a metre ruler. Place small pieces of Plasticine at the points F and 2F on both sides of the lens, as in Figure 3.2.38.

Place a small light source, such as a torch bulb, as the object supported on the ruler beyond 2F and move a white card, on the other side of the lens from the light, until a sharp image is obtained on the card.

▲ Figure 3.2.38

Note and record, in a table like the one below, the image position as 'beyond 2F', 'between 2F and F' or 'between F and lens'. Also note whether the image is enlarged or diminished compared with the actual bulb or the same size and if it is upright or *inverted*. Now repeat with the light at 2F, then between 2F and F.

Object position	Image position	Enlarged, diminished or same size?	Upright or inverted?
beyond 2F			
at 2F			
between 2F and F			
between F and lens			

So far, all the images have been real since they can be obtained on a screen. When the light is between F and the lens, the image is *virtual* and is seen by *looking through the lens* at the light. Do this. Record your findings in your table.

20 a Is the virtual image enlarged or diminished?
  b Is it upright or inverted?
21 Using your results, draw ray diagrams to locate the image for each of your object positions. Do the values you obtain agree with your measured values
  a beyond 2F
  b at 2F
  c between 2F and F
  d between F and lens?

Ray diagrams

Information about the images formed by a lens can be obtained by drawing two of the following rays.

(i) A ray parallel to the principal axis which is refracted through the principal focus, F.

(ii) A ray through the optical centre, C, which is undeviated for a thin lens.

(iii) A ray through the principal focus, F, which is refracted parallel to the principal axis.

In diagrams a thin lens is represented by a straight line at which all the refraction is considered to occur.

In each ray diagram in Figure 3.2.39, two rays are drawn from the top A of an object OA. Where these rays intersect after refraction gives the top B of the image IB. The foot I of each image is on the axis since ray OC passes through the lens undeviated.

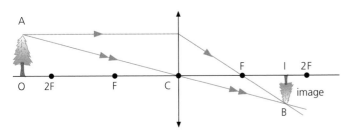

Image is between F and 2F, real, inverted, diminished

▲ **Figure 3.2.39a** Object beyond 2F

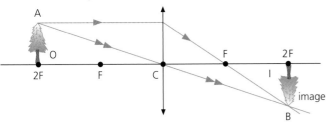

Image is at 2F, real, inverted, same size

▲ **Figure 3.2.39b** Object at 2F

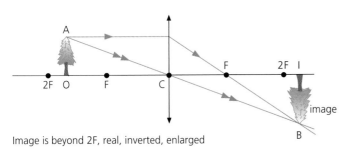

Image is beyond 2F, real, inverted, enlarged

▲ **Figure 3.2.39c** Object between 2F and F

In Figure 3.2.39d, the broken rays, and the image, are virtual. The virtual image is formed by extrapolating diverging rays backwards and cannot be formed on a screen. In all parts of Figure 3.2.39, the lens is a converging lens.

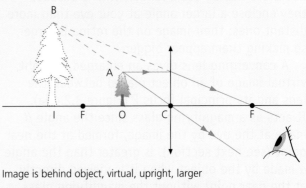

Image is behind object, virtual, upright, larger

▲ **Figure 3.2.39d** Object between F and C

Going further

Magnification

The linear magnification M is given by

$$M = \frac{\text{image size}}{\text{object size}}$$

It can be shown that in all cases

$$M = \frac{\text{distance of image from lens}}{\text{distance of object from lens}}$$

Power of a lens

The shorter the focal length of a lens, the stronger it is, i.e. the more it converges or diverges a beam of light. We define the power of a lens, P, to be 1/focal length of the lens, where the focal length is measured in metres:

$$P = \frac{1}{f}$$

Magnifying glass

The apparent size of an object depends on its actual size and on its distance from the eye. The sleepers on a railway track are all the same length but those nearby seem longer. This is because they enclose a larger angle at your eye than more distant ones: their image on the retina is larger, so making them appear bigger.

A converging lens gives an enlarged, upright, virtual image of an object placed between the lens and its principal focus F (Figure 3.2.40a). It acts as a magnifying glass since the angle β made at the eye by the image, formed at the near point (see next section), is greater than the angle α made by the object when it is viewed directly at the near point without the magnifying glass (Figure 3.2.40b).

a

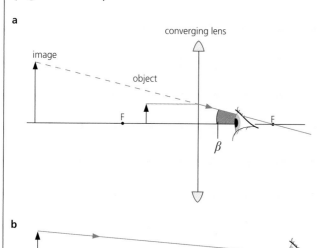

b

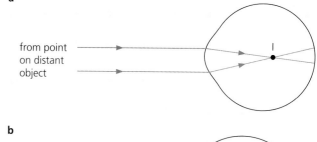

▲ **Figure 3.2.40** Magnification by a converging lens: angle β is larger than angle α

The fatter (more curved) a converging lens is, the shorter its focal length and the more it magnifies. Too much curvature, however, distorts the image.

Spectacles

From the ray diagrams shown in Figure 3.2.39 (p. 153) we would expect that the converging lens in the eye will form a real inverted image on the retina as shown in Figure 3.2.41. Since an object normally appears upright, the brain must invert the image.

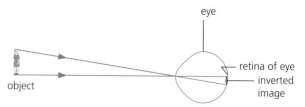

▲ **Figure 3.2.41** Inverted image on the retina

The average adult eye can focus objects comfortably from about 25 cm (the *near point*) to infinity (the *far point*). Your near point may be less than 25 cm; it gets further away with age.

Short sight

A *short-sighted* person sees near objects clearly but distant objects appear blurred. The image of a distant object is formed in front of the retina because the eyeball is too long or because the eye lens cannot be made thin enough (Figure 3.2.42a). The problem is corrected by a diverging spectacle lens (or contact lens) which diverges the light before it enters the eye, to give an image on the retina (Figure 3.2.42b).

a

b

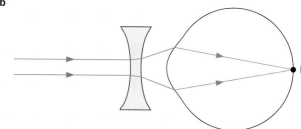

▲ **Figure 3.2.42** Short sight and its correction by a diverging lens

Long sight

A *long-sighted* person sees distant objects clearly but close objects appear blurred. The image of a near object is focused behind the retina because the eyeball is too short or because the eye lens cannot be made thick enough (Figure 3.2.43a). A converging spectacle lens (or contact lens) corrects the problem (Figure 3.2.43b).

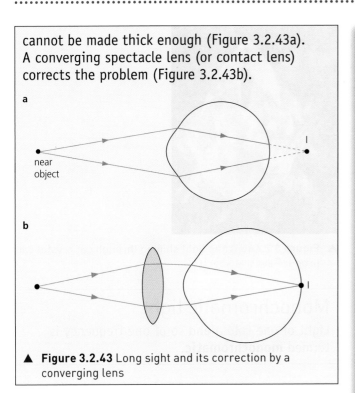

a

near object

b

▲ **Figure 3.2.43** Long sight and its correction by a converging lens

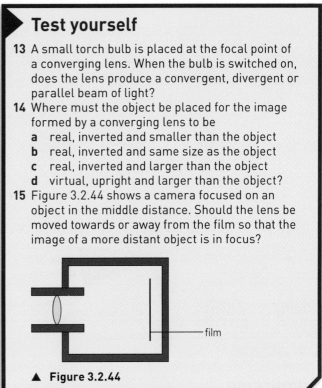

Test yourself

13 A small torch bulb is placed at the focal point of a converging lens. When the bulb is switched on, does the lens produce a convergent, divergent or parallel beam of light?

14 Where must the object be placed for the image formed by a converging lens to be
   a   real, inverted and smaller than the object
   b   real, inverted and same size as the object
   c   real, inverted and larger than the object
   d   virtual, upright and larger than the object?

15 Figure 3.2.44 shows a camera focused on an object in the middle distance. Should the lens be moved towards or away from the film so that the image of a more distant object is in focus?

film

▲ **Figure 3.2.44**

3.2.4 Dispersion of light

FOCUS POINTS

★ Describe the dispersion of light and how this can be demonstrated using a glass prism.
★ Know the seven colours of the visible spectrum in order of frequency and wavelength.

★ Understand the term monochromatic.

You can see the different colours which make up visible light in a rainbow. The sunlight is refracted by raindrops and a spectrum of colours from violet to red is produced. In this topic you will learn that the different colours which make up white light can also be separated by using a glass prism. The frequency of light determines its colour. The speed of light in a transparent medium depends on frequency, and so for a common incidence angle, the angle of refraction will vary with the colour of the light. The different colours are spread out into a spectrum and the effect is termed dispersion.

Refraction by a prism

In a triangular glass prism (Figure 3.2.45a), the bending of a ray due to refraction at the first surface is added to the bending of the ray at the second surface (Figure 3.2.45b); the overall change in direction of the ray is called the *deviation*.

The bendings of the ray do not cancel out as they do in a parallel-sided block where the emergent ray, although displaced, is parallel to the incident ray.

a

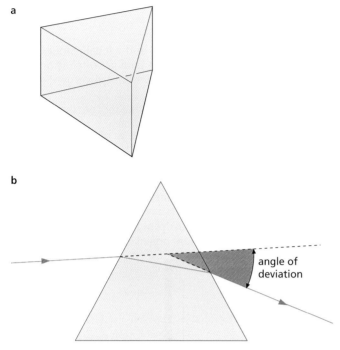

b

angle of
deviation

▲ Figure 3.2.45

▲ **Figure 3.2.46b** White light shining through cut crystal can produce several spectra

Dispersion

When sunlight (white light) falls on a triangular glass prism (Figure 3.2.46a), a band of colours called a **spectrum** is obtained (Figure 3.2.46b). The effect is termed *dispersion*. It arises because white light is a mixture of many colours; the prism separates the colours because the refractive index of glass is different for each colour (it is greatest for violet light).

The traditional colours of the visible spectrum in order of increasing wavelength are: violet, indigo, blue, green, yellow, orange and red. In order of increasing frequency, the sequence is reversed.

Red light has the longest wavelength in the optical spectrum and hence the lowest frequency and is refracted least by the prism. Violet light has the shortest wavelength and the highest frequency in the optical spectrum and is refracted most by the prism.

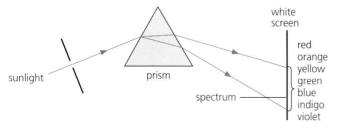

▲ **Figure 3.2.46a** Forming a spectrum with a prism

Monochromatic light

Light of one colour and so of one frequency is termed **monochromatic**.

Test yourself

16 Figure 3.2.47 shows a ray of light OP striking a glass prism and then passing through it. Which of the rays **A** to **D** is the correct representation of the emerging ray?

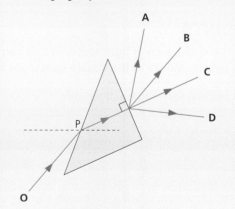

▲ **Figure 3.2.47**

17 Which diagram in Figure 3.2.48 shows the correct path of the ray through the prism?

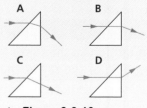

▲ **Figure 3.2.48**

Revision checklist

After studying Topic 3.2 you should know and understand:

✓ the meaning of the terms normal, angle of incidence and angle of reflection

> ✓ how to construct, calculate and measure reflections from plane mirrors

✓ the meaning of the terms refraction, critical angle, internal reflection and total internal reflection
✓ the terms used to describe images formed in lenses

> ✓ how simple converging and diverging lenses are used to correct long and short sight

✓ how a prism is used to produce a spectrum from white light
✓ the terms spectrum, dispersion

> ✓ the term monochromatic light.

After studying Topic 3.2 you should be able to:

✓ describe an experiment to show that the angle of incidence equals the angle of reflection and use the law of reflection to solve problems
✓ describe the formation of an optical image in a plane mirror and recall the properties of the image
✓ describe the passage of light through a transparent medium and an experiment to study refraction

> ✓ calculate refractive index, critical angles and describe some uses of optical fibres

✓ define the terms optical centre, principal axis, principal focus and focal length
✓ draw diagrams showing the effects of converging and diverging lenses on a beam of parallel rays and the formation of real images by a converging lens

> ✓ draw diagrams showing formation of a virtual image by a converging lens

✓ draw a diagram for the passage of a light ray through a prism and recall the seven colours of the visible spectrum in their correct order with respect to frequency and wavelength.

Exam-style questions

1 Figure 3.2.49 shows a ray of light PQ striking a mirror AB. The mirror AB and the mirror CD are at right angles to each other. QN is a normal to the mirror AB.

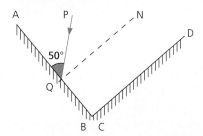

▲ **Figure 3.2.49**

a State the value of the angle of incidence of the ray PQ on the mirror AB. [2]

b Copy the diagram, and continue the ray PQ to show the path it takes after reflection at both mirrors. [3]

c State the values of the angle of reflection at AB, the angle of incidence at CD and the angle of reflection at CD. [3]

d Comment on the path of the ray PQ and the final reflected ray. [2]

[Total: 10]

2 a Which of the following statements is a true description of the image in a plane mirror?
   A Upright, real and larger
   B Upright, virtual and the same size
   C Inverted, real and smaller
   D Inverted, virtual and the same size [1]

b A person stands in front of a mirror (Figure 3.2.50). How much of the mirror is used to see from eye to toes? [2]

▲ **Figure 3.2.50**

c A girl stands 5 m away from a large plane mirror. How far must she walk to be 2 m away from her image? [4]

[Total: 7]

Alternative to Practical

3 You are asked to test the law of reflection. You are given a set of pins, a plane mirror, a protractor, a ruler, some modelling clay and a large sheet of paper attached to a cork board.
   a Describe a method you could use to test the law. Include a sketch of your proposed experimental arrangement. [4]
   b Draw up a table with headings showing what measurements you would take. [2]
   c How would you decide if your results were in agreement with the law of reflection? [2]
   d Mention two precautions you would take to ensure accurate results. [2]

[Total: 10]

4 In Figure 3.2.51 a ray of light IO changes direction as it enters glass from air.

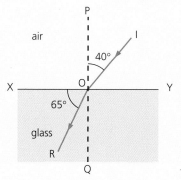

◀ **Figure 3.2.51**

a State the name given to this effect. [1]
b Identify the normal. [1]
c State whether the ray is bent towards or away from the normal in the glass. [1]
d State the value of the angle of incidence in air. [1]
e State the value of the angle of refraction in glass. [1]

[Total: 5]

5 Figure 3.2.52 shows rays of light in a semicircular glass block.

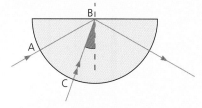

▲ **Figure 3.2.52**

a Explain why the ray entering the glass at A is not bent as it enters. [1]

b Explain why the ray AB is reflected at B and not refracted. [2]

c Ray CB does not stop at B. Copy the diagram and draw its approximate path after it leaves B. [2]

[Total: 5]

6 a Light travels up through a pond of water of critical angle 49°. Describe what happens at the surface if the angle of incidence is:

i 30° [3]

ii 60° [3]

b Calculate the critical angle for water if $n = \frac{4}{3}$. [4]

[Total: 10]

7 a Name the type of lens shown in Figure 3.2.53. [1]

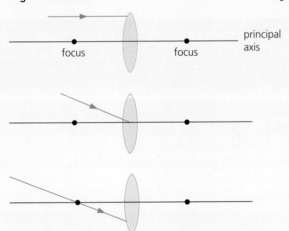

▲ **Figure 3.2.53**

b Copy the diagrams and complete them to show the path of the light after passing through the lens. [3]

c Figure 3.2.54 shows an object AB 6 cm high placed 18 cm in front of a lens of focal length 6 cm. Draw the diagram to scale and, by tracing the paths of rays from A, find the position and size of the image formed. [6]

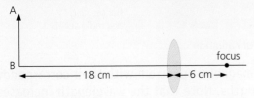

▲ **Figure 3.2.54**

[Total: 10]

8 a Three converging lenses are available, having focal lengths of 4 cm, 40 cm and 4 m, respectively. Which one would you choose as a magnifying glass? [1]

b An object 2 cm high is viewed through a converging lens of focal length 8 cm. The object is 4 cm from the lens. By means of a ray diagram find the position and nature of the image. [5]

c Calculate the ratio of the image height to object height. [2]

[Total: 8]

9 An object is placed 10 cm in front of a lens, A; the details of the image are given below. The process is repeated for a different lens, B.

Lens A: Real, inverted, magnified and at a great distance.

Lens B: Real, inverted and same size as the object.

Estimate the focal length of each lens and state whether it is converging or diverging.

[Total: 6]

10 A beam of white light strikes the face of a prism.

a Name the effect which happens to the white light when it enters the prism. [1]

b Copy Figure 3.2.55 and draw the path taken by red and blue rays of light as they pass through the prism and onto the screen AB. [5]

[Total: 6]

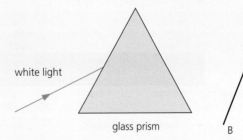

▲ **Figure 3.2.55**

3.3 Electromagnetic spectrum

FOCUS POINTS

★ Know the order, in terms of wavelength and frequency, of the main regions of the electromagnetic spectrum.

★ Understand that all electromagnetic waves travel at the same high speed in a vacuum.

★ Know the speed of electromagnetic waves in a vacuum and that it is approximately the same in air.

★ Describe typical uses of the different regions of the electromagnetic spectrum.

★ Describe harmful effects on people of prolonged exposure to electromagnetic radiation.

★ Know that microwaves are used for communication with artificial satellites.

★ Understand that many communication systems rely on electromagnetic radiation.

★ Know the difference between digital and analogue signals and the benefits of digital signalling.

Visible light forms only a small part of a very wide spectrum of electromagnetic waves, all of which travel with the speed of light in a vacuum. You will find that the properties of the different classes of waves vary with frequency from the lowest frequency radio waves through microwaves to infrared, visible and ultraviolet light, with X-rays and gamma rays at the highest frequencies. The various electromagnetic waves have a wide range of uses from communications and cooking to food sterilisation. The energy associated with an electromagnetic wave increases with frequency, so the highest frequencies are the most dangerous. Damage from over-exposure includes burns from infrared, sunburn and eye damage from ultraviolet and cell damage from X-rays and gamma rays.

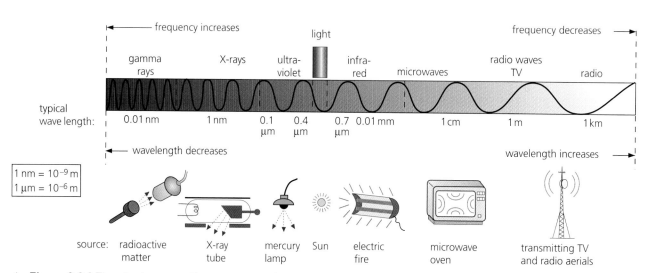

▲ **Figure 3.3.1** The electromagnetic spectrum and sources of each type of radiation

Light is one member of the family of electromagnetic radiation which forms a continuous spectrum beyond both ends of the visible (light) spectrum. Figure 3.3.1 shows the main regions of the electromagnetic spectrum with their corresponding wavelengths. Note that the wavelength increases from gamma rays to radio waves while the frequency increases in the reverse direction (radio to gamma).

While each type of radiation has a different source, all result from electrons in atoms undergoing an energy change and all have certain properties in common.

Properties of electromagnetic waves

- All types of electromagnetic radiation travel through a vacuum with the same high speed as light does.

> - The speed of electromagnetic waves in a vacuum is 3×10^8 m/s (300 000 km/s) and is approximately the same in air.

- They exhibit reflection, refraction and diffraction and have a transverse wave nature.
- They obey the wave equation, $v = f\lambda$, where v is the speed of light, f is the frequency of the waves and λ is the wavelength. Since v is constant in a particular medium, it follows that large f means small λ.
- They carry energy from one place to another and can be absorbed by matter to cause heating and other effects. The higher the frequency and the smaller the wavelength of the radiation, the greater is the energy carried, i.e. gamma rays are more 'energetic' than radio waves.

Because of its electrical origin, its ability to travel in a vacuum (e.g. from the Sun to the Earth) and its wave-like properties, electromagnetic radiation is regarded as a *progressive transverse wave*. The wave is a combination of travelling electric and magnetic fields. The fields vary in value and are directed at right angles to each other and to the direction of travel of the wave, as shown by the representation in Figure 3.3.2.

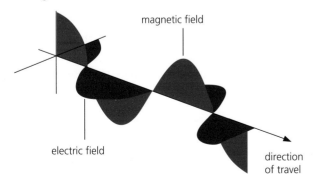

magnetic field

electric field

direction of travel

▲ **Figure 3.3.2** An electromagnetic wave

Light waves

For the visible light part of the electromagnetic spectrum, red light has the longest wavelength, which is about 0.0007 mm (7×10^{-7} m = 0.7 μm), while violet light has the shortest wavelength of about 0.0004 mm (4×10^{-7} m = 0.4 μm). Colours between these in the spectrum of white light have intermediate values. Light of one colour and so of only one frequency is called **monochromatic** light.

Since $v = f\lambda$ for all waves including light, it follows that red light has a lower frequency, f, than violet light since **(i)** the wavelength, λ, of red light is greater and **(ii)** all colours travel with the same speed 3×10^8 m/s (strictly, in a vacuum). It is the frequency of light which decides its colour, rather than its wavelength which is different in different media, as is the speed (Topic 3.2.2).

Different frequencies of light travel at different speeds through a transparent medium and so are refracted by different amounts. This explains dispersion (Topic 3.2.4), in other words why the refractive index of a material depends on the wavelength of the light.

The *amplitude* of a light (or any other) wave is greater the higher the **intensity** of the source; in the case of light the greater the intensity the brighter it is.

Visible light is the type of electromagnetic radiation used by our eyes to form images of the world around us. In addition to vision, light is used for illumination and photography (see Topic 3.2.1). Cameras and optical instruments, from microscopes to telescopes, make use of the properties of light to form images of near and distant objects.

> **?** **Worked example**
>
> The wavelength of a beam of light in air is 5×10^{-7} m. Calculate its frequency.
>
> Rearrange the equation $v = f\lambda$ to give $f = \dfrac{v}{\lambda}$.
>
> Taking $v = 3 \times 10^8$ m/s, then
>
> $$f = 3 \times 10^8 \text{ m/s} / 5 \times 10^{-7} \text{ m} = 6 \times 10^{14} \text{ Hz}$$
>
> ## Now put this into practice
> 1. The wavelength of a beam of light in air is 6×10^{-7} m. Calculate its frequency.
> 2. The frequency of a beam of infrared radiation in air is 4.0×10^{14} Hz. Calculate its wavelength.

Test yourself

1 Give the approximate wavelength in micrometres (μm) of
   a red light
   b violet light.
2 Which of the following types of radiation has
   a the longest wavelength
   b the highest frequency?
      A ultraviolet
      B radio waves
      C light
      D X-rays

Infrared radiation

Our bodies detect *infrared* radiation (IR) by its heating effect on the skin. It is sometimes called 'radiant heat' or 'heat radiation'.

Anything which is hot but not glowing, i.e. below 500°C, emits IR alone. At about 500°C a body becomes red hot and emits red light as well as IR – the heating element of an electric fire, a toaster or an electric grill are examples. At about 1500°C, things such as lamp filaments are white hot and radiate IR and white light, i.e. all the colours of the visible spectrum.

Infrared is also used in thermal imaging cameras, which show hot spots and allow images to be taken in the dark. Infrared sensors are used on satellites and aircraft for weather forecasting, monitoring of land use (Figure 3.3.3), assessing heat loss from buildings, intruder alarms and locating victims of earthquakes.

Infrared lamps are used to dry the paint on cars during manufacture and in the treatment of muscular complaints. The remote control for an electronic device contains a small infrared transmitter to send signals to the device, such as a television or DVD player. These are short range communication applications. Infrared is also used to carry data in long range optical fibre communication systems (Topic 3.2.2).

Infrared radiation can cause burns to the skin and eye damage if the intensity is high.

▲ **Figure 3.3.3** Infrared aerial photograph for monitoring land use

Ultraviolet radiation

Ultraviolet (UV) rays have shorter wavelengths than light. They cause sun tan and produce vitamins in the skin but too high an exposure can be harmful.

Ultraviolet radiation causes fluorescent paints and clothes washed in some detergents to fluoresce. They glow by re-radiating as light the energy they absorb as UV. This effect may be used in security marking to verify 'invisible' signatures on bank documents and to detect fake bank notes. Water treatment plants often use UV radiation to sterilise water.

A UV lamp used for scientific or medical purposes contains mercury vapour and this emits UV when an electric current passes through it. Fluorescent tubes also contain mercury vapour and their inner surfaces are coated with special powders called phosphors which radiate light.

Radio waves

Radio waves have the longest wavelengths in the electromagnetic spectrum. They are radiated from aerials and used to carry sound, pictures and other information over long distances.

Long, medium and short waves (wavelengths of 2 km to 10 m)

These diffract around obstacles so can be received even when hills are in their way (Figure 3.3.4a).

They are also reflected by layers of electrically charged particles in the upper atmosphere (the **ionosphere**), which makes long-distance radio reception possible (Figure 3.3.4b).

a Diffraction of radio waves

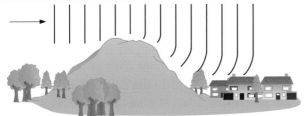

b Reflection of radio waves

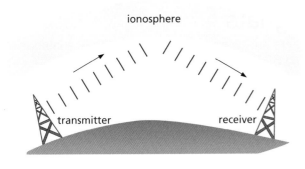

▲ **Figure 3.3.4**

VHF (very high frequency) and UHF (ultra high frequency) waves (wavelengths of 10 m to 10 cm)

These shorter wavelength radio waves need a clear, straight-line path to the receiver. They are not reflected by the ionosphere. They are used for local radio and for television transmissions.

They are also widely used in radio frequency identification (RFID) systems to track anything (from parcels to passports) to which a tag (a tiny radio transmitter and receiver) has been attached. Astronomers use radio telescopes to pick up radio signals from stars and galaxies to gain information about the Universe which cannot be obtained from optical telescopes.

Microwaves

Microwaves have wavelengths of a few cm. They are used for international telecommunications and direct broadcast satellite television relay via geostationary satellites and for mobile phone networks via microwave aerial towers and low-**orbit**

satellites (Topic 1.5.1). The microwave signals are transmitted through the ionosphere by dish aerials, amplified by the satellite and sent back to a dish aerial in another part of the world. Some satellite phones also use low-orbit artificial satellites.

Microwaves are also used for **radar** detection of ships and aircraft, and in police speed traps.

Microwaves can be used for cooking in a microwave oven since they cause water molecules in the moisture of the food to vibrate vigorously at the frequency of the microwaves. As a result, heating occurs inside the food which cooks itself.

X-rays

X-rays are produced when high-speed electrons are stopped by a metal target in an X-ray tube. X-rays have smaller wavelengths than UV.

They are absorbed to some extent by living cells but can penetrate certain solid objects and affect a photographic film. With materials such as bones, teeth and metals which they do not pass through easily, shadow pictures can be taken, like that in Figure 3.3.5 of a hand on an alarm clock. They are widely used in dentistry and in medicine, for example to detect broken bones. X-rays are also used in security machines at airports for scanning luggage; some body scanners, now being introduced to screen passengers, use very low doses of X-rays. In industry, X-ray photography is used to inspect welded joints. For safety reasons, X-ray machines need to be shielded with lead.

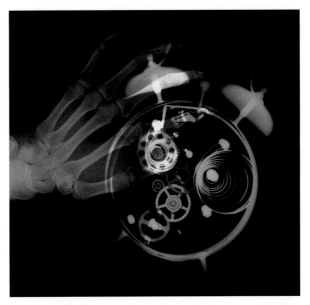

▲ **Figure 3.3.5** X-rays cannot penetrate bone and metal.

Gamma rays

Gamma rays (Topic 5.2) are more penetrating and dangerous than X-rays. They are used to both diagnose and treat cancer and also to kill harmful bacteria in food and on surgical instruments. They can also be used in engineering applications to detect flaws in metals.

Dangers of electromagnetic radiation

High exposures to electromagnetic radiation can cause harmful effects.

Microwaves

Microwaves produce heat when they are absorbed by the water in living cells. This can damage or kill the cells. There is some debate at present as to whether their use in mobile phones is harmful; 'hands-free' mode, where separate earphones are used, may be safer.

Infrared

Infrared radiation of high intensity can cause burns to the skin.

Ultraviolet

Ultraviolet radiation absorbed in high doses can cause skin cancer and eye damage. Dark skin is able to absorb more UV in the surface layers than skin of a lighter colour, so reducing the amount reaching deeper tissues. Exposure to the harmful UV rays present in sunlight can be reduced by wearing protective clothing such as a hat or by using sunscreen lotion.

X-rays and gamma rays

X-rays and gamma rays cause mutation or damage to cells in the body which can lead to cancer. X-ray and gamma ray machines are shielded by lead to reduce unnecessary exposure.

Communication systems

Electromagnetic waves are the basis of many communication systems. Microwaves only need short aerials for transmission and reception of signals, and can penetrate the walls of most buildings, so are used for mobile phones and wireless internet. Radio waves are weakened when passing through walls, but are suitable for short-range applications such as Bluetooth.

High-speed broadband internet and cable television signals are carried by optical fibres (Topic 3.2.2) because glass is transparent to visible light and some infrared. Visible light and shortwave infrared have higher frequencies than radio waves so can transmit data at a higher rate.

Information is sent in digital form as a set of pulses, rather than in analogue form as a smoothly varying signal.

Analogue and digital signals

There are two main types of signals – analogue and digital.

In **analogue** signals, voltages (and currents) can have any value within a certain range over which they can be varied smoothly and continuously, as shown in Figure 3.3.6a.

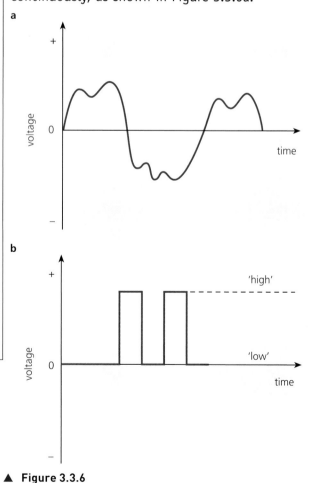

▲ Figure 3.3.6

In **digital** signals, voltages have only one of two values, either 'high' (e.g. 5 V) or 'low' (e.g. near 0 V), as shown in Figure 3.3.6b.

The advantages of using digital rather than analogue signals to transmit information are that they can be transmitted at a faster rate, can be sent over longer distances since they can be intermittently regenerated more accurately, and are less affected by noise. Analogue signals, such as sound waves, can be converted to digital signals electronically before transmission.

Test yourself

3 Name one type of electromagnetic radiation which
 a causes sun tan
 b is used for satellite communication
 c is used to sterilise surgical instruments
 d is used in a TV remote control
 e is used to cook food
 f is used to detect a break in a bone.
4 Name one type of electromagnetic radiation which causes
 a damage to surface cells in the body
 b internal heating of body cells
 c burns to the skin
 d mutation or damage to cells in the body.

5 Name one type of electromagnetic radiation used for
 a wireless internet
 b Bluetooth
 c cable television
 d mobile phones.

Revision checklist

After studying Topic 3.3 you should know and understand:
✓ that all electromagnetic waves travel with the same high speed in a vacuum

✓ that the speed of electromagnetic waves is 3×10^8 m/s in a vacuum and approximately the same in air

✓ the use of artificial satellites in communication systems

✓ the use of specific types of electromagnetic waves for different communication applications.

After studying Topic 3.3 you should be able to:
✓ describe the different types of electromagnetic radiation and distinguish between them in terms of their wavelengths, properties and uses
✓ describe the harmful effects of different types of electromagnetic radiation and how exposure to them can be reduced

✓ distinguish between analogue and digital signals and recall the benefits of digital signalling.

Exam-style questions

1 The chart below shows the complete electromagnetic spectrum.

radio waves	A	infrared	visible light	B	X-rays	gamma rays

a Name the radiation found at
  i A [1]
  ii B. [1]
b State which of the radiations marked on the diagram would have
  i the highest frequency [1]
  ii the longest wavelength. [1]
[Total: 4]

2 A VHF radio station transmits on a frequency of 100 MHz (1 MHz = 10^6 Hz). If the speed of radio waves is 3×10^8 m/s,
a calculate the wavelength of the waves [4]
b calculate the time the transmission takes to travel 60 km. [4]
[Total: 8]

3 In the diagram in Figure 3.3.7 light waves are incident on an air–glass boundary. Some are reflected and some are refracted in the glass. Select one of the following that is the same for the incident wave and the refracted wave inside the glass.

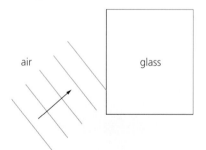

▲ **Figure 3.3.7**

A speed
B wavelength
C direction
D frequency [Total: 1]

4 a Describe the characteristics of
  i an analogue signal [3]
  ii a digital signal. [2]
b Describe the benefits of using digital signals. [3]
[Total: 8]

5 Optical fibres are used to carry cable television and high-speed broadband signals.
a Name the material used in optical fibres. [1]
b State two types of electromagnetic radiation that can be used to carry signals in optical fibres and give reasons for their choice. [5]
c Describe how sound and image information is transmitted in an optical fibre. [1]
d Give three advantages of using optical fibres to transmit data. [3]
[Total: 10]

3.4 Sound

FOCUS POINTS

★ Describe how sound is produced and the longitudinal nature of sound waves.

   ★ Understand the terms compression and rarefaction.

★ Know the approximate range of frequencies audible to humans.
★ Know the approximate speed of sound in air and describe an experiment to determine this.
★ Know that sound waves need a medium to be transmitted.

   ★ Know that sound travels fastest in solids and slowest in gases.

★ Describe how loudness and pitch of sound waves are determined by amplitude and frequency.
★ Define the terms echo and ultrasound.

   ★ Describe the uses of ultrasound.

Sound waves are produced by vibrating sources. Whether from the vibrations in your vocal cords, a violin string or a loudspeaker, all the speech, music and noise we hear around us are transmitted by sound waves. Like waves on a spring, sound waves are longitudinal waves and they require a medium in which to travel. In this topic you will learn that sound waves progress by local compression and expansion of the transmitting medium. The speed of sound varies strongly between materials from around 340 m/s in air to 1500 m/s in water. Sound waves have many of the same properties as transverse waves and can be reflected (as in an echo), refracted and diffracted. Reflected sound waves, used in sonar depth sounding at sea, can locate the position of a shoal of fish. Ultrasound, with frequencies above the range of human hearing (around 20 kHz), is used for material testing and medical imaging. Do you have some jewellery you would like cleaned? An ultrasonic cleaner could help remove the dirt.

Origin and transmission of sound

Sources of sound all have some part that *vibrates*. A guitar has strings (Figure 3.4.1), a drum has a stretched skin and the human voice has vocal cords. The sound travels through the air to our ears and we hear it. That the air is necessary may be shown by pumping the air out of a glass jar containing a ringing electric bell (Figure 3.4.2); the sound disappears though the striker can still be seen hitting the gong. Evidently sound cannot travel in a vacuum as light can. A medium is needed to transmit sound waves. In addition to air, solids and liquids are also able to transmit sound waves.

▲ **Figure 3.4.1** A guitar string vibrating. The sound waves produced are amplified when they pass through the circular hole into the guitar's sound box.

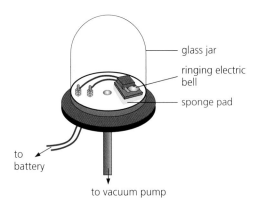

▲ **Figure 3.4.2** Sound cannot travel through a vacuum.

Sound also gives *interference* and *diffraction* effects. Because of this and its other properties, we believe sound transfers energy (as the damage from supersonic booms shows) which travels as a progressive longitudinal wave. As with transverse waves, sound waves obey the wave equation $v = f\lambda$.

Longitudinal nature of sound waves

As we saw in Topic 3.1, in a progressive longitudinal wave the particles of the transmitting medium vibrate back and forth along the same line as that in which the wave is travelling. In a sound wave, the air molecules move repeatedly closer together and then further apart in the direction of travel of the wave as can be seen in Figure 3.4.3.

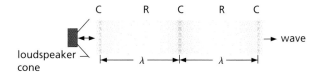

▲ **Figure 3.4.3** Sound travels as a longitudinal wave.

Compression and rarefaction

A sound wave, produced for example by a loudspeaker, consists of a train of *compressions C* (where the air molecules are closer together) and *rarefactions R* (where the air molecules are further apart) in the air (Figure 3.4.3).

The speaker has a cone which is made to vibrate in and out by an electric current. When the cone moves out, the air in front is compressed; when it moves in, the air is rarefied (becomes 'thinner'). The wave progresses through the air but the air as a whole does not move. The air particles (molecules) vibrate backwards and forwards a little as the wave passes. When the wave enters your ear the compressions and rarefactions cause small, rapid pressure changes on the eardrum and you experience the sensation of sound.

The number of compressions produced per second is the frequency f of the sound wave (and equals the frequency of the vibrating loudspeaker cone); the distance between successive compressions is the wavelength λ. As with transverse waves, the speed $v = f\lambda$.

> ### ▶ Test yourself
> 1 Describe how a sound wave can be produced.
> 2 Describe how the air molecules vibrate when a sound wave is transmitted in air.
>
> 3 How far apart are the compressions in a sound wave?

Limits of audibility

Humans hear only sounds with frequencies from about 20 Hz to 20 000 Hz. These are the **limits of audibility** in a healthy human ear; the upper limit decreases with age.

> **Key definition**
>
> **Limits of audibility** the approximate range of frequencies audible to humans, 20 Hz to 20 000 Hz

Reflection and echoes

Sound waves are reflected well from hard, flat surfaces such as walls or cliffs and obey the same laws of reflection as light. The reflected sound forms an **echo**; an echo is the reflection of a sound wave.

If the reflecting surface is nearer than 15 m from the source of sound, the echo joins up with the original sound which then seems to be prolonged. This is called **reverberation**. Some is desirable in a concert hall to stop it sounding 'dead', but too much causes 'confusion'. Modern concert halls are designed for the optimal amount of reverberation. Seats and some wall surfaces are covered with sound-absorbing material.

Speed of sound

In air

The speed of sound in air is approximately 330–350 m/s.

> ## In other materials
>
> The *speed of sound* depends on the material through which it is passing. It is faster in solids than in liquids and faster in liquids than in gases because the molecules in a solid are closer together than in a liquid and those in a liquid are closer together than in a gas. Some values are given in Table 3.4.1.
>
> ▼ **Table 3.4.1** Speed of sound in different materials
>
> | Material | air (0°C) | water | concrete | steel |
> |---|---|---|---|---|
> | **Speed/m/s** | 330–350 | 1400 | 5000 | 6000 |
>
> In air the speed *increases with temperature* and at high altitudes, where the temperature is lower, it is less than at sea level. Changes of atmospheric pressure do not affect it.

Measurement of the speed of sound

Echo method

An estimate of the speed of sound in air can be made directly if you stand about 100 metres from a high wall or building and clap your hands. Echoes are produced. When the clapping rate is such that each clap coincides with the echo of the previous one, the sound has travelled to the wall and back in the time between two claps, i.e. one interval. By timing 30 intervals with a stopwatch, the time t for one interval can be found. Also, knowing the distance d to the wall, a rough value is obtained from

$$\text{speed of sound in air} = \frac{2d}{t}$$

Direct method

The speed of sound in air can be found directly by measuring the time t taken for a sound to travel past two microphones separated by a distance d:

$$\text{speed of sound in air} = \frac{\text{distance travelled by the sound}}{\text{time taken}}$$
$$= \frac{d}{t}$$

This method can also be used to find the speed of sound in different materials. For example, placing a metal bar between the microphones would allow the speed of sound in the metal to be determined.

Practical work

Speed of sound in air

For safe experiments/demonstrations related to this topic, please refer to the *Cambridge IGCSE Physics Practical Skills Workbook* that is also part of this series.

Set two microphones about a metre apart, and attach one to the 'start' terminal and the other to the 'stop' terminal of a digital timer, as shown in Figure 3.4.4. The timer should have millisecond accuracy. Measure and record the distance d between the centres of the microphones with a metre ruler. With the small hammer and metal plate to one side of the 'start' microphone, produce a sharp sound. When the sound reaches the 'start' microphone, the timer should start; when it reaches the 'stop' microphone, the timer should stop. The time displayed is then the time t taken for the sound to travel the distance d. Record the time and then reset the timer; repeat the experiment a few times and work out an *average* value for t.

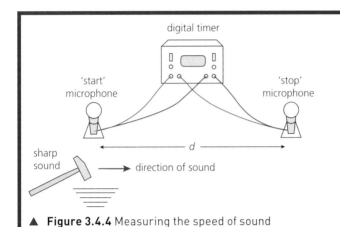

Calculate the speed of sound in air from d/t. Check how your value compares with that given in Table 3.4.1.

1 How could you improve the accuracy of this experiment?
2 If the distance between the microphones is 1.2 m and the timer records an average time interval of 3.6 ms between the start and stop signals, calculate a value for the speed of sound in air.

▲ **Figure 3.4.4** Measuring the speed of sound

Musical notes

Irregular vibrations, such as those of motor engines, cause **noise**; regular vibrations, such as occur in the instruments of a brass band (Figure 3.4.5), produce **musical notes**, which have three properties – **pitch**, *loudness* and **quality**.

▲ **Figure 3.4.5** Musical instruments produce regular sound vibrations.

Pitch

The pitch of a note depends on the frequency of the sound wave reaching the ear, i.e. on the frequency of the source of sound. A high-pitched note has a high frequency and a short wavelength. The frequency of middle C is 256 vibrations per second or 256 Hz and that of upper C is 512 Hz. Notes are an **octave** apart if the frequency of one is twice that of the other. Pitch is like colour in light; both depend on the frequency.

Notes of known frequency can be produced in the laboratory by a signal generator supplying alternating electric current (a.c.) to a loudspeaker. The cone of the speaker vibrates at the frequency of the a.c. which can be varied and read off a scale on the generator.

A set of tuning forks with frequencies marked on them can also be used. A tuning fork (Figure 3.4.6) has two steel prongs which vibrate when struck; the prongs move in and out *together*, generating compressions and rarefactions.

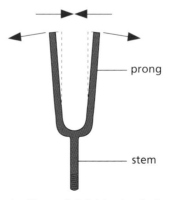

prong

stem

▲ **Figure 3.4.6** A tuning fork

Loudness

A note becomes louder when more sound energy enters our ears per second than before. This will happen when the source is vibrating with a larger amplitude. If a violin string is bowed more strongly, its amplitude of vibration increases, as does that of the resulting sound wave, and the note heard is louder because more energy has been used to produce it.

Going further

Quality

The same note on different instruments sounds different; we say the notes differ in *quality* or *timbre*. The difference arises because no instrument (except a tuning fork and a signal generator) emits a 'pure' note, i.e. of one frequency. Notes consist of a main or fundamental frequency mixed with others, called overtones, which are usually weaker and have frequencies that are exact multiples of the fundamental. The number and strength of the overtones decides the quality of a note. A violin has more and stronger higher overtones than a piano. Overtones of 256 Hz (middle C) are 512 Hz, 768 Hz and so on.

The waveform of a note played near a microphone connected to a cathode ray oscilloscope (CRO) can be displayed on the CRO screen. Those for the *same* note on three instruments are shown in Figure 3.4.7. Their different shapes show that while they have the same fundamental frequency, their quality differs. The 'pure' note of a tuning fork has a *sine* waveform and is the simplest kind of sound wave.

Note that although the waveform on the CRO screen is transverse, it represents a longitudinal sound wave.

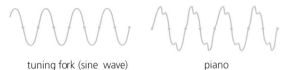

tuning fork (sine wave)   piano

violin

▲ **Figure 3.4.7** Notes of the same frequency (pitch) but different quality

Worked example

The frequency of middle C is 256 Hz. If the speed of sound in air is 340 m/s, calculate the wavelength in air.

Rearrange the equation $v = f\lambda$ to give $\lambda = v/f$.

Taking $v = 340$ m/s, then

$$\lambda = \frac{340\,\text{m/s}}{256\,\text{Hz}} = 1.3\,\text{m}$$

Now put this into practice

1 The frequency of upper C is 512 Hz. If the speed of sound in air is 340 m/s, calculate the wavelength in air.
2 The wavelength of a sound wave in air is 1.0 m. Taking the speed of sound in air to be 340 m/s, calculate the frequency of the sound wave.

Test yourself

4 Tuning forks of frequencies 128 Hz and 256 Hz are sounded. Which tuning fork has the highest pitch?
5 Sound waves of wavelengths 0.8 m and 1.2 m are produced in air. Which of the waves has the highest pitch?
6 The amplitude of the sound wave produced by tuning fork B is twice the amplitude of the sound wave produced by tuning fork A. Which tuning fork produces the loudest sound?

Ultrasound

Ultrasound is defined as sound with a frequency higher than 20 kHz. The frequency of ultrasound is too high to be detected by the human ear but can be detected electronically and displayed on a **cathode ray** oscilloscope (CRO).

Ultrasound waves are produced by a quartz crystal which is made to vibrate electrically at the required frequency; they are emitted in a narrow beam in the direction in which the crystal oscillates. An ultrasound receiver also consists of a quartz crystal, but it works in reverse, i.e. when it is set into vibration by ultrasound waves it generates an electrical signal which is then amplified. The same quartz crystal can act as both a transmitter and a receiver.

Key definition
Ultrasound sound with a frequency higher than 20 kHz

Ultrasound echo techniques

Ultrasound waves are partially or totally reflected from surfaces at which the density of the medium changes. This property is exploited in techniques such as the *non-destructive testing of materials*, sonar and medical ultrasound imaging. A bat emitting ultrasound waves can judge the distance of an object from the time taken by the reflected wave or 'echo' to return.

Ships with **sonar** can determine the depth of a shoal of fish or the seabed (Figure 3.4.8) in the same way; motion sensors also work on this principle.

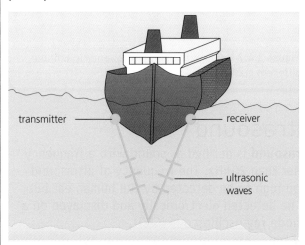

▲ **Figure 3.4.8** A ship using sonar

❓ Worked example

A research ship is using sonar to map the seabed.

How deep is the water if an ultrasound pulse reflected from the seabed takes 1.5 s to return to the ship? Take the speed of sound in water to be 1400 m/s.

Rearrange the equation for speed $v = \dfrac{s}{t}$ to give $s = v \times t$.

If the depth of the seabed is d, the ultrasound must travel a distance of $2d$ during time t between the transmission and reception of the signal, then

$$2d = vt \text{ and } d = \frac{vt}{2} = \frac{1400\,\text{m/s} \times 1.5\,\text{s}}{2} = 1050\,\text{m}$$

Now put this into practice

1   A fishing vessel is using sonar to monitor the position of shoals of fish. Calculate the depth of a shoal if an ultrasound pulse reflected from the shoal takes 0.5 s to return to the ship. Take the speed of sound in water to be 1400 m/s.

2   Sonar is being used on a research vessel to measure the height of an underground sea mount. Calculate the distance from the ship to the summit of the sea mount if an ultrasound pulse reflected from the summit takes 2 s to return to the ship. Take the speed of sound in water to be 1400 m/s.

In **medical ultrasound imaging**, used in antenatal clinics to monitor the health and sometimes to determine the sex of an unborn baby, an ultrasound transmitter/receiver is scanned over the mother's abdomen and a detailed image of the fetus is built up (Figure 3.4.9). Reflection of the ultrasound pulses occurs from boundaries of soft tissue, in addition to bone, so images can be obtained of internal organs that cannot be seen by using X-rays. Less detail of bone structure is seen than with X-rays, as the wavelength of ultrasound waves is larger, typically about 1 mm, but ultrasound has no harmful effects on human tissue.

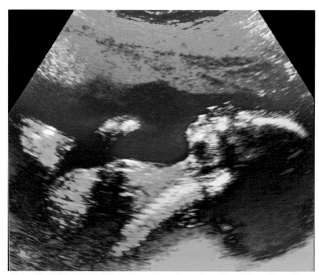

▲ **Figure 3.4.9** Checking the development of a fetus using ultrasound imaging

Other uses

Ultrasound can also be used in ultrasound drills to cut holes of any shape or size in hard materials such as glass and steel. Jewellery, or more mundane objects such as street lamp covers, can be cleaned by immersion in a tank of solvent which has an ultrasound vibrator in the base.

Going further

Seismic waves

Earthquakes produce both longitudinal waves (P-waves) and transverse waves (S-waves) that are known as seismic waves. These travel through the Earth at speeds of up to 13 000 m/s, with frequencies less than 100 Hz.

When seismic waves pass under buildings, severe structural damage may occur. If the earthquake occurs under the sea, the seismic energy can be transmitted to the water and produce tsunami waves that may travel for very large distances across the ocean. As a tsunami wave approaches shallow coastal waters, it slows down (see Topic 3.1) and its amplitude increases, which can lead to massive coastal destruction. This happened in Sri Lanka (see Figure 3.4.10) and Thailand after the great 2004 Sumatra–Andaman earthquake. The time of arrival of a tsunami wave can be predicted if its speed of travel and the distance from the epicentre of the earthquake are known; it took about 2 hours for tsunami waves to cross the ocean to Sri Lanka from Indonesia. A similar time was needed for the tsunami waves

to travel the shorter distance to Thailand. This was because the route was through shallower water and the waves travelled more slowly. If an early-warning system had been in place, many lives could have been saved.

▲ **Figure 3.4.10** This satellite image shows the tsunami that hit the south-western coast of Sri Lanka on 26 December 2004 as it pulled back out to sea, having caused utter devastation in coastal areas.

▶ Test yourself

7 a State the approximate range of frequencies of the human ear.
 b State an approximate value for the speed of sound in air.
8 If 5 seconds elapse between a lightning flash and the clap of thunder, how far away is the storm? (Speed of sound = 330 m/s)

9 a What properties of sound suggest it is a wave motion?
 b How does a progressive transverse wave differ from a longitudinal one? Which type of wave is a sound wave?

Revision checklist

After studying Topic 3.4 you should know and understand:
✔ that sound is produced by vibrating sources and echoes are produced by reflection of sound waves
✔ the term ultrasound and know its frequency

> ✔ some uses of ultrasound.

After studying Topic 3.4 you should be able to:
✔ describe the longitudinal nature of sound waves

> ✔ describe compression and rarefaction

✔ describe experiments to show that sound is not transmitted through a vacuum and measure the speed of sound in air
✔ recall the value of the speed of sound in air

> ✔ recall that in general sound travels faster in solids that in liquids and faster in liquids than in gases

✔ state the limits of audibility for the normal human ear
✔ relate the loudness and pitch of sound waves to amplitude and frequency.

Exam-style questions

1 a A girl stands 160 m away from a high wall and claps her hands at a steady rate so that each clap coincides with the echo of the one before. If her clapping rate is 60 per minute, state the value this gives for the speed of sound. [3]

b If she moves 40 m closer to the wall she finds the clapping rate has to be 80 per minute. Calculate the value these measurements give for the speed of sound. [3]

c She moves again and finds the clapping rate becomes 30 per minute. Calculate how far she is from the wall if the wave speed of sound is the value you found in **a**. [4]

[Total: 10]

2 a Draw the waveform of
  i a loud, low-pitched note [2]
  ii a soft, high-pitched note. [2]

b If the speed of sound is 340 m/s what is the wavelength of a note of frequency
  i 340 Hz [3]
  ii 170 Hz? [2]

[Total: 9]

3 a Explain with reference to a sound wave what is meant by the terms
  i compression [3]
  ii rarefaction. [3]

b State how far a compression and the nearest rarefaction are apart in terms of the wavelength of a sound wave. [1]

c A sound wave has a frequency of 220 Hz and travels in air with a speed of 330 m/s. Calculate the distance between consecutive rarefactions. [3]

[Total: 10]

4 a Name the state of matter in which sound waves travel
  i fastest [1]
  ii slowest. [1]

b Describe how sound waves are used in sonar. [4]

c Name two uses of ultrasound other than sonar. [2]

[Total: 8]

SECTION 4

Electricity and magnetism

Topics

FOCUS POINTS

★ Describe forces between magnets and magnetic materials and between magnetic poles and understand the meaning of various terms associated with magnetism.

★ Explain that interactions between magnetic fields create magnetic forces.

★ State the differences between temporary and permanent magnets and between magnetic and non-magnetic materials.
★ Describe, draw and state the direction of magnetic fields.

★ Know that the spacing of the magnetic field lines represents the relative strength of a magnetic field.

★ Describe how magnetic field lines can be plotted using a compass or iron filings.
★ Know the different uses of permanent magnets and electromagnets.

A familiar example of a magnet is a compass needle with one north-seeking pole. You will find that all magnets have two poles: like poles repel, unlike poles attract. A magnet can induce magnetism in certain materials such as iron and steel and is surrounded by a magnetic field which exerts a force on another magnet. The pattern of magnetic field lines can be made visible with the aid of iron filings. Electromagnets are formed from coils of wire through which an electrical current is passed that allows the strength of the magnet to be varied and turned on and off easily. They are used in many electrical devices from doorbells to motors. You will learn that permanent magnets and electromagnets have differing properties and uses.

In a magnetic field, the closer the field lines are at a point, the stronger is the magnetic field.

Properties of magnets
Magnetic materials

Some materials, known as ferromagnets, can be magnetised to form a magnet. In their unmagnetised form they are attracted to a magnet.

Magnetic poles

The *poles* are the places in a magnet to which magnetic materials, such as iron filings, are attracted. They are near the ends of a bar magnet and occur in pairs of equal strength.

North and south poles

A magnet has two poles; a north pole (N pole) and a south pole (S pole). If a magnet is supported so that it can swing in a horizontal plane it comes to rest with one pole, the N pole, always pointing roughly towards the Earth's north pole. A magnet can therefore be used as a *compass*.

Law of magnetic poles

If the N pole of a magnet is brought near the N pole of another magnet, repulsion occurs. Two S (south-seeking) poles also repel. By contrast, N and S poles always attract.

The law of magnetic poles summarises these facts and states:

Like poles repel, unlike poles attract.

The force between magnetic poles decreases as their separation increases.

Induced magnetism

When a piece of unmagnetised magnetic material touches or is brought near to the pole of a permanent magnet, it becomes a magnet itself. The material is said to have magnetism induced in it. Figure 4.1.1 shows that a N pole in the permanent magnet induces a N pole in the right-hand end of the magnetic material.

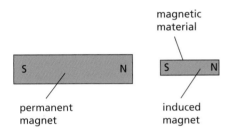

▲ **Figure 4.1.1** Induced magnetism

This can be checked by hanging two iron nails from the N pole of a magnet. Their lower ends repel each other (Figure 4.1.2a) and both are repelled further from each other when the N pole of another magnet is brought close (Figure 4.1.2b).

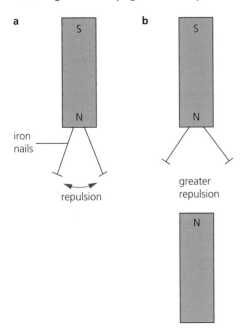

▲ **Figure 4.1.2** Magnetic repulsion

Magnetisation of iron and steel

Chains of small iron nails and steel paper clips can be hung from a magnet (Figure 4.1.3). Each nail or clip magnetises the one below it and the unlike poles so formed attract.

If the iron chain is removed by pulling the top nail away from the magnet, the chain collapses, showing that magnetism induced in iron is **temporary**. When the same is done with the steel chain, it does not collapse; magnetism induced in steel is **permanent**.

> **Key definitions**
>
> **Temporary magnets** made of soft iron, lose their magnetism easily
>
> **Permanent magnets** made of steel, retain their magnetism

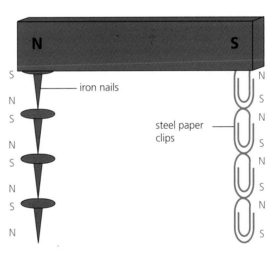

▲ **Figure 4.1.3** Investigating the magnetisation of iron and steel

Magnetic materials such as iron that magnetise easily but readily lose their magnetism (are easily demagnetised) are said to be **soft**. Those such as steel that are harder to magnetise than iron but stay magnetised are **hard**. Both types have their uses; very hard ones are used to make permanent magnets.

Magnetic and non-magnetic materials

Magnetic materials such as iron, steel, nickel and cobalt are attracted by a magnet and can be magnetised temporarily or permanently. **Non-magnetic materials** such as aluminium and wood are not attracted by a magnet and cannot be magnetised.

> **Key definitions**
>
> **Magnetic materials** materials that can be magnetised by a magnet; in their unmagnetised state they are attracted by a magnet
>
> **Non-magnetic materials** materials that cannot be magnetised and are not attracted by a magnet

Magnetic fields

The space surrounding a magnet where it produces a magnetic force is called a **magnetic field**. The force around a bar magnet can be detected and shown to vary in direction, using the apparatus in Figure 4.1.4. If the floating magnet is released near the N pole of the bar magnet, it is repelled to the S pole and moves along a curved path known as a **line of force** or a **field line**. It moves in the opposite direction if its south pole is uppermost.

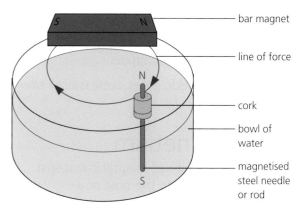

▲ **Figure 4.1.4** Detecting magnetic force

It is useful to consider that a magnetic field has a direction and to represent the field by lines of force. It has been decided that the **direction of a magnetic field at a point** should be the direction of the force on a N pole. To show the direction, arrows are put on the lines of force and point away from a N pole towards a S pole.

> **Key definition**
>
> **Direction of a magnetic field at a point** the direction of the force on the N pole of a magnet at that point

Strength and interaction of magnetic fields

A magnetic field is stronger in regions where the field lines are close together than where they are further apart.

The force between two magnets is a result of the interaction of their magnetic fields as can be seen in Figure 4.1.8 on the next page.

▶ **Test yourself**

1 Which one of these statements is true?
  A magnet attracts
  A plastics
  B any metal
  C iron and steel
  D aluminium.

2 Two bar magnets are positioned side by side as shown in Figure 4.1.5. The north pole is marked on one of the magnets.

	N

▲ **Figure 4.1.5**

Copy the diagram and mark on the position of all the poles if the magnets
  a attract each other
  b repel each other.

3 In Figure 4.1.8a on the next page, is the magnetic field stronger or weaker at X than at a point closer to one of the magnets? Explain your answer.

Practical work

Plotting lines of force

For safe experiments/demonstrations related to this topic, please refer to the *Cambridge IGCSE Physics Practical Skills Workbook* that is also part of this series.

Plotting compass method

A plotting compass is a small pivoted magnet in a glass case with non-magnetic metal walls (Figure 4.1.6a).

a

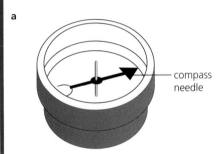

b

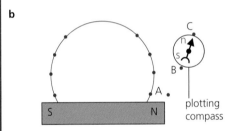

▲ **Figure 4.1.6**

Lay a bar magnet on a sheet of paper. Place the plotting compass at a point such as A (Figure 4.1.6b), near one pole of the magnet. In Figure 4.1.6b it is the N pole. Mark the position of the poles (n, s) of the compass by pencil dots B, A. Move the compass so that pole s is exactly over B, mark the new position of n by dot C.

Continue this process until the other pole of the bar magnet is reached (in Figure 4.1.6b it is the S pole). Join the dots to give one line of force and show its direction by putting an arrow on it. Plot other lines by starting at different points round the magnet.

A typical field pattern is shown in Figure 4.1.7.

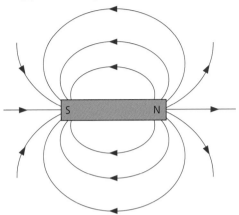

▲ **Figure 4.1.7** Magnetic field lines around a bar magnet

The combined field due to two neighbouring magnets can also be plotted to give patterns like those in Figure 4.1.8. In part a, where two like poles are facing each other, the point X is called a **neutral point**. At X, the field due to one magnet cancels out that due to the other and there are no lines of force.

a

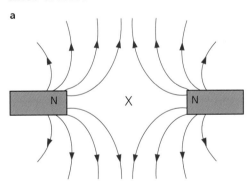

b

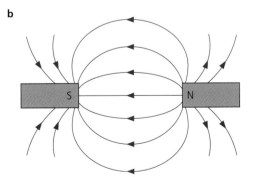

▲ **Figure 4.1.8** Field lines due to two neighbouring magnets

Iron filings method

Place a sheet of paper *on top of* a bar magnet and sprinkle iron filings *thinly and evenly* onto the paper from a 'pepper pot'.

Tap the paper gently with a pencil and the filings should form patterns showing the lines of force. Each filing turns in the direction of the field when the paper is tapped.

This method is quick but no use for weak fields.

1 Sketch the field lines around a bar magnet marking on the N and S poles and the direction of the field lines.
2 Figure 4.1.9 shows typical iron filings patterns obtained with two magnets. Why are the patterns different?
3 What combination of poles would give the observed patterns in Figure 4.1.9 a and b?

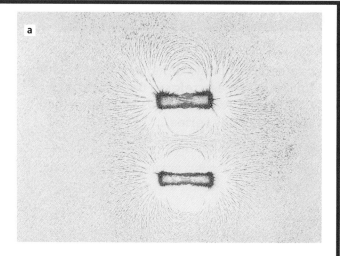

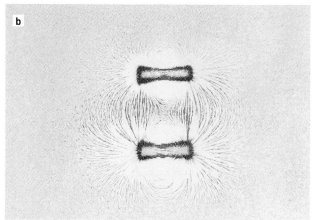

▲ **Figure 4.1.9** Field lines round two bar magnets shown by iron filings

Going further

Magnetisation and demagnetisation

A ferromagnetic material can be magnetised by placing it inside a solenoid and gradually increasing the direct current (d.c.). This increases the magnetic field strength in the solenoid (the density of the field lines increases), and the material becomes magnetised. Reversing the direction of current flow reverses the direction of the magnetic field and reverses the polarity of the magnetisation. A magnet can be demagnetised by placing it inside a solenoid through which an alternating current (a.c.) is passed and gradually reduced.

Solenoids (see Topic 4.5) can be used to magnetise and demagnetise magnetic materials; dropping or heating a magnet also causes demagnetisation. Hammering a magnetic material in a magnetic field causes magnetisation but in the absence of a field it causes demagnetisation. 'Stroking' a magnetic material several times in the same direction with one pole of a magnet will also cause it to become magnetised.

Practical work

Simple electromagnet

For safe experiments/demonstrations related to this topic, please refer to the *Cambridge IGCSE Physics Practical Skills Workbook* that is also part of this series.

An **electromagnet** is a coil of wire wound on a soft iron core. A 5 cm iron nail and 3 m of PVC-covered copper wire (SWG 26) are needed.

a Leave about 25 cm at one end of the wire (for connecting to the circuit) and then wind about 50 cm as a single layer on the nail. *Keep the turns close together and always wind in the same direction.* Connect the circuit of Figure 4.1.10, setting the rheostat (variable resistor, see p. 199) at its maximum resistance.
Find the number of paper clips the electromagnet can support when the current is varied between 0.2 A and 2.0 A. Record the results in a table.
Deduce how the strength of the electromagnet changes when the current is increased.

b Add another two layers of wire to the nail, winding in the *same direction* as the first layer. Repeat the experiment.
Deduce how the strength of the electromagnet has been changed by increasing the number of turns of wire.

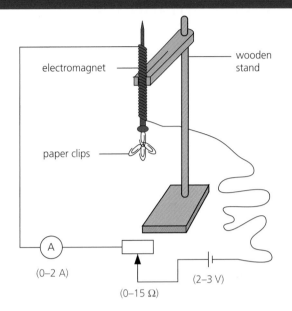

▲ **Figure 4.1.10**

c Place the electromagnet on the bench and under a sheet of paper. Sprinkle iron filings on the paper, tap it gently and observe the field pattern. Compare the pattern with that given by a bar magnet.

d Use a plotting compass to find which end of the electromagnet is a N pole.

4 Name two variables which you think could affect the strength of an electromagnet.

5 How could you use a compass to determine which end of the current-carrying coil is a north pole?

Electromagnets

An electromagnet is formed from a coil of wire through which an electrical current is passed that allows the strength of the magnet to be varied. The magnetism of an electromagnet is *temporary* and can be switched on and off, unlike that of a permanent magnet. It has a core of soft iron which is magnetised only when there is current in the surrounding coil.

The strength of an electromagnet increases if
(i) the *current* in the coil increases
(ii) the *number of turns* on the coil increases
(iii) the poles are moved *closer together*.

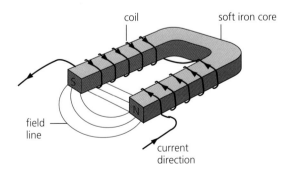

▲ **Figure 4.1.11** C-core or horseshoe electromagnet

In C-core (or horseshoe) electromagnets, condition (iii) is achieved (Figure 4.1.11). Note that the coil on each limb of the core is wound in *opposite* directions.

Uses of permanent magnets and electromagnets

Permanent magnets made from magnetic materials such as steel retain their magnetism, so can be used in applications where the magnetic field does not need to be varied. These include a compass, computer hard disk, electric motor (see Topic 4.5.5) electricity generator (see Topic 4.5.2), microphone, loudspeaker and many more everyday devices such as credit and debit cards.

An advantage over an electromagnet is that it does not require a current to maintain its magnetism.

Electromagnets are temporary and are used where one wants to be able to vary the strength of the magnetic field (by varying the current) and switch it on and off. As well as being used in cranes to lift iron objects, scrap iron, etc. (Figure 4.1.12), electromagnets are an essential part of many electrical devices such as electric bells, magnetic locks, relays and practical motors and generators (see Topic 4.5.3).

▲ **Figure 4.1.12** Electromagnet being used to lift scrap metal

---

 Going further

Magnetic shielding

Any ferromagnetic material can be used for magnetic screening of sensitive electronic equipment. Steel is often used as it is cheap, readily available and works well in strong magnetic fields. Mu-metal, a nickel-iron soft ferromagnetic material, is more effective for weaker magnetic fields but is more expensive.

Earth's magnetic field

If lines of force are plotted on a sheet of paper with no magnets nearby, a set of parallel straight lines is obtained. They run roughly from S to N geographically (Figure 4.1.13), and represent a small part of the Earth's magnetic field in a horizontal plane.

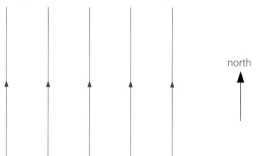

▲ **Figure 4.1.13** Lines of force due to the Earth's field

At most places on the Earth's surface a magnetic compass points slightly east or west of true north, i.e. the Earth's geographical and magnetic north poles do not coincide. The angle between magnetic north and true north is called the declination (Figure 4.1.14). In Hong Kong in 2020 it was about 3° W of N and changing slowly.

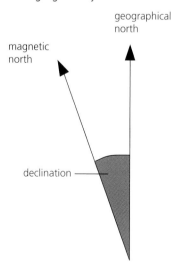

▲ **Figure 4.1.14** The Earth's geographical and magnetic poles do not coincide.

Revision checklist

After studying Topic 4.1 you should know and understand:

✔ like magnetic poles repel, unlike magnetic poles attract

✔ the difference between magnetic and non-magnetic materials, and permanent and electromagnets

✔ how to map the magnetic field around a bar magnet, by the plotting compass and iron filings methods.

After studying Topic 4.1 you should be able to:

✔ state the properties of magnets, describe induced magnetism and distinguish between the magnetic properties of iron and steel

✔ recall that a magnetic field is the region round a magnet where a magnetic force is exerted and is represented by lines of force whose direction at any point is the direction of the force on a N pole

✔ recall that the magnetic field is strongest in regions where the field lines are closest together and that magnetic forces result from the interaction of magnetic fields.

Exam-style questions

1 Copy Figure 4.1.15 which shows a plotting compass and a magnet.
  a Label the N pole of the magnet. [1]
  b Draw the magnetic field line on which the compass lies. [2]
  c State the direction of the magnetic field line. [1]

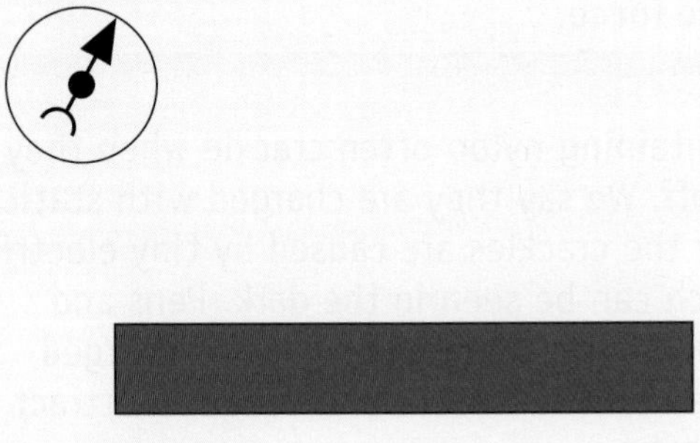

▲ Figure 4.1.15

[Total: 4]

2 a Describe an experiment using a plotting compass to map the magnetic field lines around a bar magnet. [4]
  b Explain why permanent magnets are used in some applications and electromagnets in others. [4]
  c Give two uses of a permanent magnet. [2]
  [Total: 10]

3 a Explain how magnetic forces arise. [2]
  b Where are the magnetic field lines strongest around a bar magnet? [2]
  c State how you would recognise from a pattern of magnetic field lines where the field is
    i strongest
    ii weakest. [2]
  [Total: 6]

4.2.1 Electric charge

FOCUS POINTS

★ Understand that there are positive and negative charges and that opposite charges attract and like charges repel.
★ Explain the charging of solids by friction.
★ Describe an experiment to determine whether a material is an electrical conductor or an insulator.
★ Explain the difference between electrical conductors and insulators using a simple electron model, and give examples of each.

★ Know that charge is measured in coulombs.
★ Describe an electric field, explain its direction and describe simple electric field patterns.

Electrostatic charges arise when electrons are transferred between objects by rubbing. Sparks can fly after you comb your hair or walk across a synthetic carpet when you touch an earthed object, through which the charge can be neutralised; the discharge can lead you to feel a small electric shock. A flash of lightning is nature's most spectacular static electricity effect. There are two types of electrostatic charge. Like charges repel while opposite charges attract. Charges build up on an insulator such as plastic and remain static, but for conductors like metals, charges flow away to try to neutralise charge. Both electrical conductors and insulators have their uses.

Electric charges are surrounded by an electric field which exerts a force on a nearby charge. This effect is made use of in applications from ink-jet printers to crop sprayers. As with a magnetic field, an electric field exerts an action-at-a-distance force.

▲ **Figure 4.2.1** A flash of lightning

Clothes containing nylon often crackle when they are taken off. We say they are charged with static electricity; the crackles are caused by tiny electric sparks which can be seen in the dark. Pens and combs made of certain plastics become charged when rubbed on your sleeve and can then attract scraps of paper.

Positive and negative charges

When a strip of polythene is rubbed with a cloth it becomes charged. If it is hung up and another rubbed polythene strip is brought near, repulsion occurs (Figure 4.2.2). Attraction occurs when a rubbed strip of cellulose acetate is brought near.

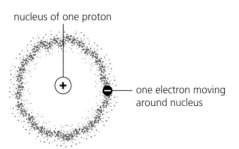

nucleus of one proton

one electron moving around nucleus

▲ **Figure 4.2.3** Hydrogen atom

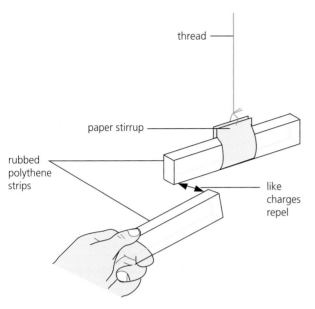

thread

paper stirrup

rubbed polythene strips

like charges repel

▲ **Figure 4.2.2** Investigating charges

This shows there are two kinds of electric charge. That on cellulose acetate is taken as **positive** (+) and that on polythene is **negative** (−). It also shows that:

Like charges (+ and +, or − and −) repel, while unlike charges (+ and −) attract.

The force between electric charges decreases as their separation increases.

> **Key definitions**
>
> **Positive charges** repel other positive charges, but positive charges attract negative charges
>
> **Negative charges** repel other negative charges, but negative charges attract positive charges

Charges, atoms and electrons

There is evidence (Topic 5.1) that we can picture an atom as being made up of a small central nucleus containing positively charged particles called **protons**, surrounded by an equal number of negatively charged **electrons**. The charges on a proton and an electron are equal and opposite so an atom as a whole is normally electrically neutral, i.e. has no net charge.

Hydrogen is the simplest atom with one proton and one electron (Figure 4.2.3). A copper atom has 29 protons in the nucleus and 29 surrounding electrons. Every nucleus except hydrogen also contains uncharged particles called **neutrons**.

The production of charges by rubbing can be explained by supposing that friction causes electrons to be transferred from one material to the other. For example, when cellulose acetate is rubbed with a cloth, electrons go from the acetate to the cloth, leaving the acetate short of electrons, i.e. positively charged. The cloth now has more electrons than protons and becomes negatively charged. Note that it is only electrons which move; the protons remain fixed in the nucleus.

> **Test yourself**
>
> 1 Two identical conducting balls, suspended on nylon threads, come to rest with the threads making equal angles with the vertical, as shown in Figure 4.2.4.
> Which of these statements is true?
> This shows that
> A the balls are equally and oppositely charged
> B the balls are oppositely charged but not necessarily equally charged
> C one ball is charged and the other is uncharged
> D the balls both carry the same type of charge.
>
>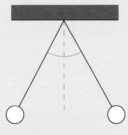
>
> ▲ **Figure 4.2.4**
>
> 2 Explain in terms of electron movement what happens when a polythene rod becomes charged negatively by being rubbed with a cloth.
> 3 Two electrostatic charges are brought close together.
> a When one charge is positive and the other is negative, are they attracted or repelled from each other?
> b When both charges are negative, are they attracted or repelled?

Units of charge

Charge is measured in **coulombs** (C) and is defined in terms of the ampere (see Topic 4.2.2).

The charge on an electron e = 1.6×10^{-19} C.

> **Key definition**
> **Coulomb** (C) unit of charge

Practical work

Gold-leaf electroscope

For safe experiments/demonstrations related to this topic, please refer to the *Cambridge IGCSE Physics Practical Skills Workbook* that is also part of this series.

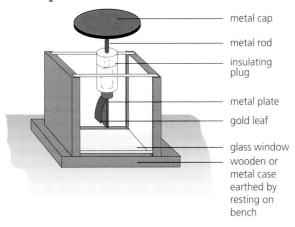

▲ **Figure 4.2.5** Gold-leaf electroscope

A gold-leaf electroscope consists of a metal cap on a metal rod at the foot of which is a metal plate with a leaf of gold foil attached (Figure 4.2.5). The rod is held by an insulating plastic plug in a case with glass sides to protect the leaf from draughts.

Detecting a charge

Bring a charged polythene strip towards the cap: the leaf rises away from the plate. When you remove the charged strip, the leaf falls again. Repeat with a charged acetate strip.

Charging by contact

Draw a charged polythene strip *firmly across the edge of the cap*. The leaf should rise and stay up when the strip is removed. If it does not, repeat the process but press harder. The electroscope has now become negatively charged by contact with the polythene strip, from which electrons have been transferred.

Insulators and conductors

Touch the cap of the charged electroscope with different things, such as a piece of paper, a wire, your finger, a comb, a cotton handkerchief, a piece of wood, a glass rod, a plastic pen, rubber tubing. Record your results.

When the leaf falls, charge is passing to or from the ground through you and the material touching the cap. If the fall is rapid the material is a *good conductor*; if the leaf falls slowly, the material is a poor conductor. If the leaf does not alter, the material is a *good insulator*.

The gold-leaf electroscope used in this experiment could be replaced by an electronic instrument capable of measuring electric charge – an electrometer.

1 How could you charge a polythene rod?
2 How could you transfer charge from a polythene rod to a gold-leaf electroscope?
3 Why does the leaf of the electroscope rise when it gains charge?
4 How can you discharge the electroscope?

Electrons, insulators and conductors

In an insulator all electrons are bound firmly to their atoms; in a conductor some electrons can move freely from atom to atom. An insulator can be charged by rubbing because the charge produced cannot move from where the rubbing occurs, i.e. the electric charge is *static*. A conductor will become charged only if it is held with an insulating handle; otherwise electrons are transferred between the conductor and the ground via the person's body.

Good insulators include plastics such as polythene, cellulose acetate, Perspex and nylon. All metals and carbon are good conductors. In between are materials that are both poor conductors and (because they conduct to some extent) poor insulators. Examples are wood, paper, cotton, the human body and the Earth. Water conducts and if it were not present in materials such as wood and on the surface of, for example, glass, these would be good insulators. Dry air insulates well.

Electric fields

When an electric charge is placed near to another electric charge it experiences a force. The electric force does not require contact between the two charges so we call it an 'action-at-a-distance force' – it acts through space. The region of space where an electric charge experiences a force due to other charges is called an **electric field**. If the electric force felt by a charge is the same everywhere in a region, the field is uniform; a uniform electric field is produced between two oppositely charged parallel metal plates (Figure 4.2.6). It can be represented by evenly spaced parallel lines drawn perpendicular to the metal surfaces. The **direction of an electric field at a point**, denoted by arrows, is the direction of the force on a small *positive* charge placed in the field (negative charges experience a force in the opposite direction to the field). An electric field is a vector quantity as it has both magnitude (strength) and direction.

> **Key definition**
> **Direction of an electric field at a point** the direction of the force on a positive charge at that point

Moving charges are deflected by an electric field due to the electric force exerted on them.

The electric field lines radiating from an isolated positively charged conducting sphere and a point charge are shown in Figures 4.2.7a and b: the field lines again emerge at right angles to the conducting surface.

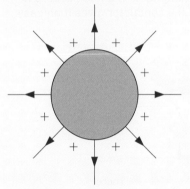

▲ **Figure 4.2.7a** Electric field around a charged conducting sphere

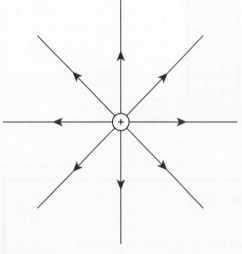

▲ **Figure 4.2.7b** Electric field around a point charge

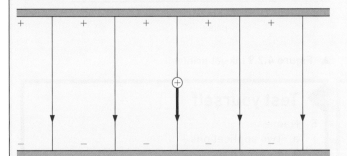

▲ **Figure 4.2.6** Uniform electric field

Going further

Dangers of static electricity

Sparks occur between electrostatic charges when the electric field is strong enough. Damage can be reduced by providing an easy path for electrons to flow safely to and from the Earth. For example, a tall building is protected by a lightning conductor consisting of a thick copper strip fixed on the outside of the building connecting metal spikes at the top to a metal plate in the ground (Figure 4.2.8).

Thunderclouds carry charges: a negatively charged cloud passing overhead repels electrons from the spikes to the Earth. The points of the spikes are left with a large positive charge (charge concentrates on sharp points) which removes electrons from nearby air molecules, so charging them positively and causing them to be repelled from the spike. This effect, called action at points, results in an 'electric wind' of positive air molecules streaming upwards which can neutralise electrons discharging from the thundercloud in a lightning flash. If a flash occurs it is now less violent and the conductor gives it an easy path to ground.

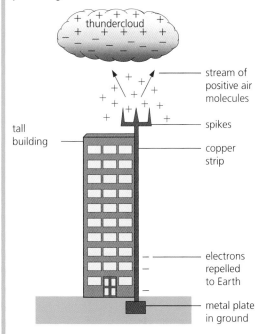

▲ **Figure 4.2.8** Lightning conductor

Sparks from static electricity can be particularly dangerous when flammable vapour is present. Fuel flowing in a pipeline (particularly a plastic pipe) experiences friction, which may lead to a build-up of static charge. During refuelling, aircraft, fuel tanker and pipeline hoses are all earthed to avoid sparks which could ignite the fuel and cause an explosion.

Computers and sensitive electronic equipment should also be earthed to avoid electrostatic damage.

Uses of static electricity

There are many uses of static electricity in applications from flue-ash precipitation in coal-burning power stations, paint and crop spraying to photocopiers and ink-jet printers.

In an ink-jet printer tiny drops of ink are forced out of a fine nozzle, charged electrostatically and then passed between two oppositely charged plates; a negatively charged drop will be attracted towards the positive plate causing it to be deflected as shown in Figure 4.2.9. The amount of deflection and hence the position at which the ink strikes the page is determined by the charge on the drop and the p.d. between the plates; both of these are controlled by a computer. About 100 precisely located drops are needed to make up an individual letter but very fast printing speeds can be achieved.

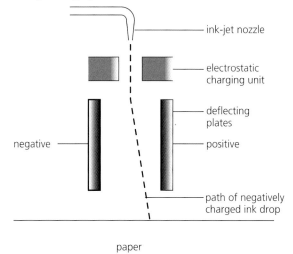

▲ **Figure 4.2.9** Ink-jet printer

> ### Test yourself
>
> 4 Describe the electric field around a negatively charged conducting sphere.

> ### Test yourself
>
> 5 Name
>   a two applications
>   b two dangers
>   of static electricity.

4.2.2 Electric current

FOCUS POINTS

★ Understand that an electric current consists of moving electric charges.

★ Define electric current and use the correct equation in calculations.

★ Describe the use of analogue and digital ammeters and the difference between alternating current (a.c.) and direct current (d.c.).

★ Describe the role of free electrons in electrical conduction in metals.

★ Know that the flow of electrons in a circuit is in the opposite direction to that of the conventional current flow.

In the previous topic you learnt about positive and negative static charges and how they were produced on conductors and insulators. In this topic you will discover that moving charges in a conductor produce an electric current which is proportional to the rate of flow of charge. Every electrical appliance you use, from hair dryer to computer, relies on the flow of an electric current. In a metal the current is produced by the movement of electrons. By convention, electric current is linked to the flow of positive charge, which is in the opposite direction to the way electrons move. You will find out how to connect an ammeter to a circuit to measure the size of an electric current and learn about the different types of current.

An **electric current** consists of moving electric charges. In Figure 4.2.10, when the van de Graaff machine is working, it produces a continuous supply of charge which produces an electric field between the metal plates to which it is connected. The table-tennis ball shuttles rapidly backwards and forwards between the plates and the very sensitive meter records a small current. As the ball touches each plate it becomes charged and is repelled to the other plate. In this way charge is carried across the gap. This also shows that static charges, produced by friction in the van de Graaff machine, cause a deflection on a meter just as current electricity produced by a battery does.

In a metal, each atom has one or more loosely held electrons that are free to move. When a van de Graaff or a battery is connected across the ends of such a conductor, the free electrons drift slowly along it in the direction from the negative to the positive terminal of a battery. There is then a current of negative charge. This is how electrical conduction occurs in a metal.

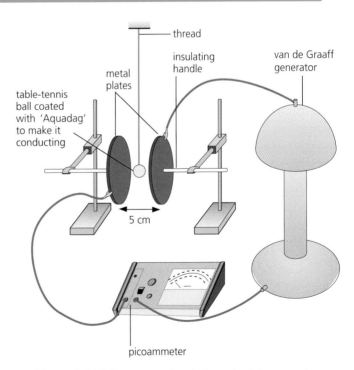

▲ **Figure 4.2.10** Demonstrating that an electric current consists of moving charges

Effects of a current

An electric current has three effects that reveal its existence and which can be shown with the circuit of Figure 4.2.11.

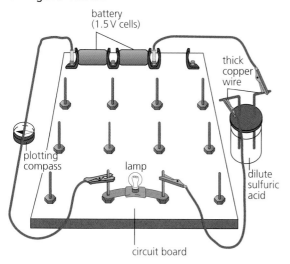

▲ **Figure 4.2.11** Investigating the effects of a current

Heating and lighting

The lamp lights because the small wire inside (the filament) is made white hot by the current.

Magnetic

The plotting compass is deflected when it is placed near the wire because a magnetic field is produced around any wire carrying a current.

Chemical

Bubbles of gas are given off at the wires in the acid because of the chemical action of the current.

The ampere and the coulomb

An **electric current** is defined as the charge passing a point per unit time and can be written in symbols as

$$I = \frac{Q}{t}$$

where I is the current when charge Q passes any point in a circuit in time t.

It shows that current is the rate of flow of charge in a circuit.

> **Key definition**
>
> **Electric current** the charge passing a point per unit time
>
> Current $I = \frac{Q}{t}$ where Q is the charge flowing past a particular point in time t.

The unit of current is the **ampere** (A) which is defined using the magnetic effect. One milliampere (mA) is one-thousandth of an ampere. Current is measured by an **ammeter.**

The unit of charge, the **coulomb** (C), is defined in terms of the ampere.

One coulomb is the charge passing any point in a circuit when a steady current of 1 ampere flows for 1 second. That is, $1\,C = 1\,A\,s$.

In general, if a steady current I (amperes) flows for time t (seconds) the charge Q (coulombs) passing any point is given by

$$Q = I \times t$$

Current must have a complete path (a circuit) of conductors if it is to flow. When drawing circuit diagrams, components are represented by symbols. Some commonly used symbols are represented in Topic 4.3.1.

❓ Worked example

Current flows in an electrical circuit.

a A charge of 2 C passes a point in the circuit in 5 s, calculate the current flowing past that point.

$$I = Q/t = 2\,C/5\,s = 0.4\,A$$

b A current of 3 A flows past another point in the circuit in 10 seconds. How much charge passes the point in this time?

$$Q = I \times t = 3\,A \times 10\,s = 30\,C$$

Now put this into practice

1 A current of 2 A flows past a point in an electrical circuit in 20 s. How much charge passes the point in this time?
2 A charge of 3 C passes a point in an electrical circuit in 7 s. Calculate the current flowing past that point.

Conventional current

Before the electron was discovered scientists agreed to think of current as positive charges moving round a circuit in the direction from positive to negative of a battery. This agreement still stands. Arrows on circuit diagrams show the direction of what we call the **conventional current**, i.e. the direction in which *positive* charges would flow. Electrons flow in the opposite direction to the conventional current.

> **Key definition**
>
> **Conventional current** flows from positive to negative; the flow of free electrons is from negative to positive

Ammeters

An ammeter is used to measure currents. It should always be placed in series in a circuit with the positive terminal on the ammeter connected to the positive terminal of the supply, as described in the practical work below (see Figure 4.2.13 overleaf). A simple moving coil ammeter will read d.c. currents only on an analogue display. It may have two ranges and two scales in the display.

A **multimeter** can have either a digital or analogue display (see Figure 4.1.12a and b) and be used to measure a.c. and d.c. currents (or voltages and also resistance). The required function is first selected, say d.c. current.

When making a measurement on either type of ammeter a suitable range must be chosen. For example, if a current of a few milliamps is expected, the 10 mA range might be selected and the value of the current (in mA) read from the display; if the reading is off-scale, the **sensitivity** should be reduced by changing to the higher, perhaps 100 mA, range.

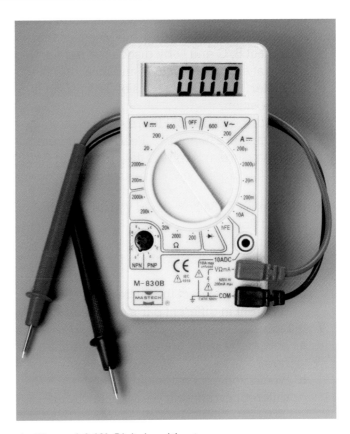

▲ **Figure 4.2.12b** Digital multimeter

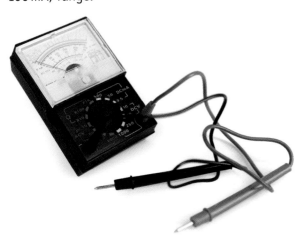

▲ **Figure 4.2.12a** Analogue multimeter

Test yourself

6   Explain how electrical conduction occurs in a metal.
7   Explain how you would connect an ammeter into a circuit.

8   What is the current in a circuit if the charge passing each point is
    a   10 C in 2 s
    b   20 C in 40 s
    c   240 C in 2 minutes?
9   How long does it take a charge of 5 C to pass a point in an electrical circuit where the current flowing is 2 A?

 Practical work

Measuring current

For safe experiments/demonstrations related to this topic, please refer to the *Cambridge IGCSE Physics Practical Skills Workbook* that is also part of this series.

a   Connect the circuit of Figure 4.2.13a (on a circuit board if possible), ensuring that the + of the cell (the metal stud) goes to the + of the ammeter (marked red). Note the current.
b   Connect the circuit of Figure 4.2.13b. The cells are in *series* (+ of one to − of the other), as are

the lamps. Record the current. Measure the current at B, C and D by disconnecting the circuit at each point in turn and inserting the ammeter. Record the values of the current in each position.

c Connect the circuit of Figure 4.2.13c. The lamps are in *parallel*. Read the ammeter. Also measure and record the currents at P, Q and R. Comment on your results.

5 In Figure 4.2.13a how could you tell when current flows?

6 In Figure 4.2.13b
   a how many paths are there for current to flow?
   b would you expect the current to be different in different parts of the circuit?

7 In Figure 4.2.13c
   a how many paths are there for current to flow?
   b would you expect the current to be different in different parts of the circuit?

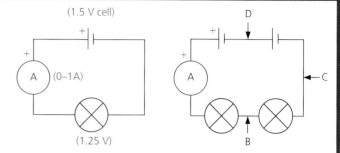

▲ **Figure 4.2.13a**   ▲ **Figure 4.2.13b**

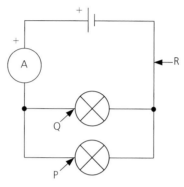

▲ **Figure 4.2.13c**

Direct and alternating current

Difference

In a **direct current** (d.c.) the electrons flow in one direction only. Graphs for steady and varying d.c. are shown in Figure 4.2.14.

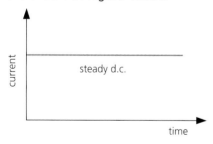

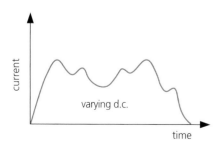

▲ **Figure 4.2.14** Direct current (d.c.)

In an **alternating current** (a.c.) the direction of flow reverses regularly, as shown in the graph in Figure 4.2.15. The circuit sign for a.c. is given in Figure 4.2.16.

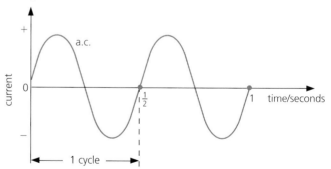

▲ **Figure 4.2.15** Alternating current (a.c.)

▲ **Figure 4.2.16** Symbol for alternating current

The pointer of an ammeter for measuring d.c. is deflected one way by the direct current. Alternating current makes the pointer move back and forth about the zero if the changes are slow enough; otherwise no deflection can be seen.

Batteries give d.c.; generators can produce either d.c. or a.c.

Frequency of a.c.

The number of complete alternations or cycles in 1 second is the *frequency* of the alternating current. The unit of frequency is the *hertz* (Hz). The frequency of the a.c. in Figure 4.2.15 is 2 Hz, which means there are two cycles per second, or one cycle lasts $1/2 = 0.5\,s$. The mains supply in many countries is a.c. of frequency 50 Hz; each cycle lasts 1/50th of a second. This regularity was used in the tickertape timer (Topic 1.2) and is relied upon in mains-operated clocks.

> ### ▶ Test yourself
> 10 Sketch
> **a** a d.c. current
> **b** an a.c. current
> **c** the circuit symbol used for a.c.
> 11 An a.c. current has a frequency of 1000 Hz. How long does each cycle last?

4.2.3 Electromotive force and potential difference

FOCUS POINTS

★ Define electromotive force.
★ Describe the use of analogue and digital voltmeters.
★ Define potential difference and know that it is measured in volts.

★ Use the correct equations for electromagnetic force and potential difference.

As you will have seen in the previous topic, a complete circuit of conductors is needed for a current to flow. In this topic you will learn that it is the electromotive force of a supply which provides the energy needed to move charge around a complete circuit. The supply may vary from a simple torch battery to your mains electricity supply. There are usually several components in a circuit, for example lamps, motors or other electrical devices, from which energy is transferred to the surroundings. The energy transferred from a device can be calculated by introducing the concept of potential difference. Previously you used an ammeter to measure the current in an electrical circuit; now you will learn how to use a voltmeter to measure potential difference.

The chemical action inside a battery produces a surplus of electrons at one of its terminals (the negative) and creates a shortage at the other (the positive). It is then able to maintain a flow of electrons, i.e. an *electric current*, in any circuit connected across its terminals for as long as the chemical action lasts. Work is done by the battery in moving charge around the circuit.

Electromotive force (e.m.f.) is defined as the electrical work done by a source in moving a unit charge around a complete circuit.

> **Key definitions**
> **Electromotive force** e.m.f. the electrical work done by a source in moving unit charge around a complete circuit
>
> **Potential difference** p.d. the work done by a unit charge passing through a component

Electromotive force is measured in **volts** (V). The e.m.f. of a car battery is 12 V and the domestic mains supply in many countries is 240 V.

There are usually a number of components in an electrical circuit through which charge flows. **Potential difference** (p.d.) is defined as the work done by a unit of charge passing through a component.

Like e.m.f., potential difference between two points is measured in volts (V). The term **voltage** is sometimes used instead of p.d.

Energy transfers and p.d.

In an electric circuit, an electric current transfers energy from an energy store, such as a battery, to components in the circuit which then transfer energy into the surroundings. In the case of a lamp, energy is transferred to the surroundings by light and by heating.

When each of the circuits shown in Figure 4.2.17 is connected up, it will be found from the ammeter readings that the current is about the same (0.4 A) in each lamp. However, the mains lamp with a potential difference of 230 V applied across it gives much more light and heat than the car lamp with 12 V across it.

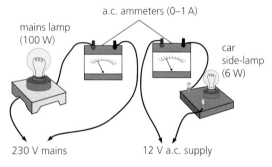

▲ **Figure 4.2.17** Investigating the effect of p.d. (potential difference) on energy transfer

Evidently the p.d. across a device affects the rate at which it transfers energy. This gives us a way of defining the unit of potential difference: the volt.

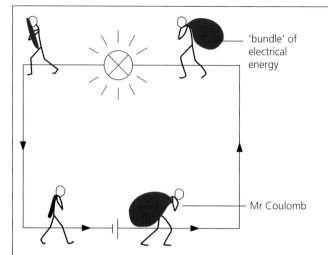

▲ **Figure 4.2.18** Model of a circuit

In our imaginary representation, Mr Coulomb travels round the circuit and unloads energy as he goes, most of it in the lamp. We think of him receiving a fresh bundle every time he passes through the battery, which suggests he must be travelling very fast. In fact, as we found earlier (Topic 4.2.2), the electrons drift along quite slowly. As soon as the circuit is complete, energy is delivered at once to the lamp, not by electrons directly from the battery but from electrons that were in the connecting wires. The model is helpful but is not an exact representation.

The volt

The demonstrations of Figure 4.2.17 show that the greater the e.m.f. of a supply, the larger is the bundle of energy given to each coulomb and the greater is the rate at which energy is transferred from a lamp.

In general, if W (joules) is the energy transferred (i.e. the work done) when charge Q (coulombs) moves around a complete circuit, the e.m.f. E (volts) of the supply is given by

$$E = W/Q$$

or $W = Q \times E$

Model of a circuit

It may help you to understand the definition of the volt, i.e. what a volt is, if you *imagine* that the current in a circuit is formed by 'drops' of electricity, each having a charge of 1 coulomb and carrying equal-sized bundles of electrical energy. In Figure 4.2.18, Mr Coulomb represents one such drop. As a drop moves around the circuit it gives up all its energy which is transferred to other energy stores. Note that *electrical energy, not charge or current, is used up.*

The p.d. between two points in a circuit is 1 volt if 1 joule of energy is transferred when 1 coulomb passes from one point to the other.

That is, 1 volt = 1 joule per coulomb (1 V = 1 J/C). If 2 J is transferred by each coulomb, the p.d. is 2 V.

In general, if W (joules) is the work done when charge Q (coulombs) passes between two points, the p.d. V (volts) between the points is given by

$$V = W/Q$$

or $W = Q \times V$

If Q is in the form of a steady current I (amperes) flowing for time t (seconds) then $Q = I \times t$ (Topic 4.2.2) and

$$W = I \times t \times V$$

? Worked example

A lamp is connected to a battery in a circuit and a current flows.

a Calculate the p.d. across the lamp if 6 J of work are done when 2 C of charge pass through the lamp.
   From the equation $V = W/Q$

   the p.d. across the lamp = $W/Q = 6\,J/\,2\,C = 3\,V$

b If the p.d. across the lamp is increased to 5 V calculate the energy transferred to the lamp when a current of 2 A flows in the lamp for 5 seconds.

   $$Q = I \times t = 2\,A \times 5\,s = 10\,C$$

   Rearranging the equation $V = W/Q$ gives

   $$W = Q \times V = 10\,C \times 5\,V = 50\,J$$

Now put this into practice

1 Calculate the p.d. across a lamp in an electric circuit when 8 J of work are done when a charge of 4 C passes through the lamp.
2 The p.d. across a lamp is 6 V.
   How many joules of energy are transferred when a charge of 2 C passes through it?
3 The p.d. across a lamp is 6 V. Find the work done when a current of 3 A flows in the lamp for 10 s.

Voltmeters

A **voltmeter** is used to measure potential differences; it should always be placed *in parallel* with the component across which the p.d. is to be measured. The positive terminal on the voltmeter should be connected to the side of the component into which current flows as is shown in the practical work below (see Figure 4.2.21 overleaf). A simple moving-coil voltmeter will read d.c. voltages only on an analogue display.

The face of an analogue voltmeter is represented in Figure 4.2.19. The voltmeter has two scales. The 0–5 scale has a full-scale deflection of 5.0 V. Each small division on the 0–5 scale represents 0.1 V. This voltmeter scale can be read to the nearest 0.1 V. The human eye is very good at judging a half division, so we are able to estimate the voltmeter reading to the nearest 0.05 V with considerable precision. The 0–10 scale has a full-scale deflection of 10.0 V; each small division on this scale represents 0.2 V so the precision of a reading is less than on the 0–5 V scale.

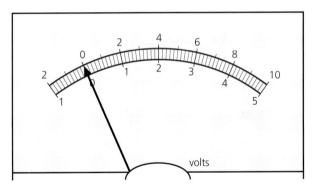

volts

▲ **Figure 4.2.19** An analogue voltmeter scale

Analogue voltmeters or multimeters are adapted moving-coil **galvanometers** (Topic 4.5.5). Digital multimeters are constructed from integrated circuits. On the voltage setting they have a very high input resistance (10 MΩ); this means they affect most circuits very little and so give very accurate readings.

When making a measurement on either an analogue or digital voltmeter a suitable range must first be chosen. For example, if a voltage of a few millivolts is expected, the 10 mV range might be selected and the value of the voltage (in mV) read from the display; if the reading is off-scale, the sensitivity should be reduced by changing to the higher, perhaps 100 mV, range.

Every measuring instrument has a calibrated scale. When you write an account of an experiment (see p. vii, *Scientific enquiry*) you should include details about each scale that you use.

? Worked example

The scales of an analogue voltmeter are shown in Figure 4.2.20.

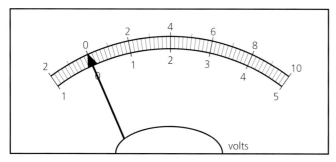

▲ Figure 4.2.20

a What are the two ranges available when using the voltmeter?
   The lower scale reads 0–5 V and the upper scale reads 0–10 V.
b What do the small divisions between the numbers 3 and 4 represent?
   0.1 V
c Which scale would you use to measure a voltage of 4.6 V?
   The lower scale 0–5 V will give a more accurate reading.
d When the voltmeter reads 4.0 V where should you position your eye to make the reading?
   Above the 4 to reduce parallax error.

Now put this into practice

1 Use the scales of the voltmeter shown in Figure 4.2.20.
   a What do the small divisions between the numbers 6 and 8 represent?
   b Which scale would you use to measure a voltage of 5.4 V?
   c When making the reading for 4.0 V an observer's eye is over the 0 V mark. Explain why the value obtained by this observer is higher than 4.0 V.

Practical work

Measuring voltage

A voltmeter is an instrument for measuring voltage or p.d. It looks like an ammeter but has a scale marked in volts. Whereas an ammeter is inserted in *series* in a circuit to measure the current, a voltmeter is connected across that part of the circuit where the voltage is required, i.e. in *parallel*.

To prevent damage to the voltmeter make sure that the + terminal (marked red) is connected to the point nearest the + of the battery.

a Connect the circuit of Figure 4.2.21a. The voltmeter gives the p.d across the lamp. Read it.

a

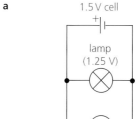

voltmeter (0–5 V)

b

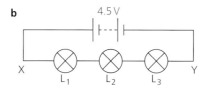

c

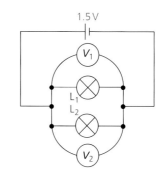

▲ Figure 4.2.21

b Connect the circuit of Figure 4.2.21b. Measure:
  i the p.d V between X and Y
  ii the p.d V_1 across lamp L_1
  iii the p.d V_2 across lamp L_2
  iv the p.d V_3 across lamp L_3.

c Connect the circuit of Figure 4.2.21c, so that two lamps L_1 and L_2 are in parallel across one 1.5 V cell. Measure the p.d.s, V_1 and V_2, across each lamp in turn.

8 For step **b** above, calculate the value of $V_1 + V_2 + V_3$ and compare with the value of V.

9 For step **c** above, compare the values of V_1 and V_2.

10 If all the lamps shown in Figure 4.2.21b are identical, what would you expect the p.d. across each to be?

11 a Explain where you would connect and how you would use a voltmeter to measure the p.d. across a device.
  b In the circuit shown in Figure 4.2.21c, if V_1 measures 1.5 V, what would you expect the value of V_2 to be?

Test yourself

12 a Define electromotive force.

  b Define potential difference.
13 The p.d. across the lamp in Figure 4.2.22 is 12 V. How many joules of electrical energy are transferred into light and heat when
  a a charge of 1 C passes through it
  b a charge of 5 C passes through it
  c a current of 2 A flows in it for 10 s?

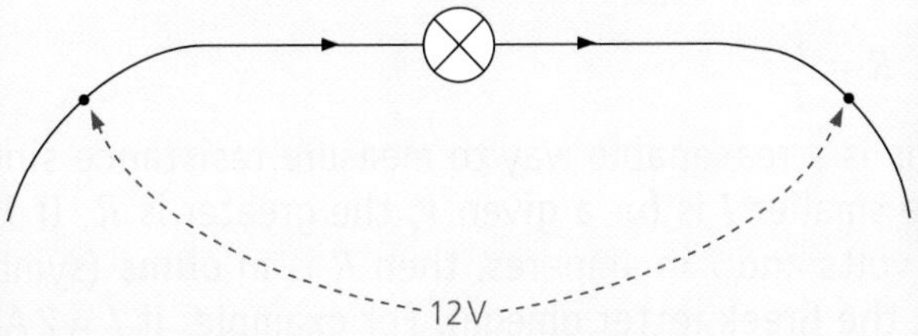

▲ **Figure 4.2.22**

4.2.4 Resistance

FOCUS POINTS

★ Know the correct equation for resistance and use it correctly to determine resistance using a voltmeter and an ammeter.

★ Draw and interpret current–voltage graphs.

★ Understand the dependence of the resistance of a metal wire on its length and cross-sectional area.

★ Know that resistance is directly proportional to length and inversely proportional to cross-sectional area in a metallic electrical conductor.

In this topic you will learn that the ease of passage of electrons depends on the nature of the material. This effect is measured by resistance. More work has to be done to drive a current through a high resistance than a low resistance. For the element in an electric fire, a high-resistance wire is needed so that a large amount of energy is transferred. The opposite is required for the connecting wires in a circuit, where low-resistance wires are used to reduce energy losses. Current flow is easier in a wire with a large cross-sectional area so thick wires are used where large currents are needed, for example in the starter motor in a car or a kitchen oven. The longer a wire, the harder it is for current to flow; energy loss is reduced by using short connecting wires.

Electrons move more easily through some conductors than others when a p.d. is applied. The opposition of a conductor to current is called its **resistance**. A good conductor has a low resistance and a poor conductor has a high resistance.

The ohm

If the current in a conductor is I when the voltage across it is V, as shown in Figure 4.2.23a, its resistance R is defined by

$$R = \frac{V}{I}$$

This is a reasonable way to measure resistance since the smaller I is for a given V, the greater is R. If V is in volts and I in amperes, then R is in **ohms** (symbol Ω, the Greek letter omega). For example, if $I = 2\,A$ when $V = 12\,V$, then $R = 12\,V/2\,A$, that is, $R = 6\,\Omega$.

The ohm is the resistance of a conductor in which the current is 1 ampere when a voltage of 1 volt is applied across it.

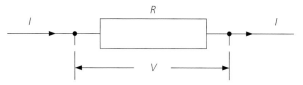

▲ **Figure 4.2.23a**

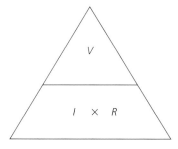

▲ **Figure 4.2.23b**

Alternatively, if R and I are known, V can be found from

$$V = IR$$

Also, when V and R are known, I can be calculated from

$$I = \frac{V}{R}$$

The triangle in Figure 4.2.23b is an aid to remembering the three equations. It is used in the same way as the 'density triangle' in Topic 1.4.

? **Worked example**

a If a p.d. of 4.5V is applied across a lamp, the current flowing through the lamp is 1.5A. Calculate the resistance of the lamp.

$$R = \frac{V}{I}$$

so $R = \dfrac{4.5\,V}{1.5\,A} = 3\,\Omega$

b A current of 0.5A flows through a resistance of 5Ω. Calculate the p.d. across the lamp.

$$V = IR = 0.5\,A \times 5\,\Omega = 2.5\,V$$

Now put this into practice

1 The current flowing through a resistor is 0.30A when a p.d. of 4.5V is applied across it.
   Calculate the value of the resistor.
2 A current of 0.2A flows through a resistor of 10Ω.
   Calculate the p.d. across the resistor.
3 A p.d. of 12.0V is applied across a lamp of 24Ω and the lamp lights up. Calculate the current passing through the lamp.

Resistors

Conductors intended to have resistance are called **resistors** (Figure 4.2.24a) and are made either from wires of special alloys or from carbon. Those used in radio and television sets have values from a few ohms up to millions of ohms (Figure 4.2.24b).

▲ **Figure 4.2.24a** Circuit symbol for a resistor

▲ **Figure 4.2.24b** Resistor

▲ **Figure 4.2.24c** Variable resistor (potentiometer)

Variable resistors are used in electronics (and are then called **potentiometers**) as volume and other controls (Figure 4.2.24c). Variable resistors that take larger currents, like the one shown in Figure 4.2.25, are useful in laboratory experiments. These consist of a coil of constantan wire (an alloy of 60% copper, 40% nickel) wound on a tube with a sliding contact on a metal bar above the tube.

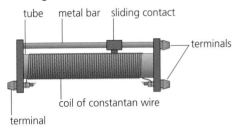

▲ **Figure 4.2.25** Large variable resistor

There are two ways of using such a variable resistor. It may be used as a **rheostat** for changing the current in a circuit; only one end connection and the sliding contact are then required. In Figure 4.2.26a moving the sliding contact to the left reduces the resistance and increases the current. This variable resistor can also act as a **potential divider** for changing the p.d. applied to a device; all three connections are then used. In Figure 4.2.26b any fraction from the total p.d. of the battery to zero can be 'tapped off' by moving the sliding contact down. Figure 4.2.27 shows the circuit diagram symbol for a variable resistor being used in rheostat mode.

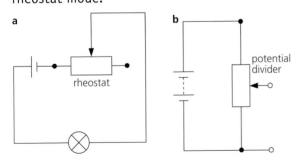

▲ **Figure 4.2.26** A variable resistor can be used as a rheostat or as a potential divider.

▲ **Figure 4.2.27** Circuit symbol for a variable resistor used as a rheostat

> **Test yourself**
>
> 14 What is the resistance of a lamp when a voltage of 12 V across it causes a current of 4 A?
> 15 Calculate the p.d. across a 10 Ω resistor carrying a current of 2 A.
> 16 The p.d. across a 3 Ω resistor is 6 V. Calculate the current flowing (in ampere).
>
> 17 Calculate the number of coulombs per second passing through a 4 Ω resistor connected across the terminals of a 12 V battery.

 Practical work

Measuring resistance

For safe experiments/demonstrations related to this topic, please refer to the *Cambridge IGCSE Physics Practical Skills Workbook* that is also part of this series.

The resistance R of a conductor can be found by measuring the current I in it when a p.d. V is applied across it and then using $R = V/I$. This is called the *ammeter–voltmeter* method.

Set up the circuit of Figure 4.2.28 in which the unknown resistance R is 1 metre of SWG 34 constantan wire. Altering the rheostat changes both the p.d. V and the current I. Record in a table, with three columns, five values of I (e.g. 0.10, 0.15, 0.20, 0.25 and 0.3 A) and the corresponding values of V.

Repeat the experiment, but instead of the wire use
i   a lamp (e.g. 2.5 V, 0.3 A),
ii a **semiconductor diode** (e.g. 1 N4001) connected first one way then the other way around and
iii a thermistor (e.g. TH 7).

(Semiconductor diodes and thermistors are considered in Topic 4.3 in more detail.)

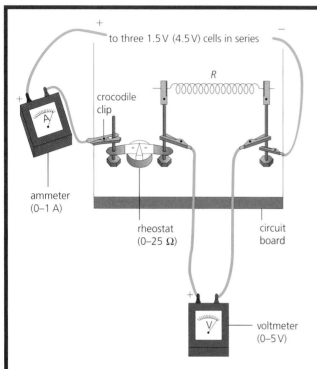

to three 1.5 V (4.5 V) cells in series

R

crocodile clip

ammeter (0–1 A)

rheostat (0–25 Ω)

circuit board

voltmeter (0–5 V)

▲ **Figure 4.2.28**

12 Work out R for each pair of readings from the equation $R = V/I$.

13 Draw the symbols for **a** a resistor and **b** a variable resistor.

14 List the equipment you would need to measure the resistance of a wire.

15 Calculate the resistance of a wire that has a current of 0.15 A passing through it when the p.d. across it is 4.5 V.

Resistance of a metal wire

The **resistance of a metallic wire**
(i) increases as its length increases
(ii) increases as its cross-sectional area decreases
(iii) depends on the material.
A long thin wire has more resistance than a short thick one of the same material. Silver is the best conductor, but copper, the next best, is cheaper and is used for connecting wires and for domestic electric cables.

Key definition

Resistance of a metallic wire directly proportional to its length and inversely proportional to its cross-sectional area

Resistivity

Experiments show that the resistance R of a wire of a given material is

(i) directly proportional to its length l, i.e. $R \propto l$

(ii) inversely proportional to its cross-sectional area A, i.e. $R \propto 1/A$ (doubling A halves R).

? **Worked example**

A copper wire has a diameter of 0.50 mm, a length of 1 km and a resistance of 84 Ω.

a Calculate the resistance of a wire of the same material and diameter with a length of 500 m.
  Let $R_1 = 84\,\Omega$, length $l_1 = 1.0\,\text{km} = 1000\,\text{m}$, length $l_2 = 500\,\text{m}$ and R_2 the required resistance.
  Then since $R \propto 1/A$ and A is constant

$$\frac{R_2}{R_1} = \frac{l_2}{l_1}$$

  and $R_2 = R_1 \times \dfrac{l_2}{l_1} = 84\,\Omega \times \dfrac{500\,\text{m}}{1000\,\text{m}} = 42\,\Omega$

  The resistance is halved when the length of the wire is halved.

b Calculate the resistance of a wire of the same material with a diameter of 1.0 mm and a length of 1 km.
  Let $R_1 = 84\,\Omega$, diameter $d_1 = 0.50\,\text{mm}$, diameter $d_2 = 1.0\,\text{mm}$ and R_2 the required resistance.
  If r is the radius of the wire, the cross-sectional area $A = \pi r^2 = \pi(d/2)^2 = (\pi/4)\,d^2$, so

$$\frac{A_1}{A_2} = \frac{(d_1)^2}{(d_2)^2} = \frac{(0.50\,\text{mm})^2}{(1.0\,\text{mm})^2} = 0.25$$

  Then since $R \propto \dfrac{l}{A}$ and l is constant

$$\frac{R_2}{R_1} = \frac{A_1}{A_2}$$

  and $R_2 = R_1 \times \dfrac{A_1}{A_2} = 84\,\Omega \times 0.25 = 21\,\Omega$

Now put this into practice

1 A certain wire has a length of 10 m and a resistance of 60 Ω.
  Calculate the resistance of 20 m of the wire.

2 A certain wire has diameter of 0.20 mm and a resistance of 60 Ω. Calculate the resistance of a wire of the same material with a diameter of 0.40 mm.

I–V graphs: Ohm's law

The variation of current with voltage is shown for various conductors in Figure 4.2.29.

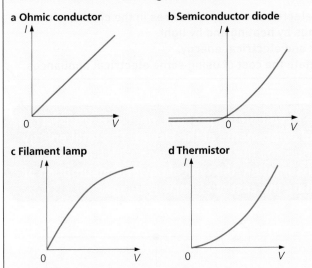

▲ **Figure 4.2.29** *I–V* graphs

Metallic conductors

Metals and some alloys give *I–V* graphs that are a straight line through the origin, as in Figure 4.2.29a, provided that their temperature is constant. *I* is directly proportional to *V*, i.e. *I* ∝ *V*. Doubling *V* doubles *I*, etc. Such conductors obey **Ohm's law**, stated as follows:

> The current in a metallic conductor is directly proportional to the p.d. across its ends if the temperature and other conditions are constant.

They are called **ohmic** or **linear conductors** and since *I* ∝ *V*, it follows that *V/I* = a constant (obtained from the slope of the *I–V* graph). The resistance of an ohmic conductor therefore does not change when the p.d. does.

Semiconductor diode

The typical *I–V* graph in Figure 4.2.29b shows that current passes when the p.d. is applied in one direction but is almost zero when the p.d. is applied in the opposite direction. A diode has a small resistance when connected one way round but a very large resistance when the p.d. is reversed. It conducts in one direction only and is a **non-ohmic** conductor.

Filament lamp

A filament lamp is a non-ohmic conductor at high temperatures. For a filament lamp the *I–V* graph curve flattens as *V* and *I* increase (Figure 4.2.29c). That is, the resistance (*V/I*) increases as *I* increases and makes the filament hotter.

Variation of resistance with temperature

In general, an increase of temperature increases the resistance of metals, as for the filament lamp in Figure 4.2.29c, but it decreases the resistance of semiconductors. The resistance of semiconductor **thermistors** decreases if their temperature rises, i.e. their *I–V* graph bends upwards, as in Figure 4.2.29d.

If a resistor and a thermistor are connected as a potential divider (Figure 4.2.30), the p.d. across the resistor increases as the temperature of the thermistor increases; the circuit can be used to monitor temperature, for example in a car radiator.

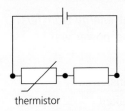

thermistor

▲ **Figure 4.2.30** Potential divider circuit for monitoring temperature

Variation of resistance with light intensity

The resistance of some semiconducting materials decreases when the intensity of light falling on them increases. This property is made use of in **light-dependent resistors** (LDRs) (see Topic 4.3.3). The *I–V* graph for an LDR is similar to that shown in Figure 4.2.29d for a thermistor. Both thermistors and LDRs are non-ohmic conductors.

> ### Test yourself
> 18 a Sketch the *I–V* graph for a resistor of constant resistance.
> b How could you obtain a value of the resistance from the graph?
> 19 a Sketch the *I–V* graph for a filament lamp.
> b Explain the shape of the graph.

4.2.5 Electrical working

FOCUS POINTS

★ Understand that energy is transferred from a source of electrical energy to devices in the circuit; in the process thermal energy is transferred to the surroundings by heating and by light.
★ Know and use the correct equations for electrical power and electrical energy.
★ Define the kilowatt-hour (kWh) and use this unit to calculate the cost of using some electrical appliances.

The e.m.f. applied to a circuit drives current around the circuit. In the process, energy is transferred from the electrical cell or mains supply to the wires and components of the circuit. The total energy transferred to a device depends on its power consumption and the time span over which it is used. In this section you will learn how to measure power consumption, the typical power consumption of some everyday household appliances and how to calculate the cost of electricity usage.

Power in electric circuits

In many circuits it is important to know the rate at which the electric current transfers energy from the source to the circuit components.

Earlier (Topic 1.7.4) we said that *energy transfers were measured by the work done* and power was defined by the equation

$$\text{power} = \frac{\text{work done}}{\text{time taken}} = \frac{\text{energy transferred}}{\text{time taken}}$$

In symbols

$$P = \frac{W}{t} \quad (1)$$

where W is in joules (J), t in seconds (s) and P is in J/s or watts (W).

From the definition of p.d. (Topic 4.2.3) we saw that if W is the work done when there is a steady current I (in amperes) for time t (in seconds) in a device (e.g. a lamp) with a p.d. V (in volts) across it, as in Figure 4.2.31, then

$$W = I \times t \times V$$

Substituting for W in (1) gives $P = \dfrac{I \times t \times V}{t}$ so

$$P = IV$$

and in time t the electrical energy transferred is $E = Pt$ so

$$E = IVt \quad (2)$$

To calculate the power P of an electrical appliance we multiply the current I in it by the p.d. V across it.

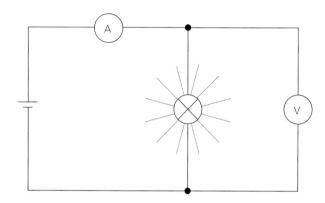

▲ Figure 4.2.31

For example if a lamp on a 240 V supply has a current of 0.25 A in it, its power

$$P = IV = 240\,\text{V} \times 0.25\,\text{A} = 60\,\text{W}$$

This means that 60 J of energy are transferred to the lamp each second. Larger units of power are the *kilowatt* (kW) and the *megawatt* (MW) where

$$1\,\text{kW} = 1000\,\text{W} \text{ and } 1\,\text{MW} = 1\,000\,000\,\text{W}$$

In units

$$\text{watts} = \text{amperes} \times \text{volts} \quad (3)$$

It follows from (3) that since

$$\text{volts} = \frac{\text{watts}}{\text{amperes}} \quad (4)$$

the volt can be defined as a *watt per ampere* and p.d. calculated from (4).

If all the energy is transferred to thermal energy in a resistor of resistance R, then $V = IR$ and the rate of transfer to thermal energy is given by

$$P = V \times I = IR \times I = I^2R$$

That is, if the current is doubled, four times as much thermal energy is produced per second. Also, $P = V^2/R$. The thermal energy can be transferred to the surroundings by light and by heating.

 Worked example

A lamp of resistance $12\,\Omega$ has a current of $0.5\,A$ flowing through it.

a   Calculate the p.d. across the lamp.

p.d. $V = IR = 0.5\,A \times 12\,\Omega = 6\,V$

b   What is the power of the lamp?

$P = IV = 0.5\,A \times 6\,V = 3\,W = 3\,J/s$

c   How much energy is transferred to the lamp in $6\,s$?

$P = E/t$ so $E = Pt = 3\,J/s \times 6\,s = 18\,J$

Now put this into practice

1   A lamp has a resistance of $12\,\Omega$ and a current of $1.0\,A$ passing through it.
   a   Calculate the p.d. across the lamp.
   b   Calculate the power of the lamp.
   c   How much energy is transferred to the lamp in $10\,s$?
2   A small electric motor attached to a $12\,V$ supply has a current of $0.3\,A$ passing through it.
   a   Calculate the power of the motor in watts.
   b   Give the power of the motor in joules/second.
   c   How much energy is transferred to the motor in 1 minute?

Practical work

Measuring electric power

For safe experiments/demonstrations related to this topic, please refer to the *Cambridge IGCSE Physics Practical Skills Workbook* that is also part of this series.

Lamp

Connect the circuit of Figure 4.2.32. Note the ammeter and voltmeter readings and work out the electric power supplied to the lamp in watts.

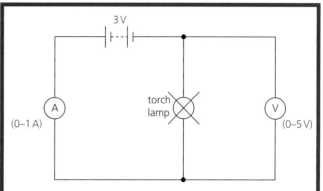

▲ **Figure 4.2.32**

Motor

Replace the lamp in Figure 4.2.32 by a small electric motor. Attach a known mass m (in kg) to the axle of the motor with a length of thin string and find the time t (in s) required to raise the mass through a known height h (in m) at a steady speed. Then the power output P_o (in W) of the motor is given by

$$P_o = \frac{\text{work done in raising mass}}{\text{time taken}} = \frac{mgh}{t}$$

If the ammeter and voltmeter readings I and V are noted while the mass is being raised, the power input P_i (in W) can be found from

$$P_i = IV$$

The efficiency of the motor is given by

$$\text{efficiency} = \frac{P_o}{P_i} \times 100\%$$

Also investigate the effect of a greater mass on: (i) the speed, (ii) the power output and (iii) the efficiency of the motor at its rated p.d.

16 When a p.d. of $30\,V$ is applied across an electric motor, a current of $0.5\,A$ flows through it. Calculate the power supplied to the motor.

17 An electric motor raises a mass of $500\,g$ through $80\,cm$ in $4\,s$. Calculate the output power of the motor.

Joulemeter

Instead of using an ammeter and a voltmeter to measure the electrical energy transferred to an appliance, a **joulemeter** can be used to measure it directly in joules. The circuit connections are shown in Figure 4.2.33.

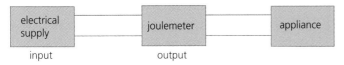

electrical supply		joulemeter		appliance
input		output		

▲ **Figure 4.2.33** Connections to a joulemeter

Paying for electricity

Electricity supply companies charge for the amount of energy they supply. A joule is a very small amount of energy and a larger unit, the **kilowatt-hour** (kWh), is used.

A kilowatt-hour is the electrical energy used by a 1 kW appliance in 1 hour.

$$1\,\text{kWh} = 1000\,\text{J/s} \times 3600\,\text{s}$$

$$= 3\,600\,000\,\text{J} = 3.6\,\text{MJ}$$

A 3 kW electric fire working for 2 hours uses 6 kWh of electrical energy – usually called 6 'units'. Electricity meters, which are joulemeters, are marked in kWh: the latest have digital readouts like the one in Figure 4.2.34.

> **Key definition**
>
> **Kilowatt-hour** (kWh) the electrical energy used by a 1 kW appliance in 1 hour
>
> $1\,\text{kWh} = 1000\,\text{J/s} \times 3600\,\text{s}$
>
> $= 3\,600\,000\,\text{J} = 3.6\,\text{MJ}$

▲ **Figure 4.2.34** Electricity meter with digital display

Typical powers of some appliances are given in Table 4.2.1.

▼ **Table 4.2.1** Power of some appliances

DVD player	20 W	iron	1 kW
laptop computer	50 W	fire	1, 2, 3 kW
light bulbs	60, 100 W	kettle	2 kW
television	100 W	immersion heater	3 kW
refrigerator	150 W	cooker	6.4 kW

Note that the current required by a 6.4 kW cooker is given by

$$I = \frac{P}{V} = \frac{6400\,\text{W}}{230\,\text{V}} = 28\,\text{A}$$

This is too large a current to draw from the ring main of a house and so a separate circuit must be used.

Worked example

If the price of 1 kWh (1 unit) of electricity is 10 cents, how much will it cost to use a 3000 W electric heater for 3 hours?

Convert watts to kilowatts: $3000\,W = 3\,kW$

Electrical energy $E = Pt = 3\,kW \times 3\,h = 9\,kWh$

Cost of using the heater $= 9\,kWh \times 10\,cents = 90\,cents$

Now put this into practice

1   If the price of 1 kWh (1 unit) of electricity is 10 cents, how much will it cost to use a 6.4 kW oven for 2 hours?

2   If the cost of 1 kWh (1 unit) of electricity is 10 cents, how much will it cost to use a 150 W refrigerator for 12 hours?

Test yourself

20 How much energy in joules is transferred to a 100 watt lamp in
    a   1 second
    b   5 seconds
    c   1 minute?
21 a   What is the power of a lamp rated at 12 V 2 A?
    b   How many joules of energy are transferred per second to a 6 V 0.5 A lamp?

Revision checklist

After studying Topic 4.2 you should know and understand:
✔   that positive and negative charges are produced by rubbing and like charges repel while unlike charges attract

✔   what is meant by an electric field and that the direction of an electric field at a point is the direction of the force on a positive charge at that point

✔   that an electric current in a metal is a flow of free electrons from the negative to the positive terminal of the battery around a circuit
✔   the difference between d.c. and a.c.
✔   the meaning of the terms electromotive force and potential difference
✔   how to use voltmeters, both analogue and digital
✔   how to solve simple problems using $R = V/I$
✔   that electric circuits transfer energy, from a battery or mains supply, to the components of the circuit and then into the surroundings.

After studying Topic 4.2 you should be able to:
✔   explain the charging of objects in terms of the motion of negatively charged electrons and describe simple experiments to show how electrostatic charges are produced and detected

✔   give examples of conductors and insulators and explain the differences between them using a simple electron model
✔   describe the use of ammeters to measure current

✔   recall the relation $I = Q/t$ and use it to solve problems
✔   distinguish between electron flow and conventional current

✔   state that e.m.f. and p.d. are measured in volts

✔   recall and use the equation $V = W/Q$

✔   describe an experiment to measure resistance and relate the resistance of a wire to its length and diameter

✔   plot and explain $I–V$ graphs for different conductors

✔   recall the relations $E = IVt$ and $P = IV$ and use them to solve simple problems on energy transfers
✔   define the kilowatt-hour and calculate of the cost of using electrical energy.

Exam-style questions

1 a Explain in terms of electron movement what happens when a piece of cellulose acetate becomes positively charged by being rubbed with a cloth. [3]

b Two positive electrostatic charges are brought close together. Will they be repelled or attracted to each other? [1]

c A positive and a negative electric charge are brought close to each other. Will they be attracted or repelled from each other? [1]

d How many types of electric charge are there? [1]

[Total: 6]

2 a Describe an experiment to distinguish between electrical conductors and insulators. [4]

b Name two good electrical conductors. [2]

c Name one electrical insulator. [1]

d Explain the difference between electrical conductors and insulators in terms of electrons. [3]

[Total: 10]

3 a Explain what is meant by an electric field. [2]

b Draw the electric field lines (including their direction) between two oppositely charged conducting parallel plates. Indicate the direction in which a positive charge would move if placed between the plates. [4]

c State the units of charge. [1]

[Total: 7]

4 a State the direction of an electric field. [3]

b Draw the field lines around a positively charged conducting sphere. [4]

[Total: 7]

5 Study the circuits in Figure 4.2.35. The switch S is open (there is a break in the circuit at this point). In which circuit would lamps Q and R light but not lamp P?

[Total: 1]

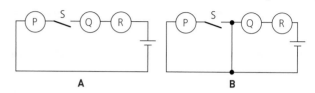

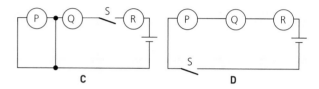

▲ **Figure 4.2.35**

6 Using the circuit in Figure 4.2.36, which of the following statements is correct?

A When S_1 and S_2 are closed, lamps A and B are lit.

B With S_1 open and S_2 closed, A is lit and B is not lit.

C With S_2 open and S_1 closed, A and B are lit.

D With S_1 open and S_2 open, A is lit and B is not lit.

[Total: 1]

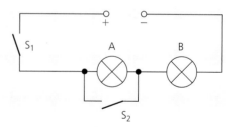

▲ **Figure 4.2.36**

7 a Name the particles which carry a current in a metal. [1]

b Distinguish between direct current (d.c.) and alternating current (a.c.). [2]

c Explain where you would connect and how you would use an ammeter to measure the d.c. current in a circuit. [3]

[Total: 6]

8 a Define electric current. [2]

b An electric current passes through a device.

i Calculate the current at a point in the circuit where 180 C of charge passes in 1 minute. [2]

ii If the current in the device is 2 A, what charge passes through it in 1 minute? [3]

[Total: 7]

9 a If the current in a floodlamp is 5 A, what charge passes in

i 10 s [2]

ii 5 minutes? [2]

b Calculate how long it will take 300 C to pass through the floodlight. [3]

[Total: 7]

10 The lamps and the cells in all the circuits of Figure 4.2.37 are the same. If the lamp in **a** has its full, normal brightness, what can you say about the brightness of the lamps in **b, c, d, e** and **f**? [Total: 5]

11 Three voltmeters V, V_1 and V_2 are connected as in Figure 4.2.38.

a If V reads 18 V and V_1 reads 12 V, what does V_2 read? [2]

b If the ammeter A reads 0.5 A, how much electrical energy is changed to heat and light in lamp L_1 in one minute? [4]

c Copy Figure 4.2.38 and mark with a + the positive terminals of the ammeter and voltmeters for correct connection. [4]

[Total: 10]

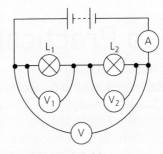

▲ **Figure 4.2.38**

12 The graph in Figure 4.2.39 illustrates how the p.d. across the ends of a conductor is related to the current in it.

a State the relationship between V and I that can be deduced from the graph, giving reasons. [4]

b Calculate the resistance of the conductor. [3]

[Total: 7]

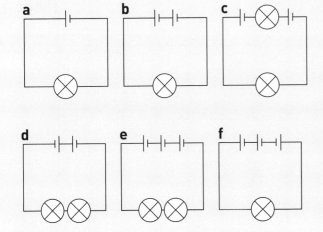

▲ **Figure 4.2.37**

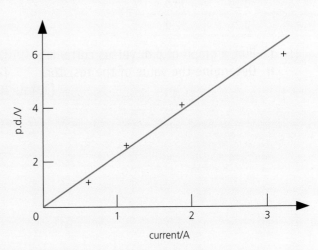

▲ **Figure 4.2.39**

13 a Describe how the resistance of a wire depends on its length and cross-sectional area. [3]

b The resistance of a wire of length 1 m is 70 Ω. Calculate the resistance of a 20 cm length of the wire. [3]

c If the 1 m length of wire is replaced with a wire of the same material and length but of half the diameter calculate its resistance. [4]

[Total: 10]

14 Sketch a current–voltage graph of

a a resistor of constant resistance [3]

b a semiconductor diode [3]

c a filament lamp. [3]

[Total: 9]

Alternative to Practical

15 a Write down an expression relating the resistance of a metal wire to the p.d. across it and the current flowing through it. [1]

b Describe how you could measure the resistance of a wire; include the equipment you would need. [4]

c In an experiment to determine the resistance of a wire the following values were obtained for the current through the wire and the p.d. across it.

Current/A	p.d./V
0.04	2.0
0.08	4.0
0.12	6.0
0.16	8.0
0.20	10.0
0.24	12.0

i Plot a graph of p.d. versus current. [3]

ii Determine the value of the resistor. [2]

[Total: 10]

16 a Calculate the energy transferred to a 6.4 kW cooker in 30 minutes. [3]

b Calculate the cost of heating a tank of water with a 3000 W immersion heater for 80 minutes if electricity costs 10 cents per kWh. [3]

[Total: 6]

17 a Below is a list of wattages of various appliances. State which is most likely to be the correct one for each of the appliances named.

60 W   250 W   850 W   2 kW   3.5 kW

i kettle [1]

ii table lamp [1]

iii iron [1]

b Calculate the current in a 920 W appliance if the supply voltage is 230 V. [4]

[Total: 7]

4.3 Electric circuits

4.3.1 Circuit diagrams and components

FOCUS POINTS

★ Draw and interpret circuit diagrams containing a variety of different components and understand how these components behave in the circuit.

You will find that electrical circuits can contain many different types of components. The circuit configuration is expressed by circuit diagrams. Conventional symbols represent the different types of components. Such diagrams are used in the design of circuits and the analysis of their behaviour.

Some of the symbols used for the various parts of an electric circuit are shown in Figure 4.3.1. So far you have encountered cells, batteries, lamps, resistors, ammeters and voltmeters. In this section you will be introduced to some more of the components frequently used in electric circuits including thermistors, light-dependent resistors (LDRs), relays, **light-emitting diodes** (LEDs) and semiconductor diodes.

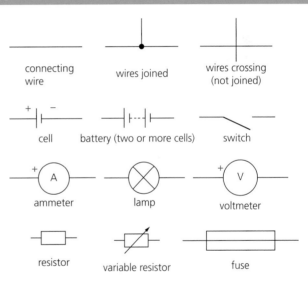

▲ **Figure 4.3.1** Circuit symbols

4.3.2 Series and parallel circuits

FOCUS POINTS

★ Understand that current in a series circuit is the same at any point.

★ Calculate the currents and p.d.s in series and parallel circuits.

★ Understand how to construct and use series and parallel circuits.
★ Calculate the combined e.m.f. and combined resistance in series circuits.
★ Know that in parallel circuits the current from the source is greater than the current in each branch and that the effective resistance of two resistors in parallel is less than that of either alone.

★ Understand that the sum of the currents into a junction equals the sum of the currents out of the junction.
★ Calculate the effective resistance of two resistors in parallel.

★ Know that in a lighting circuit there are advantages to connecting lamps in parallel.

In the preceding topic you encountered the concepts of current, p.d. and resistance and how they are related to each other in simple circuits. Electrical circuits can branch and reconnect. The net effect depends on the way the components are connected. The sum of the currents into a junction equals the sum of the currents out of the junction. This means that there are different effects when resistors follow each other (in series) from those when they lie on parallel wires. There are significant advantages in connecting lamps in parallel in a lighting circuit.

A circuit usually contains several components and the effect of connecting components together in series and parallel configurations will be now be considered.

Current in a series circuit

In a **series circuit**, such as the one shown in Figure 4.3.2, the different parts follow one after the other and there is just one path for the current to follow. The reading on an ammeter will be the same whether it is placed in the position shown or at B, C or D. That is, current is not used up as it goes around the circuit.

The current at every point in a series circuit is the same.

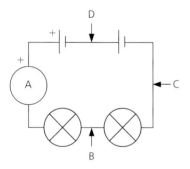

▲ **Figure 4.3.2** Current in a series circuit

Current in a parallel circuit

In a **parallel circuit**, such as the one shown in Figure 4.3.3, the lamps are side by side and there are alternative paths for the current. The current splits: some goes through one lamp and the rest through the other. The current from the source is larger than the current in each branch. For example, if the ammeter reading was 0.4 A in the position shown, then if the lamps are identical, the reading at P would be 0.2 A, and so would the reading at Q, giving a total of 0.4 A. Whether the current splits equally or not depends on the lamps; for example, if the lamps are not identical, the current might divide so that 0.3 A goes one way and 0.1 A by the other branch.

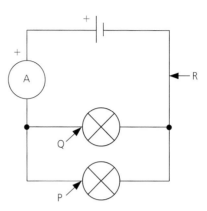

▲ **Figure 4.3.3** Currents in a parallel circuit

Practical work associated with currents in series and parallel circuits can be found in Topic 4.2.2.

> **Key definition**
>
> **Parallel circuit** components are connected side by side and the current splits into alternative paths and then recombines; current from the source is larger than the current in each branch

Current at a junction

Electric current in a circuit cannot be stored. This means that when circuits join or divide, the total current going into a junction must be equal to the total current leaving the junction. A simple example of this is provided by the splitting and re-joining of the current when it goes into and comes out of a parallel circuit.

Potential difference in a series circuit

The total p.d. across the components in a series circuit is equal to the sum of the individual p.d.s across each component. In Figure 4.3.4

$$V = V_1 + V_2 + V_3$$

where V_1 is the p.d. across L_1, V_2 is the p.d. across L_2 and V_3 is the p.d. across L_3.

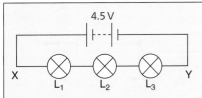

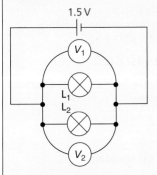

▲ Figure 4.3.4 p.d. in a series circuit

For example, if $V_1 = 1.4\,V$, $V_2 = 1.5\,V$ and

$V_3 = 1.6\,V$, then V will be $(1.4 + 1.5 + 1.6)\,V = 4.5\,V$.

Potential difference in a parallel circuit

In the circuit of Figure 4.3.5

$$V_1 = V_2$$

The p.d. across devices in parallel in a circuit are equal.

▲ Figure 4.3.5 p.d.s in a parallel circuit

The p.d. across an arrangement of parallel resistance is the same as the p.d. across one branch.

Practical work associated with voltage in series and parallel circuits can be found in Topic 4.2.3.

Cells, batteries and e.m.f.

A battery (Figure 4.3.6) consists of two or more **electric cells**. Greater e.m.f.s are obtained when cells are joined in series, i.e. + of one to − of next; the e.m.f.s of each are added together to give the combined e.m.f. In Figure 4.3.7a the two 1.5 V cells give an e.m.f. of 3 V at the terminals A, B.

▲ Figure 4.3.6 Compact batteries

The cells in Figure 4.3.7b are in opposition and the e.m.f. at X, Y is zero.

If two 1.5 V cells are connected in parallel, as in Figure 4.3.7c, the e.m.f. at terminals P, Q is still 1.5 V but the arrangement behaves like a larger cell and will last longer.

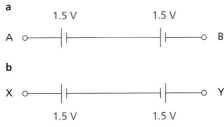

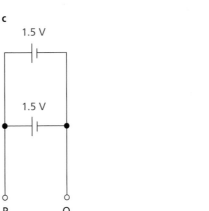

▲ Figure 4.3.7

The p.d. at the terminals of a battery decreases slightly when current is drawn from it. This effect is due to the internal resistance of the battery which transfers electrical energy to thermal energy as current flows through it. When no current is drawn from a battery it is said to be an 'open circuit' and its terminal p.d. is a maximum and equal to the e.m.f. of the battery.

In Topic 4.2.3 it was stated that if W (joules) is the work done when charge Q (coulombs) passes between two points, the p.d. V (volts) between the points is given by

$V = W/Q$

The same equation can also be used to calculate the e.m.f. of a supply.

Test yourself

1 If the lamps are both the same in Figure 4.3.8 and if ammeter A_1 reads 0.50 A, what do ammeters A_2, A_3, A_4 and A_5 read?

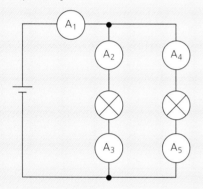

▲ **Figure 4.3.8**

2 Three 2 V cells are connected in series and used as the supply for a circuit.
   What is the p.d. at the terminals of the supply?

3 How many joules of electrical energy does 1 C gain on passing through
   a a 2 V cell
   b three 2 V cells connected in series?

Resistors in series

The resistors in Figure 4.3.9 are in series. The same current I flows through each and the total voltage V across all three is the sum of the separate voltages across them, i.e.

$V = V_1 + V_2 + V_3$

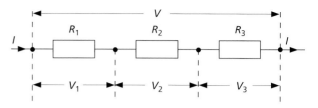

▲ **Figure 4.3.9** Resistors in series

But $V_1 = IR_1$, $V_2 = IR_2$ and $V_3 = IR_3$. Also, if R is the combined resistance, $V = IR$, and so

$IR = IR_1 + IR_2 + IR_3$

Dividing both sides by I,

$R = R_1 + R_2 + R_3$

? Worked example

A 4.5 V battery is connected across three resistors of values 3 Ω, 4 Ω and 5 Ω connected in series.

a Calculate the current flowing through the resistors.
   Combined resistance of resistors in series

   $R = R_1 + R_2 + R_3 = 3\,\Omega + 4\,\Omega + 5\,\Omega = 12\,\Omega$

   Rearrange equation $V = IR$ to give $I = V/R$ then the current flowing through the three resistors

   $$I = \frac{4.5\,\text{V}}{12\,\Omega} = 0.38\,\text{A}$$

b Calculate the p.d. across the 4 Ω resistor.
   p.d. across R_2 is given by

   $V_2 = IR_2 = 0.38\,\text{A} \times 4\,\Omega = 1.5\,\text{V}$

Now put this into practice

1 Three resistors of value 4 Ω, 6 Ω and 8 Ω are connected in series. Calculate their combined resistance.
2 A 4.5 V battery is connected across two resistors of value 3 Ω + 6 Ω. Calculate
   a the current flowing through the resistors
   b the p.d. across each.

Resistors in parallel

The resistors in Figure 4.3.10 are in parallel. The *voltage V between the ends of each is the same* and the total current I equals the sum of the currents in the separate branches, i.e.

$I = I_1 + I_2 + I_3$

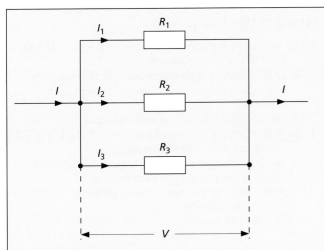

▲ **Figure 4.3.10** Resistors in parallel

But $I_1 = V/R_1$, $I_2 = V/R_2$ and $I_3 = V/R_3$.
Also, if R is the combined resistance, $I = V/R$,

$$\frac{V}{R} = \frac{V}{R_1} + \frac{V}{R_2} + \frac{V}{R_3}$$

Dividing both sides by V,

$$\frac{1}{R} = \frac{1}{R_1} + \frac{1}{R_2} + \frac{1}{R_3}$$

For the simpler case of *two* resistors in parallel

$$\frac{1}{R} = \frac{1}{R_1} + \frac{1}{R_2} = \frac{R_2}{R_1 R_2} + \frac{R_1}{R_1 R_2}$$

$$\therefore \frac{1}{R} = \frac{R_2 + R_1}{R_1 R_2}$$

Inverting both sides,

$$R = \frac{R_1 R_2}{R_1 + R_2} = \frac{\text{product of resistances}}{\text{sum of resistances}}$$

Properties of parallel circuits

We can summarise the above results for parallel circuits as follows:
(i) the current from the source is larger than the current in each branch
(ii) the **combined resistance of two resistors in parallel** is less than that of either resistor by itself.

Key definition

Combined resistance of two resistors in parallel less than that of either resistor by itself

You can check these statements are true in the Worked example below.

Lamps are connected in parallel (Figure 4.3.5) rather than in series in a lighting circuit.

The advantages are as follows:
(i) The p.d. across each lamp is fixed (at the supply p.d.), so the lamp shines with the same brightness irrespective of how many other lamps are switched on.
(ii) Each lamp can be turned on and off independently; if one lamp fails, the others can still be operated.

Practical work associated with measuring resistance can be found in Topic 4.2.4.

? Worked example

A p.d. of 24 V from a battery is applied to the network of resistors in Figure 4.3.11a.
a   What is the combined resistance of the 6 Ω and 12 Ω resistors in parallel?
Let R_1 = resistance of 6 Ω and 12 Ω in parallel.
Then

$$\frac{1}{R_1} = \frac{1}{6} + \frac{1}{12} = \frac{2}{12} + \frac{1}{12} = \frac{3}{12}$$

$$\therefore R_1 = \frac{12}{3} = 4 \, \Omega$$

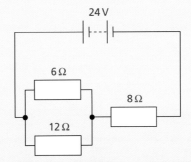

▲ **Figure 4.3.11a**

b   What is the current in the 8 Ω resistor?
Let R = total resistance of circuit = 4 Ω + 8 Ω, that is, $R = 12 \, \Omega$. The equivalent circuit is shown in Figure 4.3.11b, and if I is the current in it then, since $V = 24$ V

$$I = \frac{V}{R} = \frac{24 \, \text{V}}{12 \, \Omega} = 2 \, \text{A}$$

$\therefore$ current in 8 Ω resistor = 2 A

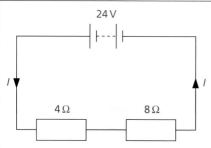

▲ **Figure 4.3.11b**

c What is the voltage across the parallel network?
Let V_1 = voltage across parallel network in Figure 4.3.11a. Then

$$V_1 = I \times R_1 = 2\,\text{A} \times 4\,\Omega = 8\,\text{V}$$

d What is the current in the $6\,\Omega$ resistor?
Let I_1 = current in $6\,\Omega$ resistor, then since $V_1 = 8\,\text{V}$

$$I_1 = \frac{V_1}{6\,\Omega} = \frac{8\,\text{V}}{6\,\Omega} = \frac{4}{3}\,\text{A}$$

Now put this into practice

1 a Calculate the combined resistance R of a $1\,\Omega$, $2\,\Omega$ and $3\,\Omega$ resistor connected in series.
  b A 12V battery is connected across the resistors. Calculate the current I flowing through each resistor.
  c What is the p.d. across each resistor?
2 a Calculate the combined resistance R of $2\,\Omega$ and $3\,\Omega$ resistors connected in parallel.
  b A 12V battery is connected across the resistors. What is the p.d. across each resistor?
  c Calculate the current I flowing through
   i the $2\,\Omega$ resistor
   ii the $3\,\Omega$ resistor.

Going further

Resistor colour code

Resistors have colour-coded bands as shown in Figure 4.3.12. In the orientation shown the first two bands on the left give digits 2 and 7; the third band gives the number of noughts (3) and the fourth band gives the resistor's 'tolerance' (or accuracy, here ±10%). So the resistor has a value of $27\,000\,\Omega$ (±10%).

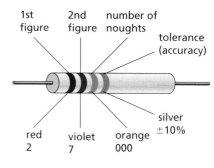

resistor value = $27\,000\,\Omega$ (±10%)
           = $27\,\text{k}\Omega$ (±10%)

Figure	Colour
0	black
1	brown
2	red
3	orange
4	yellow
5	green
6	blue
7	violet
8	grey
9	white

Tolerance	
±5%	gold
±10%	silver
±20%	no band

▲ **Figure 4.3.12** Colour code for resistors

Test yourself

4 a Write down the equation for calculating the combined resistance R of resistors R_1, R_2 and R_3 connected in series.
  b Is the current in R_1 larger, the same or smaller than in R_3?

5 a Write down the equation for calculating the combined resistance R of resistors R_1 and R_2 connected in parallel.
  b Is the current in R_1 larger, the same or smaller than in R_2 if R_1 is smaller than R_2?

4.3.3 Action and use of circuit components

> **FOCUS POINTS**
>
> ★ Know that for a constant current, as resistance increases so does the p.d. across an electrical conductor.
>
> ★ Describe how a variable potential divider works and use the correct equation for two resistors used as a potential divider.

The action of potential dividers and a range of other components, including thermistors, LDRs, relays, light-emitting diodes and semiconductor diodes, will be considered in this section. These components are widely used in electrical circuits in applications ranging from intruder and temperature alarms to indicator lamps and switching circuits.

Increase in resistance of a conductor

In a metal the current in a circuit is carried by free electrons. When the temperature of the metal increases, the atoms vibrate faster and it becomes more difficult for the electrons to move through the material. This means that the resistance of the metal increases.

From Ohm's law $V = IR$, so that if R increases then if a constant current I is to be maintained, the p.d. V across the conductor also increases.

The effect of increasing resistance can be seen in the I–V curve for a filament lamp (Figure 4.2.29c, p. 201). When the current increases, the metal filament heats up and its resistance increases as is indicated by the curvature of the graph.

Variable potential divider

The resistance of materials other than metals does not necessarily rise when their temperature increases. For example, in a semiconductor thermistor, the resistance decreases when its temperature increases.

If a thermistor is part of a potential divider circuit (see Figure 4.2.30, p. 201) then its resistance decreases when the external temperature rises. The combined resistance of the two resistors then decreases, so if the supply voltage remains constant, the current in the circuit will increase.

This means that the p.d. across the fixed resistor increases relative to that across the thermistor. The p.d. across the fixed resistor could then be used to monitor temperature.

A variable resistor can also be used as a potential divider (see Figure 4.2.26b, p. 199). Moving the contact on the resistor changes the output p.d.

Potential divider

In the circuit shown in Figure 4.3.13 overleaf, two resistors R_1 and R_2 are in series with a supply of voltage V. The current in the circuit is

$$I = \frac{\text{supply voltage}}{\text{total resistance}} = \frac{V}{(R_1 + R_2)}$$

So the voltage across R_1 is

$$V_1 = I \times R_1 = \frac{V \times R_1}{(R_1 + R_2)} = V \times \frac{R_1}{(R_1 + R_2)}$$

and the voltage across R_2 is

$$V_2 = I \times R_2 = \frac{V \times R_2}{(R_1 + R_2)} = V \times \frac{R_2}{(R_1 + R_2)}$$

Also the ratio of the voltages across the two resistors is

$$\frac{V_1}{V_2} = \frac{R_1}{R_2}$$

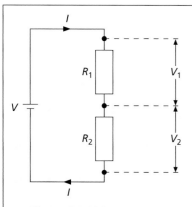

▲ **Figure 4.3.13** Potential divider circuit

? Worked example

Resistors $R_1 = 80\,\Omega$ and $R_2 = 40\,\Omega$ are connected in series and used as a potential divider.

a Calculate the ratio of the p.d.s across the resistors.
From the potential divider equation

$$V_1/V_2 = R_1/R_2 = 80\,\Omega\,/40\,\Omega = 2$$

Ratio of voltages is 2:1.

b If the supply voltage is 24 V, what is the p.d. across each resistor?
Dividing the supply voltage in the ratio 2:1 gives

$$V_1 = 2 \times \frac{24\,\text{V}}{3} = 16\,\text{V}$$

and

$$V_2 = 1 \times \frac{24\,\text{V}}{3} = 8\,\text{V}$$

Now put this into practice

1 Write down the equation relating p.d.s and resistances in a potential divider circuit.
2 Resistors $R_1 = 9\,\Omega$ and $R_2 = 6\,\Omega$ are connected in series and used as a potential divider.
   a Calculate the ratio of the p.d.s across the resistors.
   b If the supply voltage is 30 V, what is the p.d. across each resistor?

Light-dependent resistor (LDR)

The action of an LDR depends on the fact that the resistance of the semiconductor cadmium sulfide decreases as the intensity of the light falling on it increases.

An LDR and a circuit showing its action are shown in Figures 4.3.14a and b. Note the circuit symbol for an LDR, sometimes seen with a circle. When light from a lamp falls on the window of the LDR, its resistance decreases and the increased current lights the lamp.

LDRs are used in photographic exposure meters and in series with a resistor to provide an input signal in switching circuits such as a light-operated intruder alarm.

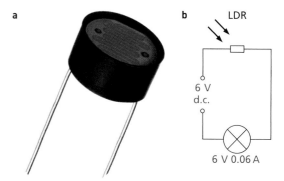

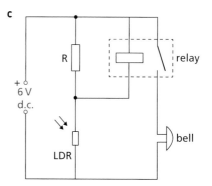

▲ **Figure 4.3.14 a** LDR; **b** LDR demonstration circuit; **c** light-operated intruder alarm

Figure 4.3.14c shows how an LDR can be used to switch a relay (Topic 4.5.3). The LDR forms part of a potential divider across the 6 V supply. When light falls on the LDR, the resistance of the LDR, and hence the voltage across it, decreases. There is a corresponding increase in the voltage across resistor R and the relay; when the voltage across the relay coil reaches a high enough p.d. (its operating p.d.) it acts as a switch and the normally open contacts close, allowing current to flow to the bell, which rings. If the light is removed, the p.d. across resistor R and the relay drops below the operating p.d. of the relay so that the relay contacts open again; power to the bell is cut and it stops ringing.

Thermistor

A negative temperature coefficient (NTC) thermistor contains semiconducting metallic oxides whose resistance decreases markedly when the temperature rises. The temperature may rise either because the thermistor is directly heated or because a current is in it.

Figure 4.3.15a shows one type of thermistor. Figure 4.3.15b shows the symbol for a thermistor in a circuit to demonstrate how the thermistor works. When the thermistor is heated with a match, the lamp lights.

A thermistor in series with a meter marked in °C can measure temperatures (Topic 4.2.4). Used in series with a resistor it can also provide an input signal to switching circuits.

Figure 4.3.15c shows how a thermistor can be used to switch a relay. The thermistor forms part of a potential divider across the d.c. source. When the temperature rises, the resistance of the thermistor falls, and so does the p.d. across it. The voltage across resistor R and the relay increases. When the voltage across the relay reaches its operating p.d. the normally open contacts close, so that the circuit to the bell is completed and it rings. If a variable resistor is used in the circuit, the temperature at which the alarm sounds can be varied.

Relays

A switching circuit cannot supply much power to an appliance so a relay is often included; this allows the small current provided by the switching circuit to control the larger current needed to operate a buzzer as in a temperature-operated switch or other device. Relays controlled by a switching circuit can also be used to switch on the mains supply for electrical appliances in the home. In Figure 4.3.16 if the output of the switching circuit is 'high' (5 V), a small current flows to the relay which closes the mains switch; the relay also isolates the low voltage circuit from the high voltage mains supply.

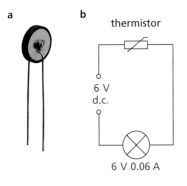

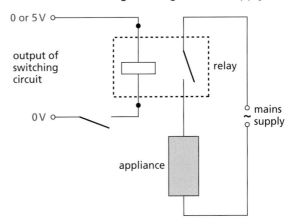

▲ **Figure 4.3.16** Use of a relay to switch mains supply

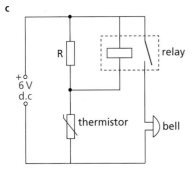

▲ **Figure 4.3.15 a** Thermistor; **b** thermistor demonstration circuit; **c** high-temperature alarm

Light-emitting diode (LED)

An LED, shown in Figure 4.3.17a, is a diode made from the semiconductor gallium arsenide phosphide. When forward biased (with the cathode C connected to the negative terminal of the voltage supply, as shown in Figure 4.3.17b), the current in it makes it emit red, yellow or green light. No light is emitted on reverse bias (when the anode A is connected to the negative terminal of the voltage supply). If the reverse bias voltage exceeds 5 V, it may cause damage.

In use, an LED must have a suitable resistor R in series with it (e.g. 300 Ω on a 5 V supply) to limit the current (typically 10 mA). Figure 4.3.17b shows the symbol for an LED in a demonstration circuit.

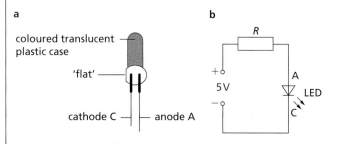

▲ **Figure 4.3.17** LED and demonstration circuit

LEDs are used as indicator lamps on computers, radios and other electronic equipment. Many clocks, calculators, video recorders and measuring instruments have seven-segment red or green numerical displays (Figure 4.3.18a). Each segment is an LED and, depending on which have a voltage across them, the display lights up the numbers 0 to 9, as in Figure 4.3.18b.

LEDs are small, reliable and have a long life; their operating speed is high and their current requirements are very low.

Diode lasers operate in a similar way to LEDs but emit coherent laser light; they are used in optical fibre communications as transmitters.

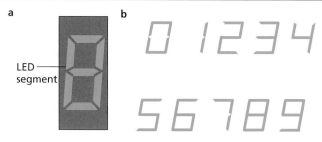

▲ **Figure 4.3.18** LED numerical display

Semiconductor diode

A diode is a device that lets current pass in one direction only. One is shown in Figure 4.3.19 with its symbol. (You will also come across the symbol without its outer circle.) The wire nearest the band is the **cathode** and the one at the other end is the **anode**.

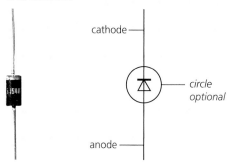

▲ **Figure 4.3.19** A diode and its symbol

The typical I–V graph is shown in Figure 4.2.29b (Topic 4.2.4). The diode conducts when the anode goes to the + terminal of the voltage supply and the cathode to the − terminal (Figure 4.3.20a). It is then **forward-biased**; its resistance is small and conventional current passes in the direction of the arrow on its symbol. If the connections are the other way around, it does not conduct; its resistance is large and it is **reverse-biased** (Figure 4.3.20b).

The lamp in the circuit shows when the diode is conducting, as the lamp lights up. It also acts as a resistor to limit the current when the diode is forward-biased. Otherwise the diode might overheat and be damaged.

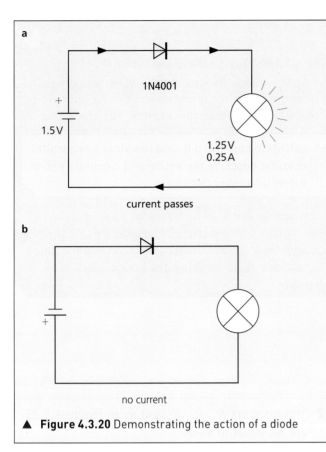

a

1N4001

+

1.5 V

1.25 V
0.25 A

current passes

b

no current

▲ **Figure 4.3.20** Demonstrating the action of a diode

A diode is a **non-ohmic** conductor. It is useful as a **rectifier** for changing alternating current (a.c.) to direct current (d.c.). Figure 4.3.21 shows the rectified output voltage obtained from a diode when it is connected to an a.c. supply.

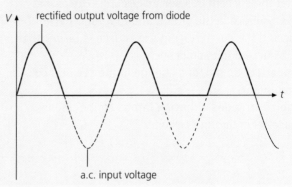

▲ **Figure 4.3.21** Rectification by a diode

Test yourself

6 Resistors $R_1 = 12\,\Omega$ and $R_2 = 36\,\Omega$ are connected in series and used as a potential divider.
   a Draw a potential divider circuit containing a battery and resistors R_1 and R_2 in series.
   b Calculate the ratio of the p.d.s across the resistors.
   c If the supply voltage is 20 V, what is the p.d. across each resistor?

7 Identify the following components from their symbols.

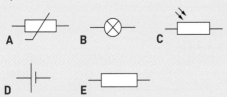

▲ **Figure 4.3.22**

8 Identify the following components from their symbols.

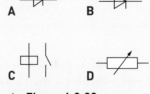

▲ **Figure 4.3.23**

9 A circuit is required to demonstrate that the resistance of a thermistor decreases when its temperature rises. Draw a circuit diagram that could be used containing a battery, a lamp and a thermistor.

Revision checklist

After studying Topic 4.3 you should know and understand:

✔ how to connect simple series and parallel circuits

✔ that the current in a series circuit is the same everywhere in the circuit and that for a parallel circuit, the current from the source is larger than the current in each branch

✔ the effect on p.d. of a change in the resistance of a conductor

✔ the advantages of having lamps connected in parallel in lighting circuits.

After studying Topic 4.3 you should be able to:

✔ use the equations for resistors in series, and recall that the combined resistance of two resistors in parallel is less than that of either resistor alone

✔ calculate current, p.d. and resistance in parallel circuits; describe the action and calculate p.d. in potential divider circuits

✔ recognise and draw symbols for a variety of components in electric circuits and be able to draw and interpret circuit diagrams incorporating those components, and explain their behaviours in a circuit.

Exam-style questions

1 Three voltmeters are connected as in Figure 4.3.24.

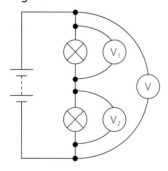

▲ **Figure 4.3.24**

What are the voltmeter readings x, y and z in the table below (which were obtained with three different batteries)?

V/V	V_1/V	V_2/V
x	12	6
6	4	y
12	z	4

x [2]

y [2]

z [2]

[Total: 6]

2 The resistors R_1, R_2, R_3 and R_4 in Figure 4.3.25 are all equal in value.
What would you expect each of the voltmeters A, B and C to read, assuming that the connecting wires in the circuit have negligible resistance?

A   [4]

B   [2]

C   [2]

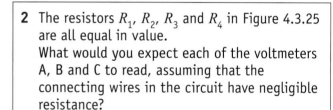
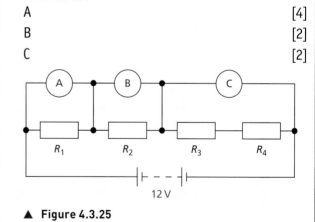

▲ **Figure 4.3.25**

[Total: 8]

3 a Calculate the effective resistance between A and B in Figure 4.3.26. [4]

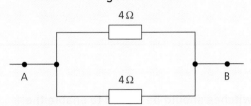

▲ **Figure 4.3.26**

b Figure 4.3.27 shows three resistors. Calculate their combined resistance in ohms. [6]

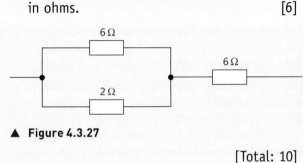

▲ **Figure 4.3.27**

[Total: 10]

4 a Resistors of value $6\,\Omega$, $7\,\Omega$ and $8\,\Omega$ are connected in series.

  i Calculate the combined resistance of the resistors. [2]

  ii The resistance of one of the resistors increases. If the current through the combination must remain unchanged does the supply voltage need to be increased or decreased? [1]

b Give two advantages of connecting lamps in parallel. [4]

c Two resistors of the same size are connected in parallel. Is the resistance of the combination greater or less than that of one of the resistors? [1]

[Total: 8]

5 What are the readings V_1 and V_2 on the high-resistance voltmeters in the potential divider circuit of Figure 4.3.28 if

a $R_1 = R_2 = 10\,\text{k}\Omega$ [2]

b $R_1 = 10\,\text{k}\Omega$, $R_2 = 50\,\text{k}\Omega$ [4]

c $R_1 = 20\,\text{k}\Omega$, $R_2 = 10\,\text{k}\Omega$? [4]

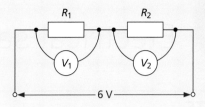

▲ **Figure 4.3.28**

[Total: 10]

6 A battery of 12 V is connected across a light-dependent resistor (LDR) in series with a resistor R.

a Draw the circuit diagram. [2]

b The value of the resistor R is $20\,\Omega$ and the resistance of the LDR is $28\,\Omega$. Calculate

  i the value of the current in the circuit [2]

  ii the p.d. across the resistor [2]

  iii the p.d. across the LDR. [2]

c The intensity of the light falling on the LDR increases. State what happens to

  i the resistance of the LDR [1]

  ii the current in the circuit [1]

  iii the p.d. across R. [1]

[Total: 11]

7 Figure 4.3.29a shows a lamp, a semiconductor diode and a cell connected in series. The lamp lights when the diode is connected in this direction. Say what happens to each of the lamps in **b**, **c** and **d**. Give reasons for your answers.

b [3]

c [4]

d [3]

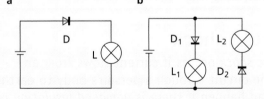

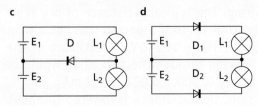

▲ **Figure 4.3.29**

[Total: 10]

4.4 Electrical safety

FOCUS POINTS

★ State various potential hazards when using a main supply.
★ Know the three wires that make up a mains circuit and where switches should be placed to enable the mains supply to be switched off safely.
★ Understand how trip switches and fuses work and choose appropriate settings and values for each.
★ Know that a fuse protects the circuit and the cabling for electrical appliances and that electrical appliances are made safer by having the outer casing non-conducting or earthed.

In the twenty-first century we would be lost without all the benefits electricity supplies bring us. Because electric circuits transfer substantial amounts of energy, use of the mains supply requires caution and electrical safety is important. You will learn that overheated wires and damaged insulation pose fire risks. Damp or wet conditions increase the risk of electric shock from faulty wiring in appliances since water reduces the electrical resistance of a person's skin. If too many appliances are connected to a circuit, the current flowing in the circuit increases and can cause cables to overheat. To prevent problems, devices such as fuses and trip switches (circuit breakers) are installed to break the circuit before the safe current level is exceeded. Safety features incorporated into appliances include double insulation and earthing of metal casing via the mains plug.

Dangers of electricity

There are a number of **hazards associated with using the mains electricity supply.**

> **Key definition**
>
> **Hazards associated with using mains electricity supply** include damaged insulation, overheated cables, damp conditions, excess current from overloaded plugs, extension leads, single and multiple sockets

Electric shock

Electric shock occurs if current flows from an electric circuit through a person's body to earth. This can happen if there is *damaged insulation* or *faulty wiring*. The typical resistance of dry skin is about $10\,000\,\Omega$, so if a person touches a wire carrying electricity at 240 V, an estimate of the current flowing through them to earth would be $I = V/R = 240/10\,000 = 0.024\,A = 24\,mA$.
For wet skin, the resistance is lowered to about $1000\,\Omega$ (since water is a good conductor of electricity) so the current would increase to around 240 mA; a lethal current.

It is the *size of the current* (not the voltage) and the *length of time* for which it acts which determine the strength of an electric shock. The path the current takes influences the effect of the shock; some parts of the body are more vulnerable than others. A current of 100 mA through the heart is likely to be fatal.

Damp conditions increase the severity of an electric shock because water lowers the resistance of the path to earth; wearing shoes with insulating rubber soles or standing on a dry insulating floor increases the resistance between a person and earth and will reduce the severity of an electric shock.

To avoid the risk of getting an electric shock:
(i) switch off the electrical supply to an appliance before starting repairs
(ii) use plugs that have an earth pin and a cord grip; a rubber or plastic case is preferred
(iii) do not allow appliances or cables to come into contact with water, for example holding a hairdryer with wet hands in a bathroom can be dangerous; keep electrical appliances well away from baths and swimming pools

(iv) do not have long cables trailing across a room, under a carpet that is walked over regularly or in other situations where the insulation can become damaged. Take particular care when using electrical cutting devices (such as hedge cutters) not to cut the supply cable.

In case of an electric shock, take the following action:

1 *Switch off the supply* if the shocked person is still touching the equipment.
2 *Send for qualified medical assistance.*
3 *If breathing or heartbeat has stopped, commence CPR* (cardiopulmonary resuscitation) by applying chest compressions at the rate of about 100 a minute until there are signs of chest movement or medical assistance arrives.

Fire risks

If flammable material is placed too close to a hot appliance such as an electric heater, it may catch fire. Similarly, if the electrical wiring in the walls of a house becomes overheated, a fire may start. Wires become hot when they carry electrical currents – the larger the current carried, the hotter a particular wire will become, since the rate of production of heat equals I^2R (see p. 202).

To reduce the risk of fire through *overheated cables*, the maximum current in a circuit should be limited by taking the following precautions:

(i) Use the correct fuse in an appliance or plug.
(ii) Do not attach too many appliances to a circuit via extension leads or single and multiple sockets.
(iii) Do not overload circuits by using too many adapters.
(iv) Appliances such as heaters use large amounts of power (and hence current), so do not connect them to a lighting circuit designed for low current use. (Thick wires have a lower resistance than thin wires so are used in circuits expected to carry high currents.)

Damaged insulation or faulty wiring which leads to a large current flowing to earth through flammable material can also start a fire.

The factors leading to fire or electric shock can be summarised as follows:

damaged insulation	→ electric shock and fire risk
overheated cables	→ fire risk
damp conditions	→ increased severity of electric shocks
overloading – plugs, extension leads or sockets	→ fire risk and electric shock

Electric lighting

LED lights

LEDs (Topic 4.3) are increasingly being used in the lighting of our homes. These semiconductor devices are 40–50% efficient in transferring electrical energy to light. The efficiency of the filament lamps used in the past was only about 10%.

Fluorescent lamps

Fluorescent strip lamps (Figure 4.4.1a) are long lasting and efficient. When one is switched on, the mercury vapour emits invisible ultraviolet radiation which makes the powder on the inside of the tube fluoresce (glow), i.e. visible light is emitted. Different powders give different colours.

Compact energy-saving fluorescent lamps (Figure 4.4.1b) are available to fit straight into normal light sockets, either bayonet or screw-in.

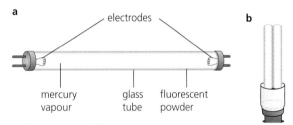

a electrodes

mercury vapour   glass tube   fluorescent powder   **b**

▲ **Figure 4.4.1** Fluorescent lamps

Going further

Electric heating

Heating elements

In domestic appliances such as electric fires, cookers, kettles and irons the 'elements' (Figure 4.4.2) are made from Nichrome wire. This is an alloy of nickel and chromium which does not oxidise (and so become brittle) when the current makes it red hot.

The elements in radiant electric fires are at red heat (about 900°C) and the radiation they emit is directed into the room by polished reflectors. In convector types the element is below red heat (about 450°C) and is designed to warm air which is drawn through the heater by natural or forced convection. In storage heaters the elements heat fire-clay bricks during the night using 'off-peak' electricity. On the following day these cool down, giving off the stored heat to warm the room.

Three-heat switch

A three-heat switch is sometimes used to control heating appliances. It has three settings and uses two identical elements. On 'high', the elements are in parallel across the supply voltage (Figure 4.4.3a); on 'medium', there is only current in one (Figure 4.4.3b); on 'low', they are in series (Figure 4.4.3c).

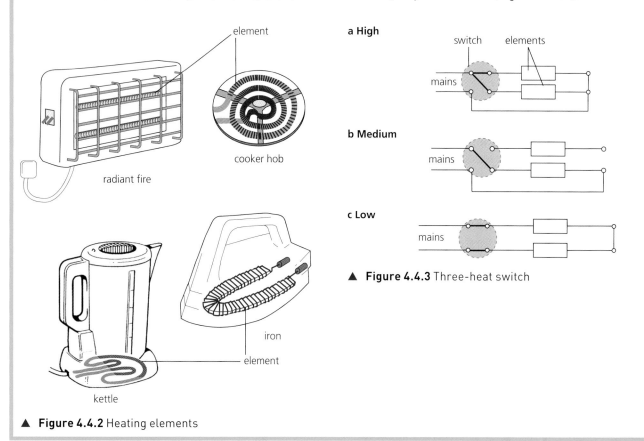

▲ **Figure 4.4.2** Heating elements

▲ **Figure 4.4.3** Three-heat switch

House circuits

Electricity usually comes to our homes by an underground cable containing two wires, the **live** (L) and the **neutral** (N). The neutral is earthed at the local sub-station and so there is no p.d. between it and earth. A third wire, the *earth* (E) also connects the top socket on the power points in the home to earth. The supply in many countries is a.c. (Topic 4.2) and the live wire is alternately positive and negative. Study the typical house circuits shown in Figure 4.4.4.

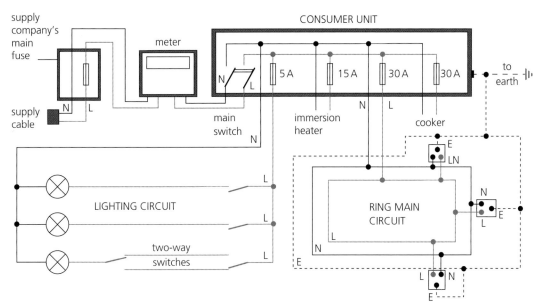

▲ **Figure 4.4.4** Electric circuits in a house

Circuits in parallel

Every circuit is connected in parallel with the supply, i.e. across the live and neutral, and receives the full mains p.d. (for example 230 V).

The advantages of having appliances connected in parallel, rather than in series, can be seen by studying the lighting circuit in Figure 4.4.4.

(i) The p.d. across each lamp is fixed (at the mains p.d.), so the lamp shines with the same brightness irrespective of how many other lamps are switched on.

(ii) Each lamp can be turned on and off independently; if one lamp fails, the others can still be operated.

In a staircase circuit, the light is controlled from two places by the two two-way switches.

Switches

Switches and fuses are always in the live wire. If they were in the neutral, light switches and power sockets would be 'live' when switches were 'off' or fuses 'blown'. A fatal shock could then be obtained by, for example, touching the element of an electric fire when it was switched off.

Ring main circuit

The live and neutral wires each run in two complete rings round the house and the power sockets, each

rated at 13 A, are tapped off from them. Thinner wires can be used since the current to each socket flows by two paths, i.e. from both directions in the ring. The ring has a 30 A fuse and if it has, say, ten sockets, then all can be used so long as the total current does not exceed 30 A, otherwise the wires overheat. A house may have several ring circuits, each serving a different area.

Fuses

A **fuse** protects a circuit; it is always placed in the live wire. It is a short length of wire of material with a low melting temperature, often 'tinned copper', which melts and breaks the circuit when the current in it exceeds a certain value. Two reasons for excessive currents are 'short circuits' due to worn insulation on connecting wires and overloaded circuits. Without a fuse the wiring would become hot in these cases and could cause a fire. *A fuse should ensure that the current-carrying capacity of the wiring is not exceeded.* In general, the thicker a cable is, the more current it can carry, but each size has a limit.

Two types of fuse are shown in Figure 4.4.5. *Always switch off before replacing a fuse,* and always replace with one of the same value as recommended by the manufacturer of the appliance. A 3 A (red) fuse will be needed for appliances with powers up to 720 W, or 13 A (brown) for those between 720 W and 3 kW.

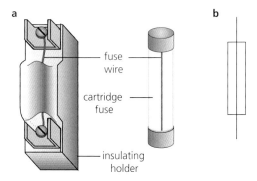

▲ **Figure 4.4.5 a** Two types of fuse; **b** the circuit symbol for a fuse

Typical power ratings for various appliances are shown in Table 4.2.1, p. 204. Calculation of the current in a device allows the correct size of fuse to be chosen.

Trip switches (circuit breakers)

Trip switches (also known as **circuit breakers**) (Figure 4.4.6) are now used instead of fuses in consumer units. They contain an electromagnet (Topic 4.1) which, when the current exceeds the rated value of the circuit breaker, becomes strong enough to separate a pair of contacts and break the circuit.

In the design shown in Figure 4.4.6, when the current is large enough in the electromagnet, the iron bolt is attracted far enough for a plunger to be released, allowing the push switch to open and contact to the rest of the circuit to be broken.

Circuit breakers operate much faster than fuses and have the advantage that they can be reset by pressing a button. As for a fuse, the trip switch setting should be chosen to be a little higher than the value of the current in the device being protected.

▲ **Figure 4.4.6** Circuit breakers

The *residual current circuit breaker* (RCCB), also called a residual current device (RCD), is an adapted circuit breaker which is used when the resistance of the earth path between the consumer and the substation is not small enough for a fault-current to blow the fuse or trip the circuit breaker. It works by detecting any difference between the currents in the live and neutral wires; when these become unequal due to an earth fault (i.e. some of the current returns to the substation via the case of the appliance and earth) it breaks the circuit before there is any danger. They have high sensitivity and a quick response.

An RCD should be plugged into a socket supplying power to a portable appliance such as an electric lawnmower or hedge trimmer. In these cases, the risk of electrocution is greater because the user is generally making a good earth connection through the feet.

? Worked example

An electric heater has a power rating of 2 kW.
a   If the supply voltage is 240 V, calculate the current in the heater.
    Power $P = IV$
    Rearrange the equation to give
    $$I = \frac{P}{V} = \frac{2000\,\text{W}}{240\,\text{V}} = 8.3\,\text{A}$$

b   Should a 3 A or 13 A fuse or trip switch setting be chosen to protect the heater?
    The fuse/trip switch setting should have a higher rating than the current in the heater, so a 13 A fuse/trip switch setting should be chosen.

Now put this into practice
1   An electric heater has a power rating of 1.5 kW.
    a   If the supply voltage is 240 V, calculate the current in the heater.
    b   Should a 3 A, 13 A or 30 A fuse be used to protect the heater?
2   A television has a power rating of 100 W.
    a   If the supply voltage is 240 V, calculate the current in the television.
    b   Should a 3 A, 13 A or 30 A fuse be chosen to protect the television?
3   An electric cooker has a power rating of 6.4 kW.
    a   If the supply voltage is 240 V, calculate the current in the cooker.
    b   Should a 3 A, 13 A or 30 A trip switch setting be chosen to protect the oven?

Earthing

A ring main has a third wire which goes to the top socket on all power points and is earthed by being connected either to a *metal* water pipe entering the house or to an earth connection on the supply cable. This third wire is a safety precaution to prevent electric shock should an appliance develop a fault.

The earth pin on a three-pin plug is connected to the metal case of the appliance which is thus joined to earth by a path of almost zero resistance. If then, for example, the element of an electric fire breaks or sags and touches the case, a large current flows to earth and 'blows' the fuse. Otherwise the case would become 'live' and anyone touching it would receive a shock which might be fatal, especially if they were 'earthed' by, say, standing in a damp environment, such as on a wet concrete floor.

Double insulation

Appliances such as vacuum cleaners, hairdryers and food mixers are usually double insulated. Connection to the supply is by a two-core insulated cable, with no earth wire, and the appliance is enclosed in a *non-conducting* plastic case. Any metal attachments that the user might touch are fitted into this case so that they do not make a direct connection with the internal electrical parts, such as a motor. There is then no risk of a shock should a fault develop.

▶ Test yourself

1 The largest number of 100 W lamps connected in parallel which can safely be run from a 230 V supply with a 5 A fuse is

   A 2
   B 5
   C 11
   D 12

2 What is the maximum power in kilowatts of the appliance(s) that can be connected safely to a 13 A 230 V mains socket?

3 a To what part of an appliance is the earth pin on a three-pin plug attached?

   b How can a two-pin appliance be designed to reduce the risk of the user receiving an electric shock if a fault develops?

Revision checklist

After studying Topic 4.4 you should know and understand
✔ why switches, fuses and circuit breakers are wired into the live wire in house circuits
✔ the benefits of earthing metal cases and double insulation.

After studying Topic 4.4 you should be able to:
✔ recall the hazards of damaged insulation, damp conditions, overheated cables and excess current from overloaded circuits
✔ state the function of a fuse and choose the appropriate fuse rating for an appliance; explain the use, choice and operation of a trip switch.

Exam-style questions

1 There are hazards in using the mains electricity supply.
   a Name two factors which can increase the risk of fire in circuits connected to the mains supply. [2]
   b Name two factors which can increase the risk of electric shock. [2]
   c Describe the steps you would take before replacing a blown fuse in an appliance. [3]
   d Explain why an electrical appliance is double insulated or the outer casing is earthed. [3]
   [Total: 10]

2 Fuses are widely used in electrical circuits connected to the mains supply.
   a Explain the function of a fuse in a circuit. [2]
   b The circuits of Figures 4.4.7a and b show 'short circuits' between the live (L) and neutral (N) wires. In both, the fuse has blown but whereas circuit a is now safe, b is still dangerous even though the lamp is out which suggests the circuit is safe. Explain. [4]

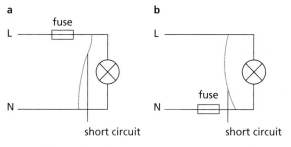

▲ Figure 4.4.7 [Total: 6]

3 a A child whose hands are damp touches a wire carrying electricity at 240 V. The resistance of the child's skin between hand and earth is 800 Ω.
   i Calculate the current which would flow through the child. [2]
   ii State whether the current you calculated in i is likely to be lethal. [1]
   iii State how the current could be reduced. [2]
   b Work out the size of fuse (3 A or 13 A) which should be used in the following appliances if the supply is 230 V
   i a 150 W television [2]
   ii a 900 W iron [2]
   iii a 2 kW kettle. [2]
   [Total: 11]

4.5 Electromagnetic effects

4.5.1 Electromagnetic induction

FOCUS POINTS

★ Know that an electromotive force (e.m.f.) is induced in a conductor when it moves across a magnetic field or a changing magnetic field links with the conductor.
★ Describe how electromagnetic induction can be demonstrated and state the factors which affect the size of an induced e.m.f.

★ Know that the direction of an induced e.m.f. is such as to oppose the change causing it.
★ Determine the relative directions of force, field and induced current.

Electricity and magnetism are closely linked. You will learn that an electrical conductor moving through a magnetic field can induce a current. Similarly, an electrical conductor in a changing magnetic field acquires an electromotive force (e.m.f.). You will find out about the factors which determine the size of the induced e.m.f. Electromagnetic induction plays an important role in many electrical applications from induction cookers and motors to electricity generators.

The effect of producing electricity from magnetism was discovered in 1831 by Faraday and is called **electromagnetic induction**. It led to the construction of generators for producing electrical energy in power stations.

Electromagnetic induction experiments

Two ways of investigating electromagnetic induction follow.

Straight wire and U-shaped magnet

First the wire is held at rest between the poles of the magnet. It is then moved in each of the six directions shown in Figure 4.5.1 and the meter observed. Only *when it is moving upwards* (direction 1) or *downwards* (direction 2) is there a deflection on the meter, indicating an induced current in the wire. The deflection is in opposite directions in these two cases and only lasts while the wire is in motion.

Bar magnet and coil

The magnet is pushed into the coil, one pole first (Figure 4.5.2 overleaf), then held still inside it. It is then withdrawn. The meter shows that current

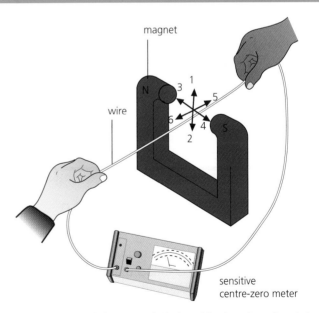

▲ **Figure 4.5.1** A current is induced in the wire when it is moved up or down between the magnet poles.

is induced in the coil in one direction as the magnet is *moved in* and in the opposite direction as it is *moved out*. There is no deflection when the magnet is at rest. The results are the same if the coil is moved instead of the magnet, i.e. only **relative motion** is needed.

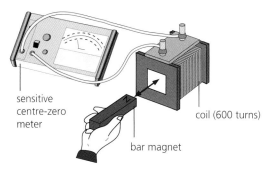

▲ **Figure 4.5.2** A current is induced in the coil when the magnet is moved in or out.

This experiment indicates that an e.m.f. is induced in a conductor when it is linked by a changing magnetic field or when it moves across a magnetic field.

Practical work

Induced currents

Connect a sensitive centre-zero meter to a 600 turn coil as shown in Figure 4.5.2.

Record the values and the direction of the current detected by the meter when you move the magnet first towards the coil and then away from the coil. Try moving the magnet faster; record your results again. Repeat the procedure by moving the coil instead of the magnet.

1 When is a current produced in the circuit?
2 How is the induced e.m.f. related to the current in the circuit?

Factors affecting the size of an induced e.m.f.

To explain electromagnetic induction Faraday suggested that an e.m.f. is induced in a conductor whenever it 'cuts' magnetic field lines, i.e. moves *across* them, but not when it moves along them or is at rest. If the conductor forms part of a complete circuit, an induced current is also produced.

Faraday identified three **factors affecting the magnitude of an induced e.m.f.** and it can be shown, with apparatus like that in Figure 4.5.2, that the induced e.m.f. increases with increases of
(i) the *speed of motion* of the magnet or coil
(ii) the *number of turns* on the coil
(iii) the *strength of the magnet*.

These facts led him to state that:

> The size of the induced e.m.f. is directly proportional to the rate at which the conductor cuts magnetic field lines.

Key definition

Factors affecting the magnitude of an induced e.m.f.
e.m.f. increases with increases of
(i) the *speed of motion* of the magnet or coil,
(ii) the *number of turns* on the coil,
(iii) the *strength of the magnet*

Direction of induced e.m.f.

> The direction of an induced e.m.f. opposes the change causing it.

In Figure 4.5.3a the magnet approaches the coil, north pole first. The induced e.m.f. and resulting current flow should be in a direction that makes the coil behave like a magnet with its top a north pole. The downward motion of the magnet will then be opposed since like poles repel.

When the magnet is withdrawn, the top of the coil should become a south pole (Figure 4.5.3b) and attract the north pole of the magnet, so hindering its removal. The induced e.m.f. and current are thus in the opposite direction to that when the magnet approaches.

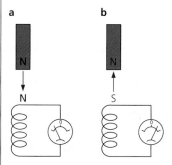

▲ **Figure 4.5.3** The induced current opposes the motion of the magnet.

This behaviour is an example of the principle of conservation of energy. If the currents produced a magnetic field of opposite polarity to those shown in Figure 4.5.3 in each coil, electrical energy would be created from nothing. As it is, work is done by whoever moves the magnet, to overcome the forces that arise.

For a straight wire moving at right angles to a magnetic field the direction of the induced current can be found from **Fleming's right-hand rule** (the '**dynamo** rule') (Figure 4.5.4).

Hold the thumb and first two fingers of the right hand at right angles to each other with the **F**irst finger pointing in the direction of the **F**ield and the thu**M**b in the direction of **M**otion of the wire, then the se**C**ond finger points in the direction of the induced **C**urrent.

Note that the direction of motion represents the direction in which the force acts on the conductor.

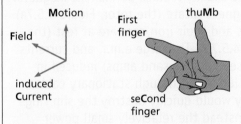

▲ **Figure 4.5.4** Fleming's right-hand (dynamo) rule

> **Key definition**
>
> **Fleming's right-hand (dynamo) rule** used to show the relative directions of force, field and induced current. When the thumb and first two fingers of the right-hand are held at right angles to each other with the first finger pointing in the direction of the magnetic field and the thumb in the direction of the motion of the wire, then the second finger points in the direction of the induced current.

> **Test yourself**
>
> 1 A magnet is pushed, N pole first, into a coil as in Figure 4.5.5. Which one of the following statements **A** to **D** is *not* true?
>   **A** An e.m.f. is induced in the coil and causes a current through the galvanometer.
>   **B** The induced e.m.f. increases if the magnet is pushed in faster and/or the coil has more turns.
>   **C** Kinetic energy is transferred to electrical energy.
>   **D** The effect produced is called electrostatic induction.

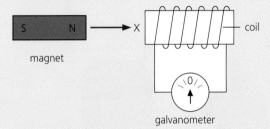

> ▲ **Figure 4.5.5**
>
> 2 A straight wire moves vertically upwards at right angles to a magnetic field acting horizontally from right to left. Make a sketch to represent the directions of the magnetic field, the force on the wire and the induced current in the wire if it is connected to a complete circuit.

4.5.2 The a.c. generator

FOCUS POINTS

★ Describe the construction and action of a simple a.c. generator.
★ Sketch and interpret a graph of e.m.f against time for an a.c. generator.

When a coil is rotated between the poles of a magnet, the conductor cuts the magnetic field lines and an e.m.f. is induced. The size of the e.m.f. generated changes with the orientation of the coil and alternates in sign during the course of each rotation. This process is used in the large generators in power stations to produce an alternating (a.c.) electricity supply.

Simple a.c. generator

The simplest alternating current (a.c.) generator (**alternator**) consists of a rectangular coil between the poles of a C-shaped magnet (Figure 4.5.6a). The ends of the coil are joined to two **slip rings** on the axle and against which *carbon brushes* press.

When the coil is rotated it cuts the field lines and an e.m.f. is induced in it. Figure 4.5.6b shows how the e.m.f. varies over one complete rotation.

As the coil moves through the vertical position with **ab** uppermost, **ab** and **cd** are moving along the lines (**bc** and **da** do so always) and no cutting occurs. The induced e.m.f. is zero.

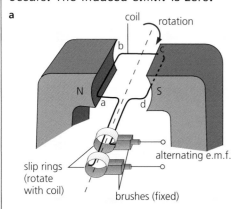

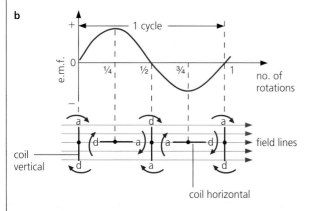

▲ **Figure 4.5.6** A simple a.c. generator and its output

During the first quarter rotation the e.m.f. increases to a maximum when the coil is horizontal. Sides **ab** and **dc** are then cutting the lines at the greatest rate.

In the second quarter rotation the e.m.f. decreases again and is zero when the coil is vertical with **dc** uppermost. After this, the direction of the e.m.f. reverses because, during the next half rotation, the motion of **ab** is directed upwards and **dc** downwards.

An alternating e.m.f. is generated which acts first in one direction and then the other; it causes a.c. to flow in a circuit connected to the brushes. The frequency of an a.c. is the number of complete cycles it makes each second and is measured in hertz (Hz), i.e. 1 cycle per second = 1 Hz. If the coil rotates twice per second, the a.c. has frequency 2 Hz. The mains supply is a.c. of frequency 50 Hz.

Practical generators

In power stations several coils are wound in evenly spaced slots in a soft iron cylinder and electromagnets usually replace permanent magnets. The electromagnets rotate (the rotor, Figure 4.5.7a) while the coils and their iron core are at rest (the stator, Figure 4.5.7b). The large e.m.f. and currents (e.g. 25 kV at several thousand amps) induced in the stator are led away through stationary cables, otherwise they would quickly destroy the slip rings by sparking. Instead the relatively small power required by the rotor is fed via the slip rings from a small generator (the **exciter**) which is driven by the same turbine as the rotor.

▲ **Figure 4.5.7** The rotor and stator of a power station alternator

In a thermal power station (Topic 1.7), the turbine is rotated by high-pressure steam obtained by heating water in a coal- or oil-fired boiler or in a nuclear reactor (or by hot gas in a gas-fired power station). A block diagram of a thermal power station is shown in Figure 4.5.8. The Sankey diagram showing energy transfer was given in Figure 1.7.5, p. 63.

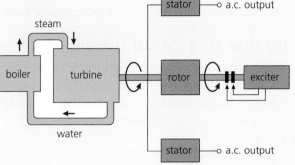

▲ **Figure 4.5.8** Block diagram of a thermal power station

> **Test yourself**
>
> 3 Which feature of the rotating coil of an a.c. generator allows the induced e.m.f. to be connected to fixed contacts?
> 4 a Sketch the output of an a.c. generator against time.
>   b At what position of the coil in an a.c. generator is the output
>   i a maximum
>   ii zero?

4.5.3 Magnetic effect of a current

FOCUS POINTS

★ Describe the pattern and direction of the magnetic field around a current-carrying straight wire and a solenoid and describe an experiment to identify the pattern.

★ Use the examples of relays and loudspeakers to describe the application of the magnetic effect of a current.

★ Describe the variation of the magnetic field strength around a current-carrying straight wire and a solenoid and recall the effect on the magnetic field of changing the current's direction and size.

A further link between electricity and magnetism comes from the presence of a magnetic field around a conductor carrying a current. The pattern and direction of the magnetic field can be found by sprinkling iron filings around a current-carrying wire and using a plotting compass. In this topic you will learn that the magnetic field can be concentrated by the geometry of the conductor. A long cylindrical coil (a solenoid) will act like a bar magnet when current is switched on. As you have seen in Topic 4.1, electromagnets have many applications. In this topic you can discover how they are also used in switches, relays, bells and loudspeakers.

Magnetic field lines are used to represent the variation in magnetic field strength around a current-carrying conductor and its dependence on the size and direction of the current.

Oersted's discovery

In 1819 Hans Oersted accidentally discovered the magnetic effect of an electric current. His experiment can be repeated by holding a wire over and parallel to a compass needle that is pointing N and S (Figure 4.5.9). The needle moves when the current is switched on. Reversing the current causes the needle to move in the opposite direction.

Evidently around a wire carrying a current there is a magnetic field. As with the field due to a permanent magnet, we represent the field due to a current by *field lines* or *lines of force*. Arrows on the lines show the direction of the field, i.e. the direction in which a N pole points.

Different field patterns are given by differently shaped conductors.

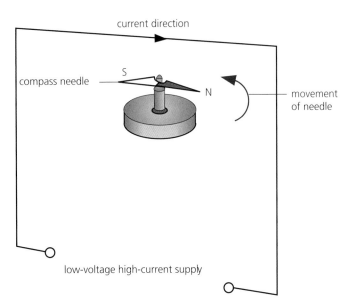

▲ **Figure 4.5.9** An electric current produces a magnetic effect.

Field due to a straight wire

If a straight vertical wire passes through the centre of a piece of card held horizontally and there is a current in the wire (Figure 4.5.10), iron filings sprinkled on the card settle in concentric circles when the card is gently tapped.

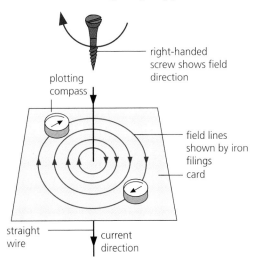

▲ **Figure 4.5.10** Field due to a straight wire

Plotting compasses placed on the card settle along the field lines and show the direction of the field at different points. When the current direction is reversed, the compasses point in the opposite direction showing that the direction of the field reverses when the current reverses.

If the current direction is known, the direction of the field can be predicted by the **right-hand screw rule**:

> If a right-handed screw moves forwards in the direction of the current (conventional), the direction of rotation of the screw gives the direction of the magnetic field.

Field due to a circular coil

The field pattern is shown in Figure 4.5.11. At the centre of the coil the field lines are straight and at right angles to the plane of the coil. The right-hand screw rule again gives the direction of the field at any point.

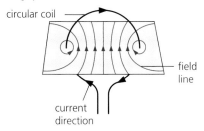

▲ **Figure 4.5.11** Field due to a circular coil

Field due to a solenoid

A **solenoid** is a long cylindrical coil. It produces a field similar to that of a bar magnet; in Figure 4.5.12a, end A behaves like a N pole and end B like a S pole. The polarity can be found as before by applying the right-hand screw rule to a short length of one turn of the solenoid. Alternatively, the **right-hand grip rule** can be used. This states that if the fingers of the right hand grip the solenoid in the direction of the current (conventional), the thumb points to the N pole (Figure 4.5.12b). Figure 4.5.12c shows how to link the end-on view of the current direction in the solenoid to the polarity. A compass could be used to plot the magnetic field lines around the solenoid (see Topic 4.1).

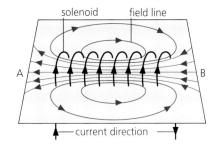

▲ **Figure 4.5.12a** Field due to a solenoid

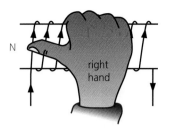

▲ **Figure 4.5.12b** The right-hand grip rule

(i) View from A   (ii) View from B

▲ **Figure 4.5.12c** End-on views

Variation of magnetic field strength

There is **variation of magnetic field strength** around a current-carrying straight wire (Figure 4.5.10) – it becomes less as the distance from the wire increases. This is shown by the magnetic field lines becoming further apart. When the current through the wire is increased, the strength of the magnetic field around the wire increases and the field lines become closer together. When the direction of the current changes, the magnetic field acts in the opposite direction.

Inside the solenoid in Figure 4.5.12a, the field lines are closer together than they are outside the solenoid. This indicates that the magnetic field is stronger inside a solenoid than outside it. When the direction of the current changes in the solenoid, the magnetic field acts in the opposite direction. The field inside a solenoid can be made very strong if it has a large number of turns or a large current. Permanent magnets can be made by allowing molten ferromagnetic metal to solidify in such fields.

> **Key definition**
>
> **Variation of magnetic field strength** the magnetic field decreases with distance from a current-carrying wire and varies around a solenoid

> **Test yourself**
>
> 5 The vertical wire in Figure 4.5.13 is at right angles to the card. In what direction will a plotting compass at A point when
>   a there is no current in the wire
>   b the current direction is upwards?
>
>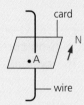
>
> ▲ **Figure 4.5.13**
>
> 6 Figure 4.5.14 shows a solenoid wound on a core of soft iron. Will the end A be a N pole or S pole when the current is in the direction shown?
>
>
>
> ▲ **Figure 4.5.14**
>
> 7 a State where the magnetic field is strongest in a current-carrying solenoid.
>   b Name two factors which affect the strength of a magnetic field around a current-carrying solenoid.

Applications of the magnetic effect of a current

Relay

A **relay** is a switch based on the principle of an electromagnet. It is useful if we want one circuit to control another, especially if the current and power are larger in the second circuit. Figure 4.5.15 overleaf shows a typical relay. When a current is in the coil from the circuit connected to AB, the soft iron core is magnetised and attracts the L-shaped iron armature. This rocks on its pivot and closes the contacts at C in the circuit connected to DE. The relay is then 'energised' or 'on'.

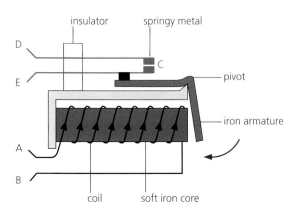

▲ **Figure 4.5.15** Relay

The current needed to operate a relay is called the *pull-on* current and the *drop-off* current is the smaller current in the coil when the relay just stops working.

If the coil resistance, R, of a relay is $185\,\Omega$ and its operating p.d. V is $12\,V$, then the pull-on current $I = V/R = 12/185 = 0.065\,A = 65\,mA$. The symbols for relays with normally open and normally closed contacts are given in Figure 4.5.16.

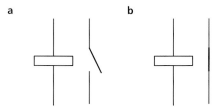

▲ **Figure 4.5.16** Symbols for a relay: **a** open; **b** closed

Reed switch

One such switch is shown in Figure 4.5.17a. When current flows in the coil, the magnetic field produced magnetises the strips (called *reeds*) of magnetic material. The ends become opposite poles and one reed is attracted to the other, so completing the circuit connected to AB. The reeds separate when the current in the coil is switched off. This type of reed switch is sometimes called a **reed relay**.

a Reed switch

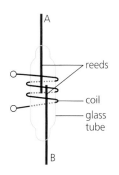

b Burglar alarm activated by a reed switch

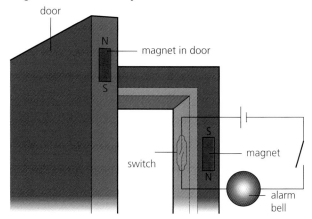

▲ **Figure 4.5.17**

Reed switches are also operated by permanent magnets. Figure 4.5.17b shows the use of a normally open reed switch as a burglar alarm. When the door is closed, the magnetic fields of the magnet in the door and door frame cancel each other and the reed switch is open. When the door is opened the magnetic field of the magnet in the door frame closes the reed switch so that current flows in the alarm circuit if it has been switched on.

Loudspeaker

Varying currents from a radio, CD player, etc. pass through a short cylindrical coil whose turns are at right angles to the magnetic field of a magnet with a central pole and a surrounding ring pole (Figure 4.5.18a).

The magnetic fields around the coil and the magnet interact and the coil vibrates with the same frequency as the a.c. of the electrical signal it receives. A paper cone attached to the coil moves with it and sets up sound waves in the surrounding air (Figure 4.5.18b).

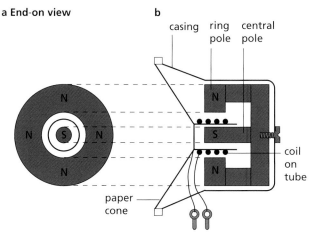

▲ **Figure 4.5.18** Moving-coil loudspeaker

Electric bell

When the circuit in Figure 4.5.19 is completed, by someone pressing the bell push, current flows in the coils of the electromagnet which becomes magnetised and attracts the soft iron bar (the armature).

The hammer hits the gong but the circuit is now broken at the point C of the contact screw.

The electromagnet loses its magnetism (becomes demagnetised) and no longer attracts the armature. The springy metal strip is then able to pull the armature back, remaking contact at C and so completing the circuit again. This cycle is repeated for as long as the bell push is depressed, and continuous ringing occurs.

> ### Test yourself
>
> 8 Explain when a relay would be used in a circuit.
> 9 The resistance, R, of the coil in a relay is $300\,\Omega$ and its operating p.d. V is $15\,V$. Calculate the pull-on current.
> 10 The pull-on current in a relay is $60\,mA$ and it operates at $12\,V$. Calculate the resistance of the coil.
> 11 Explain how an a.c. signal is converted into sound by a loudspeaker.

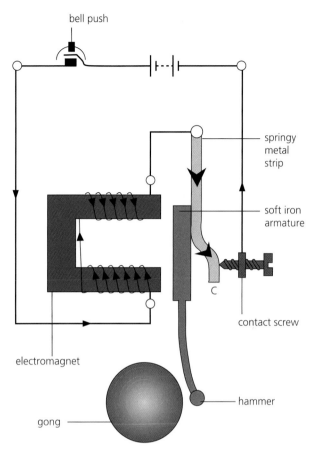

▲ **Figure 4.5.19** Electric bell

4.5.4 Force on a current-carrying conductor

> ### FOCUS POINTS
>
> ★ Describe an experiment that demonstrates that a force acts on a conductor in a magnetic field when it carries a current.
>
> ★ Know how the directions of force, magnetic field and current relate to each other.
> ★ Work out the direction of the force acting on charged particles moving in a magnetic field.

Electric motors form the heart of a whole host of electrical devices ranging from domestic appliances such as vacuum cleaners and washing machines to electric trains and lifts. In a car, the windscreen wipers are usually driven by one and the engine is started by another. All these devices rely on the fact that a current flowing in a magnetic field experiences a force. The force will cause a current-carrying conductor or beam of charged particles to move or be deflected.

The motor effect

A wire carrying a current in a magnetic field experiences a force. If the wire can move, it does so.

Demonstration

In Figure 4.5.20 the flexible wire is loosely supported in the strong magnetic field of a C-shaped magnet (permanent or electromagnet). When the switch is closed, current flows in the wire, which jumps upwards as shown. If either the direction of the current or the direction of the field is reversed, the wire moves downwards. *The force increases if the strength of the field increases and if the current increases.*

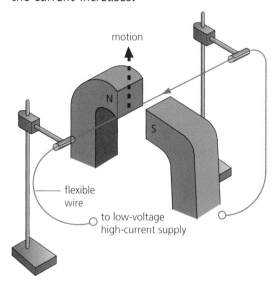

▲ **Figure 4.5.20** A wire carrying a current in a magnetic field experiences a force.

Explanation

Figure 4.5.21a is a side view of the magnetic field lines due to the wire and the magnet. Those due to the wire are circles and we will assume their direction is as shown. The dotted lines represent the field lines of the magnet and their direction is towards the right.

The resultant field obtained by combining both fields is shown in Figure 4.5.21b. There are more lines below than above the wire since both fields act in the same direction below but they are in opposition above. If we *suppose* the lines are like stretched elastic, those below will try to straighten out and in so doing will exert an upward force on the wire.

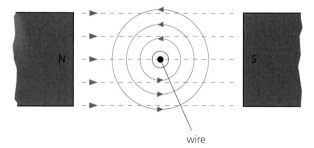

▲ **Figure 4.5.21a**

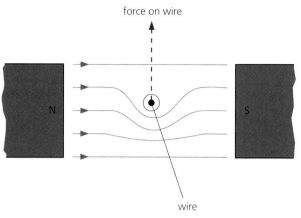

▲ **Figure 4.5.21b**

Fleming's left-hand rule

The direction of the force or thrust on the wire can be found by **Fleming's left-hand rule**, which is also called the **motor rule** (Figure 4.5.22).

Hold the thumb and first two fingers of the left hand at right angles to each other with the **F**irst finger pointing in the direction of the **F**ield and the se**C**ond finger in the direction of the **C**urrent, then the **Th**umb points in the direction of the **Th**rust (or force).

If the wire is not at right angles to the field, the force is smaller and is zero if the wire is parallel to the field.

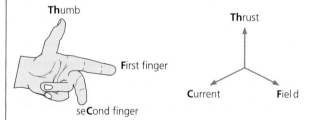

▲ **Figure 4.5.22** Fleming's left-hand (motor) rule

Force on beams of charged particles in a magnetic field

In Figure 4.5.23 the evenly spaced crosses represent a uniform magnetic field (i.e. one of the same strength throughout the area shown) acting into and perpendicular to the paper. A beam of electrons entering the field at right angles to the field experiences a force due to the motor effect whose direction is given by Fleming's left-hand rule. This indicates that the force acts at right angles to the direction of the beam and makes it follow a *circular* path as shown. The beam of negatively charged electrons is treated as being in the opposite direction to conventional current. A beam of positively charged particles would be deflected in the opposite direction to that shown (see Topic 5.2).

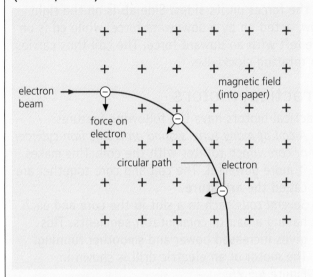

▲ **Figure 4.5.23** Path of an electron beam at right angles to a magnetic field

> **Test yourself**

12 The current direction in a wire running between the N and S poles of a magnet lying horizontally is shown in Figure 4.5.24. The force on the wire due to the magnet is directed
   A from N to S
   B from S to N
   C opposite to the current direction
   D vertically upwards.

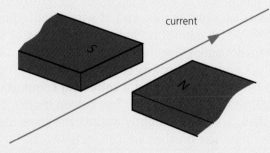

▲ **Figure 4.5.24**

13 An electron beam follows a circular path in a perpendicular magnetic field. Will the radius of the path increase or decrease if the strength of the magnetic field increases? Why?

4.5.5 The d.c. motor

FOCUS POINTS

★ Know the factors that may increase the turning effect on a current-carrying coil in a magnetic field.

★ Describe how an electric motor works.

In the previous topic you learnt that a current flowing in a magnetic field experiences a force. The force may lead to a turning effect on a current-carrying coil in a magnetic field because of a turning effect arising from the two sides of the coil. The magnitude of the turning effect is increased by increasing the number of turns on the coil, increasing the current or increasing the strength of the magnetic field. This turning effect is the basis of all electric motors from electric toothbrushes to ship propulsion.

The motor effect shows that a straight current-carrying wire in a magnetic field experiences a force. If the wire is wound into a coil, forces act on both sides of the coil and a turning effect results when the coil carries current in a magnetic field.

Turning effect on a coil

A rectangular coil of wire mounted on an axle which can rotate between the poles of a magnet may experience a turning effect when a direct current (d.c.) is passed through it.

The turning effect increases if:

(i) the number of turns on the coil increases
(ii) the current flowing in the coil increases
(iii) the strength of the magnetic field increases.

The larger the turning effect on the coil, the faster it will turn.

Simple d.c. electric motor

A simple motor to work from direct current (d.c.) consists of a rectangular coil of wire mounted on an axle which can rotate between the poles of a C-shaped magnet (Figure 4.5.25).

Each end of the coil is connected to half of a split ring of copper, called a **split-ring commutator**, which rotates with the coil. Two carbon blocks, the **brushes**, are pressed lightly against the commutator by springs. The brushes are connected to an electrical supply.

If Fleming's left-hand rule is applied to the coil in the position shown, we find that side **ab** experiences an upward force and side **cd** a downward force. (No forces act on **ad** and **bc** since they are parallel to the field.) These two forces produce a turning effect which rotates the coil in a clockwise direction until it is vertical.

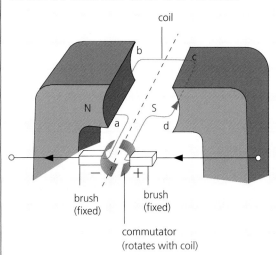

▲ **Figure 4.5.25** Simple d.c. motor

The brushes are then in line with the gaps in the commutator and the current stops. However, because of its inertia, the coil overshoots the vertical and the commutator halves change contact from one brush to the other. This reverses the current through the coil and so also the directions of the forces on its sides. Side **ab** is on the right now, acted on by a downward force, while **cd** is on the left with an upward force. The coil thus carries on rotating clockwise.

Practical motors

Practical motors have the following features:

(i) *A coil of many turns wound on a soft iron cylinder or core* which rotates with the coil. This makes it more powerful. The coil and core together are called the **armature**.

(ii) *Several coils* each in a slot in the core and each having a pair of commutator segments. This gives increased power and smoother running. The motor of an electric drill is shown in Figure 4.5.26.

(iii) *An electromagnet* (usually) to produce the field in which the armature rotates.

Most electric motors used in industry are **induction motors**. They work off a.c. (alternating current) on a different principle from the d.c. motor.

▲ **Figure 4.5.26** Motor inside an electric drill

Practical work

A model motor

The motor shown in Figure 4.5.27 is made from a kit.

a Wrap Sellotape round one end of the metal tube which passes through the wooden block.

b Cut two rings off a piece of narrow rubber tubing; slip them on to the taped end of the metal tube.

c Remove the insulation from one end of a 1.5-metre length of SWG 26 PVC-covered copper wire and fix it under both rubber rings so that it is held tight against the Sellotape. This forms one end of the coil.

d Wind 10 turns of the wire in the slot in the wooden block and finish off the second end of the coil by removing the PVC and fixing this too under the rings but on the opposite side of the tube from the first end. The bare ends act as the commutator.

e Push the axle through the metal tube of the wooden base so that the block spins freely.

f Arrange two 0.5 metre lengths of wire to act as brushes and leads to the supply, as shown. Adjust the brushes so that they are vertical and each touches one bare end of the coil when the plane of the coil is horizontal. *The motor will not work if this is not so.*

g Slide the base into the magnet with *opposite poles facing*. Connect to a 3 V battery (or other low-voltage d.c. supply) and a slight push of the coil should set it spinning at high speed.

3 List the variables in the construction of a simple d.c. motor.

4 How would the motion of the coil change if you reversed the current direction?

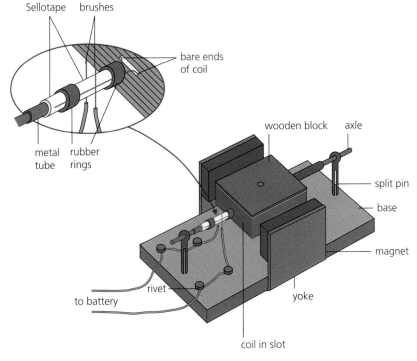

▲ **Figure 4.5.27** A model motor

Moving-coil galvanometer

A galvanometer detects small currents or small p.d.s, often of the order of milliamperes (mA) or millivolts (mV).

In the moving-coil pointer-type meter, a coil is pivoted between the poles of a permanent magnet (Figure 4.5.28a). Current enters and leaves the coil by hair springs above and below it. When there is a current, a turning effect acts on the coil (as in an electric motor), causing it to rotate until stopped by the springs. The greater the current, the greater the deflection which is shown by a pointer attached to the coil.

a

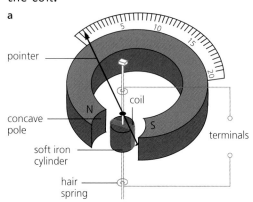

b view from above

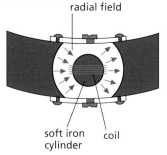

▲ **Figure 4.5.28** Moving-coil pointer-type galvanometer

The soft iron cylinder at the centre of the coil is fixed and along with the concave poles of the magnet it produces a radial field (Figure 4.5.28b), i.e. the field lines are directed to and from the centre of the cylinder. The scale on the meter is then even or linear, i.e. all divisions are the same size.

> ### Test yourself
>
> 14 How would the turning effect on a current-carrying coil in a magnetic field change if
>   a the size of the magnetic field is increased
>   b the direction of the magnetic field is reversed?
>
> 15 In the simple d.c. electric motor of Figure 4.5.29, the coil rotates anticlockwise as seen by the eye from the position X when current flows in the coil. Is the current flowing clockwise or anticlockwise around the coil when viewed from above?
>
>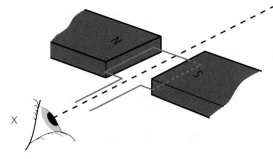
>
> ▲ **Figure 4.5.29**
>
> 16 Explain the function of the split-ring commutator and brushes in a d.c. motor.

4.5.6 The transformer

FOCUS POINTS

★ Describe the construction of a simple transformer.

★ Explain how a simple iron-cored transformer works.

★ Understand the terms primary, secondary, step-up and step-down and correctly use the transformer equation.
★ Describe how high-voltage transformers are used in the transmission of electricity and why high voltages are preferred.

★ Use the correct equations to calculate efficiency in a transformer and to explain how losses in cables are reduced by transmitting power at greater voltages.

Many household devices such as electronic keyboards, toys, lights and telephones require a lower voltage than is provided by the mains supply and a transformer is needed to reduce the mains voltage. When two coils lie in a magnetic field, variations in the current in one coil induce a current change in the other. In this section you will learn that this effect is used in a transformer to raise or lower alternating voltages. The voltage transformation depends on the ratio of the number of turns of wire in each coil. Alternating current generated in a power station is transformed into a very high voltage for long-distance electrical transmission. This reduces the size of the current flowing in the transmission cables and minimises the energy lost to heat due to the resistance of the cables.

Transformers

A **transformer** transforms (changes) an *alternating* voltage from one value to another of greater or smaller value. It has a primary coil and a secondary coil, consisting of insulation-coated wires wound on a complete soft iron core, either one on top of the other (Figure 4.5.30a) or on separate limbs of the core (Figure 4.5.30b).

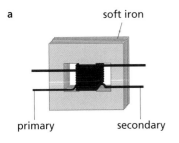

a soft iron

primary   secondary

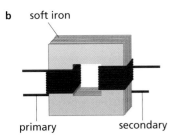

b soft iron

primary   secondary

▲ **Figure 4.5.30** Primary and secondary coils of a transformer

Mutual induction

When the current in a coil is switched on or off or changed in a simple iron-cored transformer, a voltage is induced in a neighbouring coil. The effect, called **mutual induction**, is an example of electromagnetic induction and can be shown with the arrangement of Figure 4.5.31. Coil A is the *primary* and coil B the *secondary*.

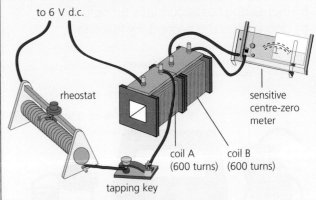

▲ **Figure 4.5.31** A changing current in a primary coil (A) induces a current in a secondary coil (B).

Switching on the current in the primary sets up a magnetic field and as its field lines grow outwards from the primary, they cut the secondary. A p.d. is induced in the secondary until the current in the primary reaches its steady value. When the current is switched off in the primary, the magnetic field dies away and we can imagine the field lines cutting the secondary as they collapse, again inducing a p.d. in it. Changing the primary current by *quickly* altering the rheostat has the same effect.

The induced p.d. is increased by having a soft iron rod in the coils or, better still, by using coils wound on a complete iron ring. More field lines then cut the secondary due to the magnetisation of the iron.

Practical work

Mutual induction with a.c.

An a.c. is changing all the time and if it flows in a primary coil, an alternating voltage and current are induced in a secondary coil.

Connect the circuit of Figure 4.5.32. All wires used should be insulated. The 1 V high current power unit supplies a.c. to the primary and the lamp detects the secondary current.

Find the effect on the brightness of the lamp of

a pulling the C-cores apart slightly
b increasing the secondary turns to 15
c decreasing the secondary turns to 5.

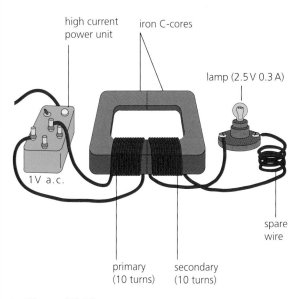

high current power unit

iron C-cores

lamp (2.5 V 0.3 A)

1 V a.c.

spare wire

primary (10 turns)   secondary (10 turns)

▲ **Figure 4.5.32**

5 In the circuit of Figure 4.5.32, if a d.c. supply were used instead of an a.c. supply would you expect the lamp to light? Explain your answer.

6 In the circuit of Figure 4.5.32 would you expect the brightness of the lamp to increase or decrease if you lowered the voltage to the primary coil?

Transformer equation

An alternating voltage applied to the primary induces an alternating voltage in the secondary. The value of the secondary voltage can be shown, for a transformer in which all the field lines cut the secondary, to be given by

$$\frac{\text{primary voltage}}{\text{secondary voltage}} = \frac{\text{primary turns}}{\text{secondary turns}}$$

In symbols

$$\frac{V_p}{V_s} = \frac{N_p}{N_s}$$

A *step-up* transformer has more turns on the secondary than the primary and V_s is greater than V_p (Figure 4.5.33a). For example, if the secondary has twice as many turns as the primary, V_s is about twice V_p. In a *step-down* transformer there are fewer turns on the secondary than the primary and V_s is less than V_p (Figure 4.5.33b).

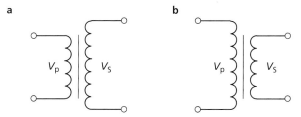

a

b

V_p V_s

V_p V_s

▲ **Figure 4.5.33** Symbols for a transformer: **a** step-up $(V_s > V_p)$; **b** step-down $(V_p > V_s)$

▶ Test yourself

17 The main function of a step-down transformer is to
  A decrease current
  B decrease voltage
  C change a.c. to d.c.
  D change d.c. to a.c.
18 A transformer has 1000 turns on the primary coil. The voltage applied to the primary coil is 230 V a.c. How many turns are on the secondary coil if the output voltage is 46 V a.c.?
  A 20   B 200
  C 2000   D 4000

Energy losses in a transformer

If the p.d. is stepped up in a transformer, the current is stepped down in proportion. This must be so if we assume that all the electrical energy given to the primary appears in the secondary, i.e. that energy is conserved and the transformer is 100% efficient or 'ideal' (many approach this efficiency). Then

power in primary = power in secondary

$$I_p V_p = I_s V_s$$

where I_p and I_s are the primary and secondary currents, respectively.

$$\therefore \frac{I_s}{I_p} = \frac{V_p}{V_s}$$

So, for the ideal transformer, if the p.d. is doubled the current is halved. In practice, it is more than halved, because of small energy losses in the transformer arising from the following three causes.

Resistance of windings

The windings of copper wire have some resistance and heat is produced by the current in them. Large transformers like those in Figure 4.5.34 have to be oil-cooled to prevent overheating.

▲ **Figure 4.5.34** Step-up transformers at a power station

Eddy currents

The iron core is in the changing magnetic field of the primary coil and currents, called **eddy currents**, are induced in it which cause heating.

These are reduced by using a **laminated** core made of sheets, insulated from one another to have a high resistance.

Leakage of field lines

All the field lines produced by the primary may not cut the secondary, especially if the core has an air gap or is badly designed.

? Worked example

A transformer steps down the mains supply from 230 V to 10 V to operate an answering machine.

a What is the turns ratio, $\frac{N_p}{N_s}$, of the transformer windings?

primary voltage, $V_p = 230$ V

secondary voltage, $V_s = 10$ V

turns ratio $= \frac{N_p}{N_s} = \frac{V_p}{V_s} = \frac{230\,\text{V}}{10\,\text{V}} = \frac{23}{1}$

b How many turns are on the primary if the secondary has 100 turns?

secondary turns, $N_s = 100$

From **a**,

$$\frac{N_p}{N_s} = \frac{23}{1}$$

$$\therefore\ N_p = 23 \times N_s = 23 \times 100$$
$$= 2300 \text{ turns}$$

---

c What is the current in the primary if the transformer is 100% efficient and the current in the answering machine is 2 A?

efficiency = 100%

$\therefore$ power in primary = power in secondary

$$I_p \times V_p = I_s \times V_s$$

$$\therefore\ I_p = \frac{V_s \times I_s}{V_p} = \frac{10\,\text{V} \times 2\,\text{A}}{230\,\text{V}} = \frac{2}{23}\,\text{A} = 0.09\,\text{A}$$

Note that in this ideal transformer the current is stepped up in the same ratio as the voltage is stepped down.

Now put this into practice

1 A transformer steps down the mains supply from 240 V to 12 V to operate a doorbell.

a What is the turns ratio $\frac{N_p}{N_s}$ of the transformer windings?

b How many turns are on the primary if the secondary has 80 turns?

2 A transformer steps up an a.c. voltage of 240 V to 960 V.
  a Calculate the turns ratio of the transformer windings.
  b How many turns are on the secondary if the primary has 500 turns?
3 A transformer is 100% efficient. The current in the primary is 0.05 A when the p.d. is 240 V.
  Calculate the current in the secondary where the p.d. is 12 V.

Transmission of electrical power

Grid system

The **National Grid** is a network of cables, mostly supported on pylons, that connects all the power stations in a country to consumers. In the largest modern stations, electricity is generated at 25 000 V (25 kilovolts = 25 kV) and stepped up in a transformer to a higher p.d. to be sent over long distances. Later, the p.d. is reduced by substation transformers for distribution to local users (Figure 4.5.35).

At the National Control Centre, engineers direct the flow of electricity and re-route it when breakdown occurs. This makes the supply more reliable and cuts costs by enabling smaller, less efficient stations to be shut down at off-peak periods.

Advantages of high-voltage transmission

The efficiency with which transformers step alternating p.d.s up and down accounts for the use of a.c. rather than d.c. in power transmission.

Higher voltages are used in the transmission of electric power so that smaller currents can be used to transfer the energy. **Advantages of high-voltage transmission of electricity** include:
(i) reducing the amount of thermal energy lost in the transmission cables
(ii) allowing wires with small cross-sectional areas to be used; these are cheaper and easier to handle than the thicker wires required to carry large currents.

High p.d.s require good insulation but are readily produced by a.c. generators.

> **Key definition**
>
> **Advantages of high-voltage transmission of electricity**
> (i) lower power loss in transmission cables
> (ii) lower currents in cables so thinner/cheaper cables can be used

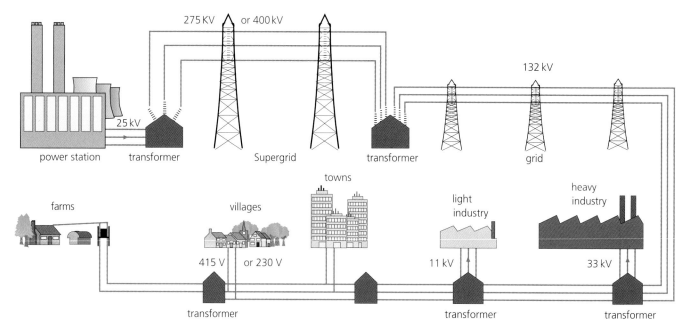

▲ **Figure 4.5.35** The National Grid transmission system in Britain

Power loss in transmission cables

Power cables have resistance, and so electrical energy is transferred to thermal energy during the transmission of electricity from the power station to the user. The power P lost as thermal energy in cables of resistance R is given by

$$P = I^2 R$$

In order to reduce energy losses the current I in the cables should be kept low.

? Worked example

A power station transfers 200 000 W of electrical power to consumers.
a Calculate the current in the transmission cables if the power is transferred at
  i 10 000 V
    power transferred $P = IV$ so
    $$I = P/V = \frac{200\,000\,\text{W}}{10\,000\,\text{V}} = 20\,\text{A}$$
  ii 400 000 V
    power transferred $P = IV$ so
    $$I = P/V = \frac{200\,000\,\text{W}}{400\,000\,\text{V}} = 0.5\,\text{A}$$
b Calculate the energy loss in cables of resistance 400 Ω if the power is transferred at
  i 10 000 V
    power lost in cables $P = I^2R = (20\,\text{A})^2 \times 400\,\Omega$
    $$= 1.6 \times 10^5\,\text{W}$$
  ii 400 000 V
    power lost in cables $P = I^2R = (0.5\,\text{A})^2 \times 400\,\Omega = 100\,\text{W}$
c Is it preferable to transfer the power at the greater or smaller voltage?
  Greater because less energy is transferred to thermal energy in the transmission lines at the smaller current and higher voltage.

Now put this into practice

1 A power station transfers 300 000 W of electrical power to consumers.
  a Calculate the current in the transmission cables if the power is transferred at
    i 20 000 V
    ii 200 000 V.
  b Calculate the energy loss in cables of resistance 400 Ω if 300 000 W of power is transferred at
    i 20 000 V
    ii 200 000 V.
2 Calculate the electrical power transferred to thermal energy in transmission lines of resistance 375 Ω when 210 000 W is transferred at 700 000 V.

➡ Going further

Applications of eddy currents

Eddy currents are the currents induced in a piece of metal when it cuts magnetic field lines. They can be quite large due to the low resistance of the metal. They have their uses as well as their disadvantages.

Car speedometer

The action depends on the eddy currents induced in a thick aluminium disc (Figure 4.5.36), when a permanent magnet, near it but *not touching* it, is rotated by a cable driven from the gearbox of the car. The eddy currents in the disc make it rotate in an attempt to reduce the relative motion between it and the magnet (see Topic 4.5.2). The extent to which the disc can turn, however, is controlled by a spring. The faster the magnet rotates, the more the disc turns before it is stopped by the spring. A pointer fixed to the disc moves over a scale marked in mph (or km/h) and gives the speed of the car.

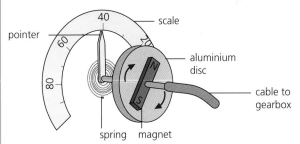

▲ **Figure 4.5.36** Car speedometer

Metal detector

The metal detector shown in Figure 4.5.37 consists of a large primary coil (A), through which an a.c. current is passed, and a smaller secondary coil (B). When the detector is swept over a buried metal object (such as a nail, coin or pipe) the fluctuating magnetic field lines associated with the alternating current in coil A 'cut' the hidden metal and induce eddy currents in it. The changing magnetic field lines associated with these eddy currents cut the secondary coil B in turn and induce a current which can be used to operate an alarm. The coils are set at right angles to each other so that their magnetic fields do not interact.

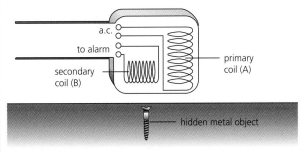

▲ **Figure 4.5.37** Metal detector

Revision checklist

After studying Topic 4.5 you should know and understand:

✔ Faraday's explanation of electromagnetic induction

✔ the right-hand screw and right-hand grip rules for relating current direction and magnetic field direction

✔ the action and applications of a relay and a loudspeaker

✔ that a rectangular current-carrying coil experiences a turning effect in a magnetic field and that the effect is increased by increasing the number of turns on the coil, the current in the coil or the strength of the magnetic field

✔ how to use Fleming's left-hand rule for relating directions of force, field and current

✔ the terms primary, secondary, step-up and step-down in relation to a transformer

✔ the use of transformers in the high voltage transmission of electrical power

✔ the reasons why greater voltage a.c. is preferred with reference to the equation $P = I^2R$.

After studying Topic 4.5 you should be able to:

✔ describe experiments to show electromagnetic induction

✔ predict the direction of induced e.m.f.s and currents and describe and explain the operation of a simple a.c. generator

✔ draw sketches and describe an experiment to identify the pattern of magnetic field lines arising from currents in straight wires and solenoids

✔ identify regions of different magnetic field strength around a solenoid and straight wire and describe the effect on their magnetic fields of changing the magnitude and direction of the current

✔ describe an experiment that demonstrates a force acts on a current-carrying conductor in a magnetic field, and recall the factors which influence the size and direction of the force

✔ explain the action of a simple d.c. electric motor

✔ describe the construction of a transformer and use the transformer equation $V_s/V_p = N_s/N_p$

✔ explain the action of a transformer and recall and use the equation $I_p \times V_p = I_s \times V_s$ for an ideal transformer.

Exam-style questions

1 a Describe an experiment to demonstrate electromagnetic induction. [4]
  b State the factors affecting the magnitude of an induced e.m.f. [3]
  [Total: 7]

2 a Describe the deflections observed on the sensitive, centre-zero galvanometer G (Figure 4.5.38) when the copper rod XY is connected to its terminals and is made to vibrate up and down (as shown by the arrows), between the poles of a U-shaped magnet, at right angles to the magnetic field. [2]
  b Explain the behaviour of the galvanometer in part **a**. [4]

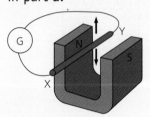

▲ **Figure 4.5.38**

[Total: 6]

3 A simple a.c. generator is shown in Figure 4.5.39.
  a Name A and B and describe their purpose. [3]
  b Describe changes that could be made to increase the e.m.f. generated. [3]
  c Sketch a graph of e.m.f. against time for the generator and relate the position of the generator coil to the peaks, troughs and zeros of the e.m.f. [4]

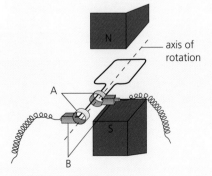

▲ **Figure 4.5.39** [Total: 10]

4 a Describe an experiment to plot the magnetic field lines around a straight current-carrying wire. [4]
  b Sketch the magnetic field lines (including their direction) around a current-carrying solenoid. [4]
  c What happens if the direction of the current in the wire is reversed? [1]
  [Total: 9]

5 Part of the electrical system of a car is shown in Figure 4.5.40. Explain why
  a connections are made to the car body [2]
  b there are *two* circuits in parallel with the battery [2]
  c wire A is thicker than wire B [1]
  d a relay is used. [2]

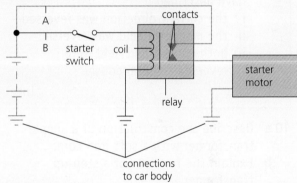

▲ **Figure 4.5.40**

[Total: 7]

6 Explain how a loudspeaker works.
  [Total: 6]

7 a Describe an experiment to show that a force acts on a current-carrying conductor in a magnetic field, including the effect of reversing the current or the direction of the magnetic field. [5]
  b A current-carrying coil experiences a turning effect in a magnetic field. State the effect on the magnitude of the turning effect of:
  i increasing the current in the coil [1]
  ii reducing the number of turns on the coil [1]
  iii increasing the strength of the magnetic field. [1]
  [Total: 8]

8 An electric motor is used to raise a weight attached to a string. Select which of the following is not used to transfer energy in the process.
  A Mechanical working
  B Heating
  C Electrical working
  D Electromagnetic waves
[Total: 1]

9 a Draw a labelled diagram of the essential components of a simple d.c. motor. [3]
  b Explain how continuous rotation is produced in a d.c. motor and show how the direction of rotation is related to the direction of the current. [4]
  c State what would happen to the direction of rotation of the motor you have described if
    i the current direction was reversed [1]
    ii the magnetic field was reversed [1]
    iii both current and field were reversed simultaneously. [1]
[Total: 10]

10 a Describe the construction of a simple transformer with a soft iron core. [4]
  b Explain the function of a step-up transformer. [2]
  c A step-up transformer is used to obtain a p.d. of 720 V from a mains supply of 240 V. Calculate the number of turns that will be needed on the secondary if there are 120 turns on the primary. [4]
[Total: 10]

11 a Calculate the number of turns on the secondary of a step-down transformer which would enable a 12 V lamp to be used with a 230 V a.c. mains power, if there are 460 turns on the primary. [4]

  b Assuming there are no energy losses, what current will flow in the secondary when the primary current is 0.10 A? [4]
[Total: 8]

12 Two coils of wire, A and B, are placed near one another (Figure 4.5.41). Coil A is connected to a switch and battery. Coil B is connected to a centre-reading moving-coil galvanometer, G.
  a If the switch connected to coil A were closed for a few seconds and then opened, the galvanometer connected to coil B would be affected. Explain and describe, step by step, what would actually happen. [7]
  b What changes would you expect if a bundle of soft iron wires was placed through the centre of the coils? Give a reason for your answer. [3]
  c What would happen if more turns of wire were wound on the coil B? [1]

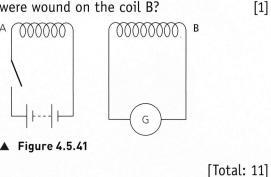

▲ **Figure 4.5.41**
[Total: 11]

13 a Explain the use of transformers in the transmission of electrical power. [3]
  b Give two reasons for the use of high voltages in the transmission of electricity. [2]
[Total: 5]

SECTION 5

Nuclear physics

Topics

5.1.1 The atom

FOCUS POINTS
★ Describe the structure of an atom in terms of negatively charged electrons orbiting a positively charged nucleus.
★ Understand that atoms may form positive or negative ions by losing or gaining electrons.

★ Describe how the nuclear model of the atom is supported by the scattering of alpha (α-) particles by a sheet of thin metal.

The discoveries of the electron and of radioactivity indicated that atoms contained negatively and positively charged particles and were not indivisible as was previously thought. The questions then were 'How are the particles arranged inside an atom?' and 'How many are there in the atom of each element?' In this topic you will learn that atoms have a small dense nucleus surrounded by a cloud of negatively charged electrons orbiting the nucleus. In a neutral atom the charge of the electrons and nucleus are equal and opposite.

The experiments by Geiger and Marsden on the scattering of α-particles from thin metal films provided evidence for the nuclear model of the atom.

An early theory, called the 'plum-pudding' model, regarded the atom as a positively charged sphere in which the negative electrons were distributed all over it (like currants in a pudding) and in sufficient numbers to make the atom electrically neutral.

Experiments carried out by physicists in the early twentieth century cast doubts about this model. The structure of an atom is now described in terms of negatively charged electrons orbiting a positively charged nucleus.

Scattering experiments

While investigating radioactivity, the physicist Ernest Rutherford noticed that not only could α-particles (Topic 5.2) pass straight through very thin metal foil as if it was not there but also that some α-particles were deflected from their initial direction. With the help of Hans Geiger (of GM tube fame, see p. 261) and Ernest Marsden, Rutherford investigated this in detail at Manchester University in the UK using the arrangement in Figure 5.1.1. The fate of the α-particles after striking the gold foil was detected by the scintillations (flashes of light) they produced on a glass screen coated with zinc sulfide and fixed to a rotatable microscope.

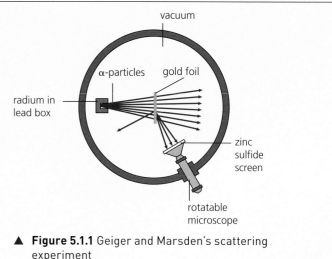

▲ **Figure 5.1.1** Geiger and Marsden's scattering experiment

They found that most of the α-particles were undeflected, some were scattered by appreciable angles and a few (about 1 in 8000) surprisingly 'bounced' back. To explain these results Rutherford proposed in 1911 a 'nuclear' model of the atom in which *all the positive charge and most of the mass of an atom* formed a dense core or **nucleus**, of very small size compared with the whole atom. The electrons surrounded the nucleus some distance away.

He derived a formula for the number of α-particles deflected at various angles, assuming that the electrostatic force of repulsion between the positive charge on an α-particle and the positive charge on the nucleus of a gold atom obeyed an inverse-square law (i.e. the force increases four times if the separation is halved). Geiger and Marsden's experimental results confirmed Rutherford's formula and supported the view that an atom is mostly empty space. In fact, the nucleus and electrons occupy about one-million-millionth of the volume of an atom. Putting it another way, the nucleus is like a sugar lump in a very large hall and the electrons a swarm of flies.

Figure 5.1.2 shows the paths of three α-particles.

Particle 1 is clear of all nuclei and passes straight through the gold atoms.

Particle 2 is deflected slightly.

Particle 3 approaches a gold nucleus so closely that it is violently repelled by it and 'rebounds', appearing to have had a head-on 'collision'.

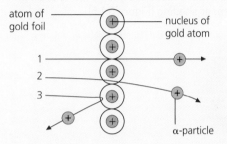

▲ **Figure 5.1.2** Electrostatic scattering of α-particles

Going further

Rutherford–Bohr model of the atom

Shortly after Rutherford proposed his nuclear model of the atom, Niels Bohr, a Danish physicist, developed it to explain how an atom emits light. He suggested that the electrons circled the nucleus at high speed, being kept in certain orbits by the electrostatic attraction of the nucleus for them. He pictured atoms as miniature solar systems. Figure 5.1.3 shows the model for three elements.

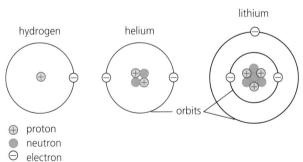

▲ **Figure 5.1.3** Electron orbits

Normally the electrons remain in their orbits but if the atom is given energy, for example by being heated, electrons may jump to an outer orbit. The atom is then said to be excited. Very soon afterwards the electrons return to an inner orbit and, as they do, energy is transferred by bursts of electromagnetic radiation (called photons), such as infrared light, ultraviolet or X-rays (Figure 5.1.4). The wavelength of the radiation emitted depends on the two orbits between which the electrons jump. If an atom gains enough energy for an electron to escape altogether, the atom becomes an ion and the energy needed to achieve this is called the ionisation energy of the atom.

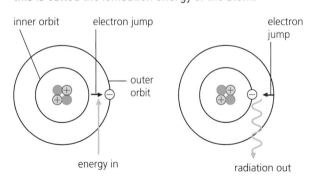

▲ **Figure 5.1.4** Bohr's explanation of energy changes in an atom

Ions

A normal atom is neutral, i.e. with no net charge. If a negatively charged electron is lost, the atom acquires a positive charge and is termed a positive ion. When an extra electron is gained, a negative ion is produced. Such electron exchange forms a way of binding atoms together in molecules and in crystal structures.

Going further

Schrödinger model of the atom

Although it remains useful for some purposes, the Rutherford–Bohr model was replaced by a mathematical model developed by Erwin Schrödinger, which is not easy to picture. The best we can do, without using advanced mathematics, is to say that in the Schrödinger model the atom consists of a nucleus surrounded by a hazy cloud of electrons. Regions of the atom where the mathematics predicts that electrons are more likely to be found are represented by denser shading (Figure 5.1.5).

▲ **Figure 5.1.5** Electron cloud

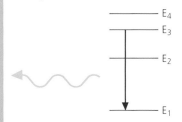

▲ **Figure 5.1.6** Energy levels of an atom

This theory does away with the idea of electrons moving in definite orbits and replaces them by *energy levels* that are different for each element. When an electron 'jumps' from one level, say E_3 in Figure 5.1.6, to a lower one E_1, a photon of electromagnetic radiation is emitted with energy equal to the difference in energy of the two levels. The frequency (and wavelength) of the radiation emitted by an atom is thus dependent on the arrangement of energy levels. For an atom emitting visible light, the resulting *spectrum* (produced for example by a prism) is a series of coloured *lines* that is unique to each element. Sodium vapour in a gas discharge tube (such as a yellow street light) gives two adjacent yellow–orange lines (Figure 5.1.7a). Light from the Sun is due to energy changes in many different atoms and the resulting spectrum is a *continuous* one with all colours (see Figure 5.1.7b).

▲ **Figure 5.1.7a** Line spectrum due to energy changes in sodium

▲ **Figure 5.1.7b** A continuous spectrum

Test yourself

1 Describe how charge and mass are distributed in an atom.

2 How can a positive ion be formed from a neutral atom?

5.1.2 The nucleus

FOCUS POINTS

★ Describe how the nucleus consists of protons and neutrons and state their relative charges.

★ Describe nuclear fission and nuclear fusion and the corresponding nuclide equations.

★ Understand what is meant by proton number (Z) and nucleon number (A) and calculate the number of neutrons in a nucleus.

★ Know how the proton and nucleon numbers relate to the charge on or mass of a nucleus.

★ Describe and explain isotopes in terms of number of protons and neutrons and use nuclide notation.

The nucleus of an atom is composed of positively charged protons and uncharged neutrons of very similar mass. The properties of an atom are determined by the number of protons, but the number of neutrons can vary to give different isotopes of the same element.

Some heavy nuclei are unstable and break up into multiple parts in nuclear fission. Light nuclei may combine in nuclear fusion. Both processes emit substantial amounts of energy.

Protons and neutrons

We now believe as a result of many experiments, in some of which α- and other high-speed particles were used as 'atomic probes', that atoms contain three basic particles – protons, neutrons and electrons.

A **proton** is a hydrogen atom minus an **electron**, i.e. a positive hydrogen ion. Its charge is equal in size but opposite in sign to that of an electron, but its mass is about 2000 times greater.

A **neutron** is uncharged with almost the same mass as a proton. The **relative charges of protons, neutrons and electrons** are +1, 0, −1, respectively.

Protons and neutrons are in the nucleus and are called **nucleons**. Together they account for the mass of the nucleus (and most of that of the atom); the protons account for its positive charge. These facts are summarised in Table 5.1.1.

Key definition

Relative charges of protons, neutrons and electrons
+1, 0, −1, respectively

▼ **Table 5.1.1**

Particle	Relative mass	Relative charge	Location
proton	1836	+1	in nucleus
neutron	1839	+0	in nucleus
electron	1	−1	outside nucleus

In a neutral atom the number of protons equals the number of electrons surrounding the nucleus. Table 5.1.2 shows the particles in some atoms. Hydrogen is simplest with one proton and one electron. Next is the inert gas helium with two protons, two neutrons and two electrons. The soft white metal lithium has three protons and four neutrons.

▼ **Table 5.1.2**

	Hydrogen	Helium	Lithium	Oxygen	Copper
protons	1	2	3	8	29
neutrons	0	2	4	8	34
electrons	1	2	3	8	29

The atomic or **proton number Z** of an atom is the number of protons in the nucleus. The proton number is also the number of electrons in the atom. The electrons determine the chemical properties of an atom and when the elements are arranged in order of atomic number in the Periodic Table, they fall into chemical families.

The **mass** or **nucleon number A** of an atom is the number of nucleons in the nucleus.

In general, the number of neutrons in the nucleus = $A - Z$.

Atomic nuclei are represented by symbols. Hydrogen is written as $_{1}^{1}\text{H}$, helium as $_{2}^{4}\text{He}$ and lithium as $_{3}^{7}\text{Li}$. In general atom X is written in nuclide notation as $_{Z}^{A}\text{X}$, where A is the nucleon number and Z the proton number.

Key definitions

Proton number Z number of protons in the nucleus

Nucleon number A number of protons and neutrons in the nucleus

---

Mass and charge on a nucleus

The relative charge on a nucleus is equal to the product of the proton number Z and the charge on a proton.

The relative mass of a nucleus is equal to the total mass of the neutrons and protons in the nucleus. For a nucleus of nucleon number A this will be approximately equal to the product of A and the mass of a proton.

Test yourself

3 An atom consists of protons, neutrons and electrons. Which one of the following statements is *not* true?
   A An atom consists of a tiny nucleus surrounded by orbiting electrons.
   B The nucleus always contains protons and neutrons, called nucleons, in equal numbers.
   C A proton has a positive charge, a neutron is uncharged and their mass is about the same.
   D An electron has a negative charge of the same size as the charge on a proton but it has a much smaller mass.

4 A lithium atom has a nucleon number of 7 and a proton number of 3.
   a State the number of
      i nucleons in the nucleus
      ii protons in the nucleus
      iii neutrons in the nucleus
      iv electrons in a neutral atom.
   b Give the symbol for the lithium atom in nuclide notation.

5 An atom of sodium has the symbol $^{23}_{11}$Na.
   a State the values of Z and A.
   b How many protons are there in the nucleus?

6 An atom of helium has the symbol $^4_2$He. State the charge on the nucleus of the atom.

7 An atom of chlorine has the symbol $^{37}_{17}$Cl.
   a State the nucleon number of the atom.
   b Explain how the nucleon number is related to the mass of a nucleus.

Isotopes and nuclides

Isotopes of an element are atoms that have the same number of protons but different numbers of neutrons. That is, their proton numbers are the same but not their nucleon numbers. An element may have more than one isotope.

Isotopes have identical chemical properties since they have the same number of electrons and occupy the same place in the Periodic Table. (In Greek, *isos* means same and *topos* means place.)

Few elements consist of identical atoms; most are mixtures of isotopes. Chlorine has two isotopes; one has 17 protons and 18 neutrons (i.e. $Z = 17$, $A = 35$) and is written $^{35}_{17}$Cl, the other has 17 protons and 20 neutrons (i.e. $Z = 17$, $A = 37$) and is written

$^{37}_{17}$Cl. They are present in ordinary chlorine in the ratio of three atoms of $^{35}_{17}$Cl to one atom of $^{37}_{17}$Cl, giving chlorine an average atomic mass of 35.5.

Hydrogen has three isotopes: $^1_1$H with one proton, **deuterium** $^2_1$H with one proton and one neutron and **tritium** $^3_1$H with one proton and two neutrons. Ordinary hydrogen consists of 99.99% $^1_1$H atoms. Water made from deuterium is called heavy water: it has a density of 1.108 g/cm^3, it freezes at 3.8°C and boils at 101.4°C.

Each form of an element is called a **nuclide**. Nuclides with the same Z number but different A numbers are isotopes. Radioactive isotopes are termed **radioisotopes** or **radionuclides**; their nuclei are unstable.

? Worked example

Three stable isotopes of oxygen are $^{16}_8$O, $^{17}_8$O and $^{18}_8$O.

Give the number of i protons, ii nucleons and iii neutrons in each isotope.

Oxygen-16
i number of protons $Z = 8$
ii number of nucleons $A = 16$
iii number of neutrons = $A - Z = 8$

Oxygen-17
i number of protons $Z = 8$
ii number of nucleons $A = 17$
iii number of neutrons = $A - Z = 9$

Oxygen-18
i number of protons $Z = 8$

ii number of nucleons $A = 18$
iii number of neutrons = $A - Z = 10$

Now put this into practice

1 Two isotopes of carbon are $^{12}_6$C and $^{14}_6$C.
   Give the number of
      a protons   b nucleons   c neutrons
   in each isotope.

2 In the nucleus of two isotopes of the same element, state whether the following numbers are the same or different.
   A Number of neutrons
   B Number of protons
   C Number of nucleons

Nuclear energy

$E = mc^2$

Einstein predicted that if the energy of a body changes by an amount E, its mass changes by an amount m given by the equation

$$E = mc^2$$

where c is the speed of light (3×10^8 m/s). The implication is that any reaction in which there is a decrease of mass, called a **mass defect**, is a source of energy. The energy and mass changes in physical and chemical changes are very small; those in some nuclear reactions, such as radioactive decay, are millions of times greater. It appears that mass (matter) is a very concentrated store of energy.

Fission

The heavy metal uranium is a mixture of isotopes of which $^{235}_{92}U$, called uranium-235, is the most important. Some atoms of this isotope decay quite naturally, emitting high-speed neutrons. If one of these hits the nucleus of a neighbouring uranium-235 atom (as it is uncharged the neutron is not repelled by the nucleus), the nucleus may break (**fission** of the nucleus) into two nearly equal radioactive nuclei, often of barium and krypton, with the production of two or three more neutrons:

$$^{235}_{92}U + ^{1}_{0}n \quad \rightarrow \quad ^{144}_{56}Ba + ^{90}_{36}Kr \quad + \quad 2^{1}_{0}n$$

$$\text{neutron} \qquad \text{fission fragments} \qquad \text{neutrons}$$

The mass defect is large and appears mostly as kinetic energy of the fission fragments. These fly apart at great speed, colliding with surrounding atoms and raising their average kinetic energy, i.e. their temperature.

Note that the total values of A and Z must balance on both sides of the equation because nucleons and charge are conserved.

On the left of the equation:
total $A = 235 + 1 = 236$ and total $Z = 92 + 0 = 92$.

On the right of the equation:
total $A = 144 + 90 + 2 = 236$ and
total $Z = 56 + 36 + 0 = 92$.

The total values of A and Z on each side of the equation are equal.

If the fission neutrons split other uranium-235 nuclei, a **chain reaction** is set up (Figure 5.1.8). In practice, some fission neutrons are lost by escaping from the surface of the uranium before this happens. The ratio of those causing fission to those escaping increases as the mass of uranium-235 increases. This must exceed a certain **critical** value to sustain the chain reaction.

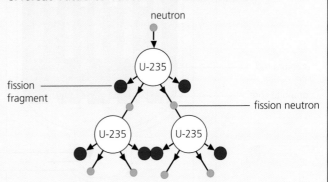

▲ **Figure 5.1.8** Chain reaction

Nuclear reactor

In a nuclear power station, heat from a nuclear reactor produces the steam for the turbines. Figure 5.1.9 (on the next page) is a simplified diagram of one type of reactor.

The chain reaction occurs at a steady rate which is controlled by inserting or withdrawing neutron-absorbing control rods of boron among the uranium rods. The graphite core is called the **moderator** and slows down the fission neutrons: fission of uranium-235 occurs more readily with slow than with fast neutrons. Carbon dioxide gas is pumped through the core as a coolant. In the **heat exchanger** the heated gas transfers energy to pipes containing cold water so that the water boils to produce steam. The concrete shield gives workers protection from γ-emissions and escaping neutrons. The radioactive fission fragments must be removed periodically if the nuclear fuel is to be used efficiently.

In an **atomic bomb**, an increasingly uncontrolled chain reaction occurs when two pieces of uranium-235 come together and exceed the critical mass.

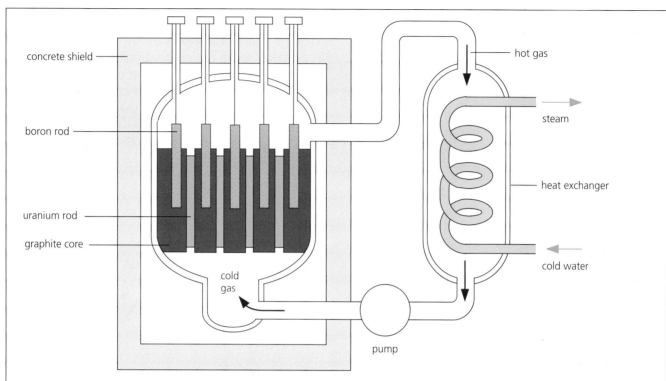

▲ **Figure 5.1.9** Nuclear reactor

Fusion

If light nuclei join together to make heavier ones, this can also lead to a loss of mass and, as a result, the release of energy. Such a reaction has been achieved in the *hydrogen bomb*. At present, research is being done on the controlled **fusion** of isotopes of hydrogen (deuterium and tritium) to give helium.

$$\underset{\text{deuterium}}{{}_{1}^{2}\text{H}} + \underset{\text{tritium}}{{}_{1}^{3}\text{H}} \rightarrow \underset{\text{helium}}{{}_{2}^{4}\text{He}} + \underset{\text{neutron}}{{}_{0}^{1}\text{n}}$$

 Going further

Thermonuclear reactions

The following thermonuclear reactions occur in the Sun:

$${}_{1}^{1}\text{H} + {}_{1}^{1}\text{H} \rightarrow {}_{1}^{2}\text{H} + \text{positron}({}_{+1}^{0}\text{e}) + \text{neutrino}(\nu)$$

$${}_{1}^{1}\text{H} + {}_{1}^{2}\text{H} \rightarrow {}_{2}^{3}\text{He} + \gamma\text{-ray}$$

$${}_{2}^{3}\text{He} + {}_{2}^{3}\text{He} \rightarrow {}_{2}^{4}\text{He} + {}_{1}^{1}\text{H} + {}_{1}^{1}\text{H}$$

Each of these fusion reactions results in a loss of mass and a release of energy. Overall, tremendous amounts of energy are created that help to maintain the very high temperature of the Sun. You can learn more about positrons and neutrinos in Topic 5.2.

Fusion can only occur if the reacting nuclei have enough energy to overcome their mutual electrostatic repulsion. This can happen if they are raised to a very high temperature (over 100 million °C) so that they collide at very high speeds. If fusion occurs, the energy released is enough to keep the reaction going; since heat is required, it is called **thermonuclear fusion.**

The source of the Sun's energy is nuclear fusion. The temperature in the Sun is high enough for the conversion of hydrogen into helium to occur, in a sequence of thermonuclear fusion reactions known as the 'hydrogen burning' sequence.

▶ Test yourself

8 a Explain what is meant by the term 'mass defect' in a nuclear reaction.
   b What happens to the mass defect in a nuclear reaction?
9 a Under what conditions does a fusion reaction occur?
   b Why is a fusion reaction self-sustaining?
10 Do fission or fusion reactions occur in
   a the reactor of a nuclear power station
   b the Sun?

Revision checklist

After studying Topic 5.1 you should know and understand:
✔ the terms proton number (Z), nucleon number (A), isotope and nuclide, and how to use the nuclide notation.

After studying Topic 5.1 you should be able to:
✔ describe the location in the atom of protons, neutrons and electrons

✔ describe the Geiger–Marsden experiment which established the nuclear model of the atom
✔ balance equations involving nuclide notation and outline the processes of nuclear fission and fusion.

Exam-style questions

1 a Define **i** proton number Z and
  ii nucleon number A. [2]
  b Explain the meaning of the term isotope. [2]
  c An isotope of helium can be written in nuclide notation as ^3_2He.
  Describe how an atom of ^3_2He differs from one of ^4_2He. [2]
  [Total: 6]

2 An atom of calcium can be written in nuclide notation as $^{40}_{20}\text{Ca}$.
  a State the values of the nucleon number A and the proton number Z for an atom of calcium. [2]
  b In a neutral atom of calcium state the number of
  i protons [1]
  ii nucleons [1]
  iii neutrons [1]
  iv electrons. [1]
  c When a calcium atom loses an electron it becomes an ion. State whether the ion is negatively or positively charged. [1]
  [Total: 7]

3 a Describe the importance of Geiger and Marsden's experiment on the scattering of α-particles by a thin metal film to understanding the structure of the atom. [5]
  b State the charge on **i** a proton, **ii** a neutron and **iii** a helium nucleus. [3]
  [Total: 8]

4 a Explain what is meant by
  i nuclear fission [2]
  ii nuclear fusion. [2]
  b Energy is released in the following nuclear fusion reaction
  $$^3_2\text{He} + {}^3_2\text{He} \rightarrow {}^4_X\text{Y} + {}^1_1\text{H} + {}^1_1\text{H}$$
  i Determine the total values of A and Z on the left side of the equation. [2]
  ii Work out X and identify Y. [2]
  [Total: 8]

5.2 Radioactivity

5.2.1 Detection of radioactivity

FOCUS POINTS
★ Understand what background radiation is and know the sources that significantly contribute to it.
★ Know that you can measure ionising nuclear radiation using a detector connected to a counter and use count-rate in calculations.

★ Determine a corrected count-rate using measurements of background radiation.

In the previous topic you learnt that an atom has a small dense nucleus containing neutrons and protons which is surrounded by a cloud of electrons. In this topic you will encounter the consequences of instability in the nucleus – radioactivity. We are all continually exposed to radiation from a range of radioactive sources, including radon gas in the air we breathe, cosmic rays from the Sun, food and drink and the rocks and buildings around us. These sources all contribute to what is called background radiation.

The discovery of radioactivity in 1896 by the French scientist Henri Becquerel was accidental. He found that uranium compounds emitted radiation that (i) affected a photographic plate even when it was wrapped in black paper and (ii) ionised a gas. Soon afterwards Marie Curie discovered the radioactive element radium. We now know that radioactivity arises from unstable nuclei which may occur naturally or be produced in reactors. Radioactive materials are widely used in industry, medicine and research.

We are all exposed to natural **background radiation,** both natural and artificial, as indicated in Figure 5.2.1.
(i) Cosmic rays (high-energy particles from the Sun) are mostly absorbed by the atmosphere but some reach the Earth's surface.
(ii) Radon gas is in the air.
(iii) Numerous homes, particularly in Scotland, are built from granite rocks that emit radioactive radon gas; this can collect in basements or well-insulated rooms if the ventilation is poor.
(iv) Radioactive potassium-40 is present in food and is absorbed by our bodies.
(v) Various radioisotopes are used in certain medical procedures.
(vi) Radiation is produced in the emissions from nuclear power stations and in fall-out from the testing of nuclear bombs; the latter produce strontium isotopes with long half-lives which are absorbed by bone.

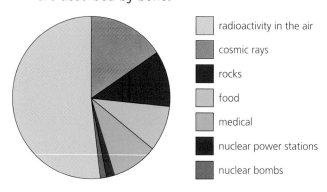

- radioactivity in the air
- cosmic rays
- rocks
- food
- medical
- nuclear power stations
- nuclear bombs

▲ **Figure 5.2.1** Radiation sources

Ionising effect of radiation

A charged electroscope discharges when a lighted match or a radium source (*held in forceps*) is brought near the cap (Figure 5.2.2).

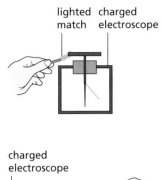

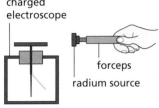

▲ **Figure 5.2.2**

On reaching the electrodes, the ions produce a current pulse which is amplified and fed either to a **scaler** or a **ratemeter**. A scaler counts the pulses and shows the total received in a certain time. A ratemeter gives the counts per second (or minute), or **count-rate**, directly. It usually has a loudspeaker which gives a 'click' for each pulse.

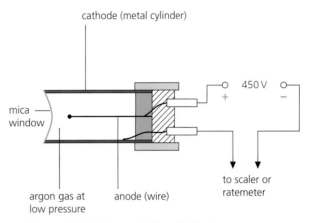

▲ **Figure 5.2.4** Geiger–Müller (GM) tube

In the first case the flame knocks electrons out of surrounding air molecules leaving them as positively charged **ions**, i.e. air molecules which have lost one or more electrons (Figure 5.2.3); in the second case radiation causes the same effect, called **ionisation**. The positive ions are attracted to the cap if it is negatively charged; if it is positively charged the electrons are attracted. As a result, in either case the charge on the electroscope is neutralised, i.e. it loses its charge. Note that if a neutral atom gains an electron it becomes a negative ion.

▲ **Figure 5.2.3** Ionisation

Geiger–Müller (GM) tube

The ionising effect is used to detect radiation.

When radiation enters a **Geiger–Müller (GM) tube** (Figure 5.2.4), either through a thin end-window made of mica, or, if the radiation is very penetrating, through the wall, it creates argon ions and electrons. These are accelerated towards the **electrodes** and cause more ionisation by colliding with other argon atoms.

Count-rate corrections

Count-rates measured from a radioactive source need to be corrected by subtracting the contribution from the background radiation from the readings. The background count-rate in a location can be obtained by recording the count-rate, well away from any radioactive source, for a few minutes. An average value for the background count-rate can then be calculated in counts/s or counts/minute as required.

> ### ▶ Test yourself
>
> 1 Name two major sources of background radiation.
> 2 What property of radiation is used to detect radioactivity?
>
> 3 A count-rate of 190 counts/s was recorded for a radioactive source. If the background count over one minute was 300, calculate
>   a the background count-rate/s
>   b the corrected count-rate for the source.

5.2.2 The three types of nuclear emission

FOCUS POINTS

★ Know that radiation emitted from a nucleus is described as spontaneous and random in direction.
★ Identify α-, β- and γ-emissions from the nucleus from their specific qualities.

★ Describe the deflection of α-particles, β-particles and γ-radiation in magnetic and electric fields.
★ Explain the ionising effects of α-particles, β-particles and γ-radiation in terms of electric charge and kinetic energy.

Certain isotopes can decay into a more stable state with emission of radiation in the form of discrete particles (α- and β-particles) or high-frequency electromagnetic waves (γ-radiation). The emission of the radiation is random in direction. You will learn about the different properties of these three types of radiation and how they interact with matter. An α-particle is a doubly positively charged helium ion and a β-particle is a negatively charged electron.

Owing to their electric charge both α-particles and β-particles are deflected by electric and magnetic field; γ-radiation is deflected by neither. Their kinetic energy and electric charge affect the ionising properties of each type of radiation.

Alpha, beta and gamma radiation

Experiments to study the penetrating power, ionising ability and behaviour of radiation in magnetic and electric fields show that a radioactive substance emits one or more of three types of radiation – called **alpha** (α-), **beta** (β-) and **gamma** (γ-) radiation. The emission of radiation from a nucleus is a spontaneous and random process in direction; we will consider this again later (see p. 269).

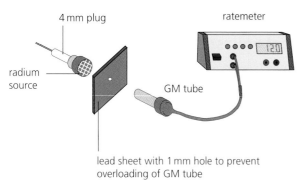

(see p. 269)

▲ **Figure 5.2.5** Investigating the penetrating power of radiation

Experiments using radiation sources are teacher demonstrations only.

Penetrating power of the different types of radiation can be investigated as shown in Figure 5.2.5 by observing the effect on the count-rate of placing one of the following in turn between the GM tube and the lead sheet:
(i) a sheet of thick paper (the radium source, lead and tube must be close together for this part)
(ii) a sheet of aluminium 2 mm thick
(iii) a further sheet of lead 2 cm thick.
Radium (Ra-226) emits α-particles, β-particles and γ-emissions. Other sources can be used, such as americium, strontium and cobalt.

α-particles

These are stopped by a thick sheet of paper and have a range in air of only a few centimetres since they cause intense ionisation in a gas due to frequent collisions with gas molecules. They are deflected by electric and strong magnetic fields in a direction and by an amount which suggests they are helium atoms minus two electrons, i.e. helium ions with a double positive charge. From a particular substance, they are all emitted with the same speed (about 1/20th of that of light).

Americium (Am-241) is a pure α-particle source, used in smoke detectors.

β-particles

These are stopped by a few millimetres of aluminium and some have a range in air of several metres. Their ionising power is much less than that of α-particles. As well as being deflected by electric fields, they are more easily deflected by magnetic fields. Measurements show that β⁻-particles are streams of *high-energy electrons*, emitted with a range of speeds up to that of light. Strontium (Sr-90) emits β⁻-particles only. Note that a β⁻-particle is often just referred to as a β-particle.

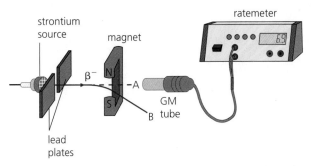

strontium source

magnet

ratemeter

lead plates

GM tube

▲ **Figure 5.2.6** Demonstrating magnetic deflection of β⁻-particles

The magnetic deflection of β⁻-particles can be shown as in Figure 5.2.6. With the GM tube at A and without the magnet, the count-rate is noted. Inserting the magnet reduces the count-rate but it increases again when the GM tube is moved sideways to B.

γ-emissions

These are the most penetrating and are stopped only by many centimetres of lead. They ionise a gas even less than β-particles and are not deflected by electric and magnetic fields. They give interference and diffraction effects and are electromagnetic radiation travelling at the speed of light. Their wavelengths are those of very short X-rays, from which they differ only because they arise in atomic nuclei whereas X-rays come from energy changes in the electrons outside the nucleus.

Cobalt (Co-60) emits γ-radiation and β⁻-particles but can be covered with aluminium to provide pure γ-radiation.

Comparing α-, β- and γ-radiation

Of α-, β- and γ-radiation, α-particles have the highest mass, charge and kinetic energy of the three (although their speed is the lowest). Their large mass and double positive charge enable them to interact strongly with matter and they have the least penetrating power. In collisions, the large positive charge attracts and easily knocks outer electrons from atoms they are passing, leading to the production of ions and giving them the greatest ionising power; they slow down and transfer kinetic energy in each collision. Lower mass β-particles interact less strongly with matter than α-particles and so have a greater penetrating power. Their single negative charge repels the outer electron from atoms and produces ions, but their kinetic energy and ionising power are less than that of α-particles. Since γ-emissions are electromagnetic radiation, they have no charge and almost no mass, so have little interaction with matter and the highest penetrating power. They still have sufficient kinetic energy to knock an outer electron from an atom and produce ionisation, but are the least likely to do so.

A GM tube detects β-particles and γ-emissions and energetic α-particles; a charged electroscope detects only α-particles. All three types of radiation cause fluorescence.

The behaviour of the three kinds of radiation in a magnetic field is summarised in Figure 5.2.7a. The deflections (not to scale) are found from Fleming's left-hand rule, taking negative charge moving to the right as equivalent to positive (conventional) current to the left.

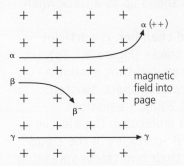

▲ **Figure 5.2.7a** Deflection of α-, β- and γ-radiation in a magnetic field

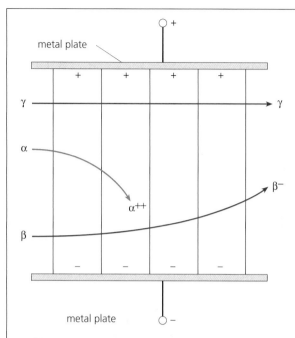

▲ **Figure 5.2.7b** Deflection of α-, β- and γ-radiation in a uniform electric field

Figure 5.2.7b shows the behaviour of α-particles, β⁻-particles and γ-radiation in a uniform electric field: α-particles are attracted towards the negatively charged metal plate, β⁻-particles are attracted towards the positively charged plate and γ-emissions pass through undeflected.

Particle tracks

The paths of particles of radiation were first shown up by the ionisation they produced in devices called *cloud chambers*. When air containing a vapour, alcohol, is cooled enough, saturation occurs. If ionising radiation passes through the air, further cooling causes the saturated vapour to condense on the ions created. The resulting white line of tiny liquid drops shows up as a track when illuminated.

In a **diffusion cloud chamber**, α-particles showed straight, thick tracks (Figure 5.2.8a). Very fast β-particles produced thin, straight tracks while slower ones gave short, twisted, thicker tracks (Figure 5.2.8b). γ-emissions eject electrons from air molecules; the ejected electrons behaved like β⁻-particles in the cloud chamber and produced their own tracks spreading out from the γ-emissions.

a α-particles

b Fast and slow β-particles

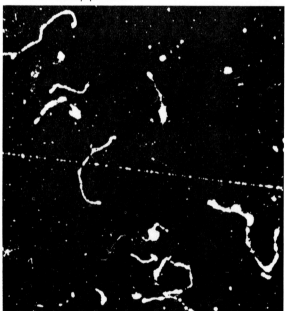

▲ **Figure 5.2.8** Tracks in a cloud chamber

The **bubble chamber**, in which the radiation leaves a trail of bubbles in liquid hydrogen, has now replaced the cloud chamber in research work. The higher density of atoms in the liquid gives better defined tracks, as shown in Figure 5.2.9 than obtained in a cloud chamber. A magnetic field is usually applied across the bubble chamber which causes charged particles to move in circular paths; the sign of the charge can be deduced from the way the path curves.

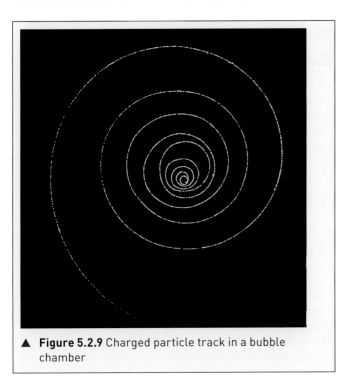

▲ **Figure 5.2.9** Charged particle track in a bubble chamber

▶ **Test yourself**

4 Which type of radiation from radioactive materials
   a has a positive charge
   b is the most penetrating
   c has the shortest range in air
   d has a negative charge?

5 Which type of radiation from radioactive materials
   a is easily deflected by a magnetic field
   b consists of waves
   c causes the most intense ionisation
   d is not deflected by an electric field?

5.2.3 Radioactive decay

FOCUS POINTS

★ Understand that radioactive decay is a spontaneous random change in an unstable nucleus and understand what happens during α- or β-decay.

★ Know why isotopes of an element may be radioactive and describe the effect of α-, β-decay and γ-emissions on the nucleus.
★ Determine the emission of α-, β-decay and γ-radiation using decay equations.

Radioactive decay occurs when an unstable nucleus emits an α- or β-particle. Radioactive decay is a random process; when and in which direction a particular atom will decay cannot be determined. When radioactive decay occurs, a different element is formed.

A nucleus is unstable if it has too many neutrons or is too heavy. Stability is increased in β-decay because the number of neutrons is decreased when a neutron turns into a proton and an electron. Heavy nuclei become lighter and more stable by emission of an α-particle. Decay equations using nuclide notation can be used to represent the changes that a nucleus undergoes in radioactive decay.

Radioactive decay

Radioactive decay is the emission of an α-particle or a β-particle from an unstable nucleus. During α- or β-decay, the nucleus changes to that of a different element, which may itself be unstable.

After a series of changes a stable element is formed. These changes are spontaneous and random and cannot be controlled; also it does not matter whether the material is pure or combined chemically with something else.

α-decay

An α-particle is a helium nucleus, having two protons and two neutrons, and when an atom decays by emission of an α-particle, its nucleon number decreases by 4 and its proton number by 2. For example, when radium of nucleon number 226 and proton number 88 emits an α-particle, it decays to radon of nucleon number 222 and proton number 86. We can write the following *radioactive decay equation* using nuclide notation:

$$^{226}_{88}\text{Ra} \rightarrow {}^{222}_{86}\text{Rn} + {}^{4}_{2}\text{He}$$

The values of A and Z must balance on both sides of the equation since nucleons and charge are conserved. In α-decay the number of nucleons in the nucleus is reduced and a heavy nucleus becomes lighter, tending to increase its stability.

β-decay

In β⁻-decay a neutron changes to a proton and an electron.

$$\text{neutron} \rightarrow \text{proton} + \text{electron}$$

The proton remains in the nucleus and the electron is emitted as a β⁻-particle. The new nucleus has the same nucleon number, but its proton number increases by one since it has one more proton. Radioactive carbon, called carbon-14, decays by β⁻-emission to nitrogen:

$$^{14}_{6}\text{C} \rightarrow {}^{14}_{7}\text{N} + {}^{0}_{-1}\text{e}$$

Note that a β⁻-decay is often just referred to as a β-decay.

In β-decay a neutron turns into a proton and an electron so that the number of neutrons in the nucleus decreases; this increases the stability of a nucleus that has an excess of neutrons. The factors affecting the stability of nuclei are considered in more detail below.

γ-emission

After emitting an α-particle, or β-particle, some nuclei are left in an 'excited' state. Rearrangement of the protons and neutrons occurs and a burst of γ-emissions is released.

 Going further

Other particles

Positrons are subatomic particles with the same mass as an electron but with opposite (positive) charge. They are emitted in some decay processes as β⁺-particles. Their tracks can be seen in bubble chamber photographs. The symbol for a positron is β⁺. In β⁺-decay a proton in a nucleus is converted to a neutron and a positron as, for example, in the reaction:

$$^{64}_{29}\text{Cu} \rightarrow {}^{64}_{28}\text{Ni} + {}^{0}_{+1}\text{e}$$

A *neutrino* (ν) is also emitted in β⁺-decay. Neutrinos are emitted from the Sun in large numbers, but they rarely interact with matter so are very difficult to detect. *Antineutrinos* and positrons are the 'antiparticles' of neutrinos and electrons, respectively. If a particle and its antiparticle collide, they annihilate each other and energy is transferred to γ-radiation.

An antineutrino ($\bar{\nu}$), with no charge and negligible mass, is also emitted in β⁻-decay.

? **Worked example**

a   Polonium $^{218}_{84}\text{Po}$ decays by emission of an α-particle to an isotope of lead (symbol Pb). Write down the decay equation.
 An α-particle consists of two protons and two neutrons so when polonium decays by α-emission, Z decreases by 2 and A decreases by 4:

$$^{218}_{84}\text{Po} \rightarrow {}^{214}_{82}\text{Pb} + {}^{4}_{2}\text{He}$$

b   Strontium $^{90}_{38}\text{Sr}$ decays by emission of a β-particle to an isotope of yttrium (symbol Y). Write down the decay equation.
 A β-particle is an electron so when strontium decays by β-emission, A remains constant but a proton changes to a neutron so Z increases by 1:

$$^{90}_{38}\text{Sr} \rightarrow {}^{90}_{39}\text{Y} + {}^{0}_{-1}\text{e}$$

Now put this into practice

1   Americium $^{241}_{95}\text{Am}$ decays by emission of an α-particle to an isotope of neptunium (symbol Np). Write down the decay equation.
2   Yttrium $^{91}_{39}\text{Y}$ decays by emission of a β-particle to an isotope of zirconium (symbol Zr). Write down the decay equation.

Nuclear stability

The stability of a nucleus depends on both the number of protons (Z) and the number of neutrons (N) it contains. Figure 5.2.10 is a plot of N against Z for all known nuclides. The blue band indicates the region over which stable nuclides occur; unstable nuclides occur outside this band. The continuous line, drawn through the centre of the band, is called the stability line.

It is found that for *stable* nuclides:

(i) $N = Z$ for the lightest
(ii) $N > Z$ for the heaviest
(iii) most nuclides have *even* N and Z, implying that the α-particle combination of two neutrons and two protons is likely to be particularly stable.

For *unstable* nuclides:

(i) disintegration tends to produce new nuclides nearer the stability line and continues until a stable nuclide is formed
(ii) a nuclide above the stability line decays by β⁻-emission (a neutron changes to a proton and electron) so that the N/Z ratio decreases

(iii) a nuclide below the stability line decays by β⁺-emission (a proton changes to a neutron and positron) so that the N/Z ratio increases
(iv) nuclei with more than 82 protons usually emit an α-particle when they decay.

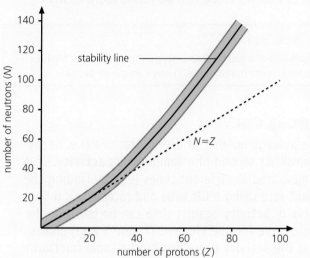

▲ **Figure 5.2.10** Stability of nuclei

Test yourself

6 Write down the change that occurs in a nucleus in β-decay.

7 How can the stability be increased of
   a a heavy nucleus
   b a nucleus with a high ratio of neutrons to protons?

5.2.4 Half-life

FOCUS POINTS

★ Know what is meant by the half-life of a particular isotope, and use this in calculations.

★ Calculate the half-life when background radiation has not been subtracted.
★ Understand why certain isotopes are used for particular applications based on the type of radiation emitted and the half-life of the isotope.

When an α- or β-particle is emitted from a nucleus, a different element is formed. The emission is random for any individual atom, but occurs at a constant rate known as the half-life, whose value depends on the isotope. Half-lives can be calculated from decay data.

Accurate calculation of the half-life of a radioactive isotope requires the background radiation to be taken into account. Radioisotopes find many applications. They are used as tracers in medicine to detect tumours, in industry to measure fluid flow, and in agriculture to detect the uptake of nutrients. Other applications range from radiotherapy to archaeology and smoke alarms.

Radioactive atoms have unstable nuclei which change or decay into atoms of a different element when they emit α- or β-particles. The *rate of decay* is unaffected by temperature but every radioactive element has its own definite decay rate, expressed by its **half-life**.

The **half-life of a particular isotope** is the time taken for half the nuclei of that isotope in a given sample to decay.

It is difficult to know when a substance has lost all its radioactivity, but the time for its activity to fall to half its value can be found more easily.

> **Key definition**
>
> **Half-life of a particular isotope** the time taken for half the nuclei of that isotope in any sample to decay

Decay curve

The average number of disintegrations (i.e. decaying atoms) per second of a sample is its **activity**. If it is measured at different times (e.g. by finding the count-rate using a GM tube and ratemeter), a decay curve of activity against time can be plotted. The ideal curve for one element (Figure 5.2.11) shows that the activity decreases by the same fraction in successive equal time intervals. It falls from 80 to 40 disintegrations per second in 10 minutes, from 40 to 20 in the next 10 minutes, from 20 to 10 in the third 10 minutes and so on. The half-life is 10 minutes.

Half-lives vary for different isotopes from millionths of a second to millions of years. For radium it is 1600 years.

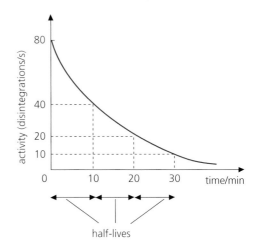

▲ **Figure 5.2.11** Decay curve

The information in Figure 5.2.11 may also be represented in table form:

Time/min	0	10	20	30
Counts/s	80	40	20	10

? Worked example

a   In an experiment to find the half-life of radioactive iodine, the following results were obtained.

Time/min	0	25	50	75
Counts/s	200	100	50	25

  i   What is the half-life of the iodine?
    The count-rate drops by half every 25 minutes.
    The half-life of iodine is 25 minutes.
  ii What fraction of the original material is left after 75 minutes?
    75 minutes corresponds to 3 half-lives.
    After 25 minutes, fraction left = 1/2
    After 50 minutes, fraction left = 1/2 × 1/2 = 1/4
    After 75 minutes, fraction left = 1/2 × 1/4 = 1/8

b   Carbon-14 has a half-life of 5700 years. A 10 g sample of wood cut recently from a living tree has an activity of 160 counts/minute. A piece of charcoal taken from a prehistoric campsite also weighs 10 g but has an activity of 40 counts/minute. Estimate the age of the charcoal.
    After 1 × 5700 years the activity will be 160/2 = 80 counts per minute.
    After 2 × 5700 years the activity will be 80/2 = 40 counts per minute.
    The age of the charcoal is 2 × 5700 = 11 400 years.

Now put this into practice

1   In an experiment to find the half-life of a radioisotope the following results were obtained.

Time/min	0	10	20	30	40	50
Counts/s	200	133	88	59	39	26

  a   Estimate the half-life of the radioisotope from the data.
  b   Determine the half-life of the radioisotope by plotting a graph.
2   A radioactive source has a half-life of 15 minutes. Find the fraction left after 1 hour.
3   In an experiment to determine the half-life of a radioactive material its count-rate falls from 140 counts/minute to 35 counts/minute in an hour. Calculate the half-life of the material.
4   Carbon-14 has a half-life of 5700 years. A 5 g sample of wood cut recently from a living tree has an activity of 80 counts/minute. A 5 g piece of wood taken from an ancient dugout canoe has an activity of 20 counts/minute. Estimate the age of the canoe.

Experiment to find the half-life of thoron

The half-life of the α-emitting gas **thoron** can be found as shown in Figure 5.2.12. The thoron bottle is squeezed three or four times to transfer some thoron to the flask (Figure 5.2.12a). The clips are then closed, the bottle removed and the stopper replaced by a GM tube so that it seals the top (Figure 5.2.12b).

When the ratemeter reading has reached its maximum and started to fall, the count-rate is noted every 15 s for 2 minutes and then every 60 s for the next few minutes. (The GM tube is left in the flask for at least 1 hour until the radioactivity has decayed.)

A measure of the background radiation is obtained by recording the counts for a period (say 10 minutes) at a position well away from the thoron equipment. The count-rates in the thoron decay experiment are then corrected by subtracting the average background count-rate from each reading. A graph of the corrected count-rate against time is plotted and the half-life (52 s) estimated from it.

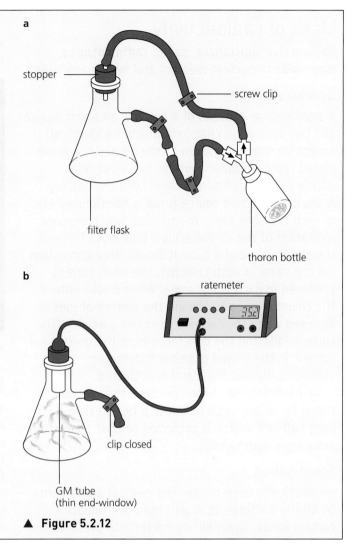

▲ **Figure 5.2.12**

Random nature of decay

During the previous experiment it becomes evident that the count-rate varies irregularly: the loudspeaker of the ratemeter 'clicks' erratically, not at a steady rate. This is because radioactive decay is a spontaneous and **random** process, in that it is a matter of pure chance whether or not a particular atom will decay during a certain period of time. All we can say is that about half the atoms in a sample will decay during the half-life. We cannot say which atoms these will be, nor can we influence the process in any way. Radioactive emissions occur spontaneously and randomly over space and time.

Test yourself

8 In an experiment to determine the half-life of thorium the following results were obtained.

Time/s	0	30	60	90	120	150	180
Counts/s	165	112	77	53	37	27	20

The background count-rate was found to be 5 counts/s.

    a Correct the count-rates for background radiation.

    b Plot a graph of corrected counts/s against time and estimate the half-life of thorium.

9 If the half-life of a radioactive gas is 2 minutes, then after 8 minutes the activity will have fallen to a fraction of its initial value. This fraction is

    A 1/4   **B** 1/6   **C** 1/8   **D** 1/16

Uses of radioactivity

Radioactive substances, called **radioisotopes**, are now made in nuclear reactors and have many uses.

Smoke alarm

A smoke alarm consists of a battery, an alarm device and two ionisation chambers which are identical, except for one being open to the air and the other closed. Each chamber contains two electrodes with a p.d. applied across them from the battery. A small radioactive source (usually americium-241) in each chamber emits α-particles, which produce ionisation of the air molecules; ions move towards the electrodes and a current flows. Since conditions are the same in each chamber, the small current produced in each is the same. When smoke enters the chamber open to the air, the motion of ions is impeded when they adhere to smoke particles. The current falls and the difference from the unchanged current in the closed chamber is detected electronically and the alarm is activated.

α-particles are chosen because they do not travel far in air so do not pose a health risk; a long half-life source is preferred so that a constant activity is maintained.

Sterilisation

γ-radiation is used to sterilise medical instruments by killing bacteria. It is also used to irradiate certain foods, again killing bacteria to preserve the food for longer. The radiation is safe to use as no radioactive material goes into the food. γ-radiation is chosen for its high penetrating power (it can pass through packaging) and a long half-life source is preferred so that a constant activity is maintained.

Thickness gauge

If a radioisotope is placed on one side of a moving sheet of material and a GM tube on the other, the count-rate decreases if the thickness increases. This technique is used to control automatically the thickness of paper, plastic and metal sheets during manufacture (Figure 5.2.13). Because of their range, β emitters are suitable sources for monitoring the thickness of thin sheets but γ emitters would be needed for thicker materials as they are more penetrating and are absorbed less. The half-life of the source should be long so that its activity remains constant over time.

Flaws in a material can be detected in a similar way; the count-rate will increase where a flaw is present.

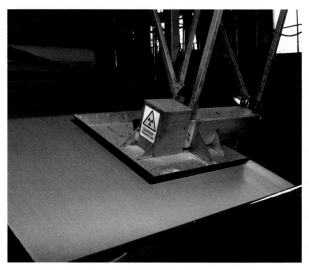

▲ **Figure 5.2.13** Quality control in the manufacture of paper using a radioactive gauge

Diagnosis and treatment of cancer

γ-radiation emitted by radioactive chemicals that are preferentially taken up by cancer cells are used to image and diagnose cancer. γ-radiation is chosen because of its high penetrating power. For use of a radioisotope inside the body, short half-life sources are preferred so that exposure time is limited. In the treatment of cancer by **radiotherapy**, high energy beams of γ-radiation are focused directly onto a tumour in order to kill the cancerous cells. The beams are rotated round the body to minimise damage to surrounding tissue. For external use, where a constant dose is required, a long half-life radioisotope will be used.

Tracers

The progress of a small amount of a weak radioisotope injected into a system can be traced by a GM tube or other detector. The method is used in medicine to detect brain tumours and internal bleeding, in agriculture to study the uptake of fertilisers by plants, and in industry to measure fluid flow in pipes.

A **tracer** should be chosen whose half-life matches the time needed for the experiment; the activity of the source is then low after it has been used and so will not pose an ongoing radiation threat. For medical purposes, where short exposures are preferable, the time needed to transfer the source from the production site to the patient also needs to be considered.

Archaeology

A radioisotope of carbon present in the air, carbon-14, is taken in by living plants and trees along with non-radioactive carbon-12. When a tree dies no fresh carbon is taken in. So as the carbon-14 continues to decay, with a half-life of 5700 years, the amount of carbon-14 compared with the amount of carbon-12 becomes smaller. By measuring the residual radioactivity of carbon-containing material such as wood, linen or charcoal, the age of archaeological remains can be estimated within the range 1000 to 50000 years (Figure 5.2.14).

The ages of rocks have been estimated in a similar way by measuring the ratio of the number of atoms of a radioactive element to those of its decay product in a sample.

> ### Test yourself
>
> **10** What type of radiation is used
>   **a** to sterilise equipment
>   **b** in a smoke alarm?
> **11** Explain how radioisotopes can be used to monitor the thickness of metal sheets in industry.

▲ **Figure 5.2.14** The year of construction of this Viking ship has been estimated by radiocarbon techniques to be 800 CE.

5.2.5 Safety precautions

FOCUS POINTS

★ Understand the effects of ionising nuclear radiation on living things.
★ Describe safe movement, usage and storage of radioactive materials.

★ Explain the main safety precautions for ionising radiation.

Although α-, β- and γ-radiations have many uses, they can also be very dangerous to humans due to their ionising properties which damage body cells. Careful safety measures have to be taken when handling radioactive materials and moving, storing and disposing of them. The symbol warning of the presence of radioactive materials is universally recognised.

Unnecessary exposure to radiation is to be avoided. Lead shielding is used to absorb radiation and protect health workers and other workers who are frequently close to radiation sources. To reduce the danger from ionising radiations, time of exposure to sources should be minimised and distance from them maximised.

Dangers of nuclear radiation

We cannot avoid exposure to radiation in small doses but large doses can be dangerous to our health. The ionising effect produced by nuclear radiation causes damage to cells and tissues in our bodies and can also lead to the mutation of genes. This damage may cause cell death and also induce cancers.

The danger from α-particles is small, unless the source enters the body, but β- and γ-radiation can cause radiation burns (i.e. redness and sores on the skin) and delayed effects such as eye cataracts and cancer. Large exposures may lead to radiation sickness and death. The symbol used to warn of the presence of radioactive material is shown in Figure 5.2.15.

▲ **Figure 5.2.15** Radiation hazard sign

The increasing use of radioisotopes in medicine and industry has made it important to find safe ways of transporting and using radioactive materials and of disposing of radioactive waste. In using and transporting radioactive materials the effects of ionising radiation need to be minimised with thick lead containers and remote handling where possible.

A variety of methods are used for radioactive storage. Waste with very low levels of radioactivity is often enclosed in steel containers which are then buried in concrete bunkers; possible leakage is a cause of public concern, as water supplies could be contaminated allowing radioactive material to enter the food chain. Waste with high levels of radioactivity is immobilised in glass or synthetic rock and stored deep underground.

The weak sources used at school should always be:
(i)   lifted with forceps
(ii) held away from the eyes and
(iii) kept in their boxes when not in use.

Safety precautions

To reduce exposure to ionising radiations:
(i)   exposure time to the radiation should be minimised
(ii) the distance between a source and a person should be kept as large as possible
(iii) people should be protected by the use of shielding which absorbs the radiation.
In industry, sources are handled by long tongs and transported in thick lead containers.

Workers are protected by lead and concrete walls, and wear radiation dose badges that keep a check on the amount of radiation they have been exposed to over a period (usually one month). The badge contains several windows which allow different types of radiation to fall onto a photographic film; when the film is developed it is darkest where the exposure to radiation was greatest.

Revision checklist

After studying Topic 5.2 you should know and understand:

✔ the meaning of the term 'background radiation' and recall its sources

✔ that radioactivity is (i) a spontaneous random process, (ii) due to nuclear instability and (iii) independent of external conditions

✔ that some isotopes of an element are unstable and that during α- or β-decay the nucleus changes to a different element

✔ discuss the dangers of radioactivity and necessary safety precautions.

After studying Topic 5.2 you should be able to:

✔ recall the nature of α-, β- and γ-radiation and describe experiments to compare their range, penetrating power and ionising effects

✔ predict how α-, β- and γ-radiation will be deflected in magnetic and electric fields and explain their relative ionising effects

✔ describe the effect of α-particles, β-particles and γ-ray emissions on the stability and the number of excess neutrons in the nucleus and interpret decay equations

✔ define the term half-life, and solve simple problems on half-life from data given in tables and decay curves

✔ recall some uses of radioactivity.

Exam-style questions

1 a Three types of radiation, X, Y and Z, are shown in Figure 5.2.16.

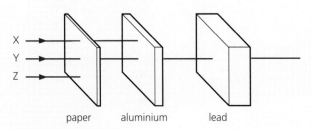

▲ Figure 5.2.16

State which of the columns **A** to **D** correctly names the radiations X, Y and Z.

	A	B	C	D
X	α	β	γ	γ
Y	β	γ	α	β
Z	γ	α	β	α

[3]

b Nuclide notation is used to represent particles in nuclear reactions. Which symbol **A** to **D**, is used in equations to represent
   i   an α-particle [1]
   ii a β-particle [1]
   iii a neutron [1]
   iv an electron? [1]

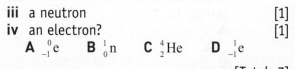
   A $_{-1}^{0}e$ **B** $_{0}^{1}n$ **C** $_{2}^{4}He$ **D** $_{-1}^{1}e$

[Total: 7]

2 a Explain what is meant by the half-life of a radioisotope. [2]

b The graph represented in Figure 5.2.17 shows the decay curve of a radioisotope.

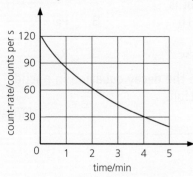

▲ Figure 5.2.17

Calculate its half-life in minutes. [4]

[Total: 6]

3 The ratio of the number of atoms of argon-40 to potassium-40 in a sample of radioactive rock is analysed to be 1:3. Assuming that there was no potassium in the rock originally and that argon-40 decays to potassium-40 with a half-life of 1500 million years, estimate the age of the rock. Assume there were N atoms of argon-40 in the rock when it was formed.

  a Calculate

    i the number of argon atoms left after 1500 million years [1]

    ii the number of potassium atoms formed after 1500 million years [1]

    iii the argon:potassium ratio after 1500 million years [1]

    iv the number of argon atoms left after 3000 million years [1]

    v the number of potassium atoms formed after 3000 million years [1]

    vi the argon:potassium ratio after 3000 million years. [1]

  b Estimate the age of the rock. [1]

    [Total: 7]

4 **a i** Radon-222 decays by emitting an α-particle to form an element whose symbol is

    A $^{216}_{85}\text{At}$ **B** $^{216}_{86}\text{Rn}$

    C $^{218}_{84}\text{Po}$ **D** $^{216}_{84}\text{Po}$ [1]

   ii Write the decay equation in nuclide notation. [3]

  b i Thorium-234 decays by emitting a β-particle to form an element whose symbol is

    A $^{235}_{90}\text{Th}$ **B** $^{234}_{91}\text{Pa}$

    C $^{234}_{89}\text{Ac}$ **D** $^{232}_{88}\text{Ra}$ [1]

   ii Write the decay equation in nuclide notation. [3]

    [Total: 8]

5 **a** Discuss the effects ionising nuclear radiation has on living things. [2]

  b α-, β- and γ-emissions have different powers of ionisation and range.
Choose the type of radiation whose power of ionisation is

    i greatest [1]

    ii least. [1]

    Choose the type of radiation whose range in air is

    iii greatest [1]

    iv least. [1]

  c Describe two safety precautions you should take when using radioactive materials. [2]

    [Total: 8]

6 Describe how radioactive sources are used in the following applications. Name and explain the choice of the radiation used in each case.

  a Smoke alarms [4]

  b Thickness measurements in industry [3]

  c Irradiation of food [3]

    [Total: 10]

SECTION 6

Space physics

Topics

6.1.1 The Earth

FOCUS POINTS

★ Use the rotations of the Earth on its axis to explain the apparent daily motion of the Sun and the periodic change between night and day.

★ Use the Earth's orbiting of the Sun to explain the seasons.

> ★ Use the correct equation to define average orbital speed.

★ Use the Moon's orbiting of the Earth to explain the Moon's phases.

Day and night and the rising and setting of the Sun can be explained by the rotation of the Earth on its axis. The repeating pattern of spring, summer, autumn and winter arises from the motion of the Earth around the Sun. In this section you will learn more about these natural occurrences. The Earth is the third planet from the Sun and travels in a nearly circular orbit controlled by the large gravitational attraction exerted by the Sun. The Earth takes one year to travel around the orbit, moving closer and further away from the Sun with the seasons. The Moon orbits the Earth under the influence of the Earth's gravitational attraction and always has the same area of its surface facing towards the Earth. The Moon is lit by the light from the Sun so that its appearance changes over the course of each month.

> You will learn how to calculate the average orbital speed of an object such as the Moon or a planet.

The Earth is a planet of the Sun travelling in a nearly circular orbit around the Sun, and the Moon orbits the Earth as a satellite. The motion of the Earth and Moon account for the occurrence of a number of natural events.

Motion of the Earth

Day and night

These are caused by the Earth spinning on its axis (i.e. about the line through its north and south poles) and making one complete revolution every 24 hours. This creates day for the half of the Earth's surface facing the Sun and night for the other half, facing away from the Sun.

Rising and setting of the Sun

The Earth's rotation on its axis causes the Sun to have an apparent daily journey from east to west. It rises exactly in the east and sets exactly in the west only at the equinoxes (around 20 March and 23 September). In the northern hemisphere in summer it rises north of east and sets north of west. In winter it rises and sets south of these points.

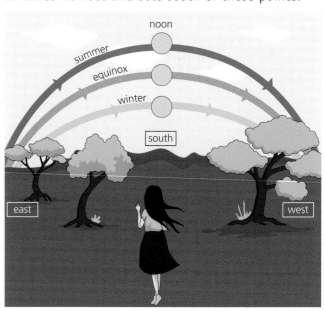

▲ **Figure 6.1.1** Rising and setting of the Sun (in the northern hemisphere)

Each day the Sun is highest above the horizon at noon and directly due south in the northern hemisphere; this height itself is greatest and the daylight hours longest about 21 June. After that the Sun's height slowly decreases and near 21 December it is lowest and the number of daylight hours is smallest, as shown in Figure 6.1.1.

The seasons

Two factors are responsible for these. The first is the motion of the Earth around the Sun once in approximately 365 days (i.e. in one year). The second is the tilt (23.5°) of the Earth's axis to the plane of its path around the Sun. Figure 6.1.2 shows the tilted Earth in four different positions of its orbit.

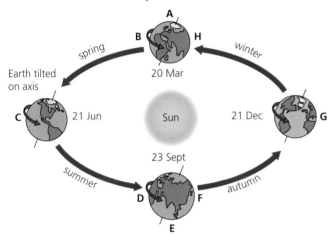

▲ **Figure 6.1.2** Seasons for the northern hemisphere

Over part BCD of the Earth's orbit, the northern hemisphere is tilted towards the Sun so that the hours of daylight are greater than those of darkness, i.e. it is spring and summer. The southern hemisphere is tilted away from the Sun and has shorter days than nights so it is autumn and winter there. The northern hemisphere receives more solar radiation and so the weather is warmer.

Over part FGH of the orbit, the situation is reversed. The southern hemisphere is tilted towards the Sun, while the northern hemisphere is tilted away from it, and experiences autumn and winter.

At C, around 21 June, the northern hemisphere has its longest day while the southern hemisphere has its shortest day. At G, around 21 December, the opposite is true.

At A and E of the orbit, night and day are equal in both hemispheres; these are the *equinoxes*, which often fall on 20 March and 23 September.

Motion of the Moon

The moon is a satellite of the Earth and travels round it in an approximately circular orbit approximately once a month at an average distance away of about 400 000 km. It also revolves on its own axis in a month and so always has the same side facing the Earth, so that we never see the 'dark side of the Moon'. We see the Moon by reflected sunlight since it does not produce its own light. It does not have an atmosphere. It does have a gravitational field due to its mass, but the field strength is only one-sixth of that on Earth. Hence the astronauts who walked on the Moon moved in a 'springy' fashion but did not fly off into space.

Phases of the Moon

The Moon's appearance from the Earth changes during its monthly journey; it has different *phases*. In Figure 6.1.3 the outer circle shows that exactly half of it is always illuminated by the Sun. How it looks from the Earth in its various positions is shown inside this. In the new Moon phase, the Moon is between the Sun and the Earth and the side facing the Earth, being unlit, is not visible from the Earth. A thin new crescent appears along one edge as it travels in its orbit, gradually increasing in size until at the first quarter phase, when half of the Moon's surface can be seen. At full Moon, the Moon is on the opposite side of the Earth from the Sun and appears as a complete circle. After that it wanes through the last quarter phase until only the old crescent can be seen. Figure 6.1.4 shows the surface of the Moon partially illuminated as seen from Earth.

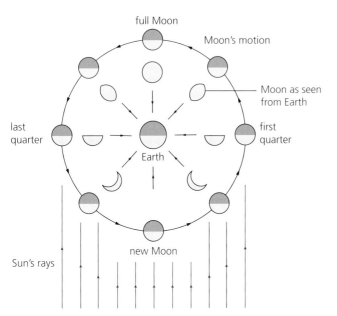

▲ **Figure 6.1.3** Phases of the Moon

▲ **Figure 6.1.4** The surface of the Moon partially illuminated as seen from Earth

Rising and setting of the Moon

Like the Sun, the Moon appears to have a daily trip across the sky from the east, where it rises, to the west, where it sets; again, this is due to the Earth's rotation on its axis.

Orbital speed

In Topic 1.2 we defined
average speed = total distance/total time.

So, for the Moon, moving in a circular orbit around the Earth with **average orbital speed** v,

$$v = \frac{\text{circumference of orbit}}{T} = \frac{2\pi r}{T}$$

where r is the average radius of the orbit, and T is the orbital period (the time for one orbit), also known as the orbital duration.

> **Key definitions**
> **Average orbital speed** $v = \dfrac{2\pi r}{T}$ where r is the average radius of the orbit and T is the orbital duration

Test yourself

1 The Earth is a planet of the Sun.
   Which one of the following statements is *not* true?
   A The Earth orbits the Sun once each year.
   B The Earth spins once on its axis every 24 hours.
   C Day and night are due to the Earth spinning on its axis.
   D At places on the Earth's hemisphere tilted towards the Sun, the day is shorter than the night.

2 Make a sketch of the phases of the Moon as seen from Earth.

3 a Name the months when there are equal hours of day and night in both the northern and southern hemispheres. State the name given to these dates.
   b Name the season in the northern hemisphere when the Earth is tilted towards the Sun.

4 Taking the average distance of the Moon from the Earth to be 380 000 km and the orbital period of the Moon to be 27 days, calculate its average orbital speed in km/s.

6.1.2 The Solar System

FOCUS POINTS

★ Describe the Solar System in terms of one star, planets, moons and smaller Solar System bodies.

★ Describe elliptical orbits in relation to planets, comets and the Sun.
★ Analyse and interpret planetary data.

★ Know which planets are rocky and small and which are large and gaseous and explain the difference with reference to an accretion model for Solar System formation.
★ Know what affects the gravitational field strength of a planet.
★ Calculate the time it takes for light to travel large distances in the Solar System.
★ Explain why planets orbit the sun and know that gravitational attraction of the Sun is what keeps an object in orbit around the Sun.

★ Know the effect of the distance from the Sun on gravitational field and orbital speed of the planets.
★ Use conservation of energy to explain the effect of the distance from the Sun on the speed of an object in an elliptical orbit.

In this topic you will learn that all the bodies in the Solar System move under the influence of the gravitational attraction of the large mass of the Sun. The orbits of the planets, dwarf planets and comets are elliptical with the Sun at one focus. Planetary orbits are nearly circular, but comets travel far away before returning close to the Sun after a considerable period of time. Moons orbit planets bound by the gravitation of the planet. The planets are thought to have condensed from a cloud of dust and gas, called a nebula, with the heavier materials drawn close to the Sun by its gravity; the inner four small planets are rocky and the much larger outer planets are gaseous with icy moons. Light from the Sun takes about 8 minutes to reach the Earth.

Space programmes have allowed much data to be collected about the planets and other objects in our Solar System. You will learn how to use of some of this data, about elliptical orbits, and why comets and other bodies travel faster when they are nearer to the Sun.

The Solar System

The Solar System consists of one star (the Sun) and eight planets moving around it in elliptical orbits (slightly flattened circles) (Figure 6.1.5). It also includes dwarf planets and asteroids which orbit the Sun, moons that orbit many of the planets and smaller Solar System bodies such as comets and natural satellites.

The four *inner* planets, Mercury, Venus, Earth and Mars, are all small, of similar size, solid and rocky, with a layered structure, and have a high density.

The four *outer* planets, Jupiter, Saturn, Uranus and Neptune, are much larger and colder and consist mainly of gases; their density is low. These outer planets have many moons and other natural satellites in the form of rings of icy materials.

Most of the **dwarf planets**, the best known of which is Pluto, orbit the Sun at average distances greater than Neptune. These are Trans-Neptunian Objects (TNOs) and, together with their satellites, number over 2000; they are generally of low density and are thought to be composed of a mixture of rock and ice.

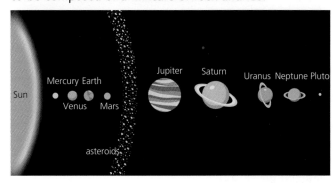

▲ **Figure 6.1.5** The Solar System (distances from the Sun not to scale)

The *asteroids* are pieces of rock of various sizes which mostly orbit the Sun between Mars and Jupiter; their density is similar to that of the inner planets. If they enter the Earth's atmosphere they burn up and fall to Earth as meteors or shooting stars. The largest asteroid, Ceres, is classed as a dwarf planet.

Asteroids are presently classified as minor planets, which are defined as any object that orbits a star that does not have a large enough mass for gravitational attraction to have pulled it into a spherical shape. Larger asteroids, such as Ceres, are classified as dwarf planets if they have enough mass and a gravitational field strength high enough to have formed a spherical shape, but not enough to attract and clear the area around them of smaller objects.

Comets consist of dust embedded in ice made from water and methane and are sometimes called 'dirty snowballs'. Their density is similar to the outer planets and they orbit the Sun in highly elliptical orbits and are much closer to it at some times than others, as shown in Figure 6.1.6. They return to the inner Solar System at regular intervals but in many cases these intervals are so long (owing to journeys far beyond Pluto) that they cannot be predicted. On approaching the Sun, the dust and gas are blown backwards by radiation pressure from the Sun and the comet develops a bright head and long tail pointing away from the Sun.

One of the most famous is Halley's comet (Figure 6.1.7) which visits the inner Solar System about every 76 years, the last occasion being 1986.

The Sun is a star and produces its own light. The planets and our moon are seen from Earth by reflected solar light.

Elliptical orbits

The planets, dwarf planets and comets orbit the Sun in an *ellipse*; the Sun is at one focus of the ellipse, not the centre, as is shown in Figure 6.1.6. For a comet with a highly elliptical path, the focus is not close to the centre of the ellipse. For the approximately circular paths of planets, the focus can be taken as the centre.

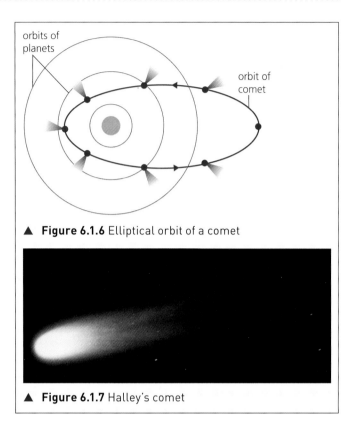

▲ **Figure 6.1.6** Elliptical orbit of a comet

▲ **Figure 6.1.7** Halley's comet

Origin of the Solar System

Our Sun is thought to have formed when gravitational attraction pulled together swirling clouds of hydrogen gas and dust (called nebulae) in a region of space where their density was high. The Solar System may well have been formed at the same time, about 4500 million years ago, with the planets being created from the disc of matter left over from the nebula that formed the Sun. As this material rotated around the Sun, gravitational attraction between small particles caused them to join together and grow in size in an accretion process. A rotating accretion disc is thought to have formed from which the planets emerged. Evidence for this **accretion model** of the formation of the Solar System is the approximate age of the Earth, which has been found by the radioactive dating of minerals in rocks (Topic 5.2). Further confirmation was provided by rocks brought back from the Moon, found to be 4500 million years old.

The view that the whole Solar System was formed at the same time from a rotating disc of accreting material is also supported by the fact that the orbits of the planets are more or less in the same plane and all revolve around the Sun in the same direction. Neither of these circumstances are likely to have happened by chance.

To account for the existence of heavier chemical elements in the Sun and inner planets, it is thought these might have come from an exploding supernova. During the lifetime of a star, atoms of hydrogen, helium and other light elements are fused into atoms of heavier elements. It is likely that matter from previous stellar explosions mixed with interstellar hydrogen before our Sun and the planets were formed, making our Sun a second-generation star. Evidence for the presence of many elements in interstellar clouds of gas and dust comes from their visible light spectra.

As the Sun grew in size, it become hotter. In the region of space where the inner planets were forming, the temperature would have been too high for light molecules such as hydrogen, helium, water and methane to exist in a solid state. As a result, the inner planets are built from materials with high melting temperatures such as metals (for example, iron) and silicates. Small nodules collided with each other to form larger bodies, and the planets began to grow. Since less than 1% of a nebula is composed of heavy elements, this means that Venus, Mercury, the Earth and Mars only grew to a small size and are solid and rocky.

Further away from the Sun, in the cooler regions of the Solar System, light molecules could exist in a solid icy form and, being more abundant than the heavy elements, the outer planets could grow to a size large enough to capture even the lightest element, hydrogen. These outer planets, Jupiter, Saturn, Uranus and Neptune, are large, gaseous and cold; together their mass constitutes 99% of the mass orbiting the Sun.

More than 99% of the total mass of the Solar System is concentrated in the Sun itself, so it exerts a very high gravitational attraction that keeps objects, such as the planets, orbiting around it.

Gravitational field strength of a planet

Newton proposed that all objects in the Universe having mass attract each other with a force called gravity. The greater the mass of each object and the smaller their distance apart the greater is the force: halving the distance quadruples the force.

For a planet the gravitational field strength at its surface depends on its mass, and is nearly uniform across its surface. The strength of the gravitational field decreases as the distance from the planet increases.

Travel times

The planets and their moons are visible from Earth only because they reflect light from the Sun. The outer regions of the Solar System are over 5000 million kilometres (5×10^{12} m) from the Sun and even light takes time to travel to such distant places.

? Worked example

Calculate the time t for light from the Sun to reach the Earth.

From the equation

average speed = total distance/total time

Rearrange the equation to give

total time = total distance/speed

If we take the distance of the Earth from the Sun to be 1.5×10^{11} m and the speed of light to be 3×10^8 m/s, then

$t = (1.5 \times 10^{11} \text{ m}) / (3 \times 10^8 \text{ m/s}) = 500 \text{ s (about 8 minutes)}$

Now put this into practice

1 Calculate the time t for light from the Sun to reach Mercury if the radius of the orbit of the planet is 5.8×10^{10} m.
2 Pluto is 5.9×10^{12} m distant from the Sun. How long does it take light to reach Pluto?

Test yourself

5 Name the planets in the Solar System in order of increasing distance from the Sun.
6 The planets orbit the Sun.
   Which of the following statements is *false*?
   A The Sun is a star which produces its own light.
   B Planets are seen by reflected solar light.
   C The planets revolve around the Earth in elliptical orbits.
   D Venus is the planet closest to the Earth.
7 The Sun exerts a large gravitational force on matter. State whether the following statements are *true* or *false*.
   A The Sun's gravitational attraction keeps the planets in orbit around it.
   B The force of gravity keeps the Moon in orbit around the Earth.
   C The Sun contains most of the mass of the Solar System.

8 A comet travels in an elliptical path around the Sun. Draw a sketch of the path of a comet and mark the position of the Sun on your sketch.

The planets

Fact and figures about the Sun, Earth, Moon and planets are listed in Tables 6.1.1 and 6.1.2. Times are given in Earth hours (h), days (d) or years (y).

As well as being of general interest, this data indicates (i) factors that affect conditions on the surface of the planets and (ii) some of the environmental problems that a visit or attempted colonisation would encounter.

▼ **Table 6.1.1** Data for the Sun, Earth and Moon

	Mass/kg	Radius/m	Density kg/m³	Surface gravity /N/kg	Orbital duration
Sun	2.0×10^{30}	7.0×10^8	1410	274	
Earth	6.0×10^{24}	6.4×10^6	5520	9.8	365 days (around Sun)
Moon	7.4×10^{22}	1.7×10^6	3340	1.7	27 days (around Earth)

▼ **Table 6.1.2** Data for the planets

Planet	Av distance from Sun /million km	Orbit time round sun /days or years	Surface temperature /°C	Density /kg/m³	Diameter /10^3 km	Mass /10^{24} kg	Surface gravity /N/kg	No. of moons
Mercury	57.9	88 d	350	5427	4.8	0.330	3.7	0
Venus	108.2	225 d	460	5243	12.1	4.87	8.9	0
Earth	149.6	365 d	20	5514	12.8	5.97	9.8	1
Mars	227.9	687 d	−23	3933	6.8	0.642	3.7	2
Jupiter	778.6	11.9 y	−120	1326	143	1898	23.1	79
Saturn	1433.5	29.5 y	−180	687	120	568	9.0	82
Uranus	2872.5	84 y	−210	1271	51	86.8	8.7	27
Neptune	4495.1	165 y	−220	1638	50	102	11.0	14

It can be seen from Table 6.1.2 that a planet's year (i.e. orbit time around the Sun) increases with distance from the Sun. The orbital speed, however, decreases with distance; Neptune travels much more slowly than Mercury. Surface temperatures decrease markedly with distance from the Sun with one exception. *Venus* has a high surface temperature (460°C) due to its dense atmosphere of carbon dioxide acting as a heat trap (i.e. the greenhouse effect). Its very slow 'spin time' of 243 Earth days means its day is longer than its year of 225 days!

Mercury, also with a slow 'spin time' (58.5 days), has practically no atmosphere and so while its noon temperature is 350°C, at night it falls to −170°C.

Mars is the Earth's nearest neighbour. It is colder than the Earth, temperatures on its equator are rarely greater than 0°C even in summer. Its atmosphere is very thin and consists mostly of carbon dioxide with traces of water vapour and oxygen. Its axis is tilted at an angle of 24° and so it has seasons, but these are longer than on Earth. There is no liquid water on the surface, but it has polar ice-caps of water, ice and solid carbon

dioxide. In some parts there are large extinct volcanoes and evidence such as gorges of torrential floods in the distant past (Figure 6.1.8). Mars is now a comparatively inactive planet although high winds do blow at times causing dust storms.

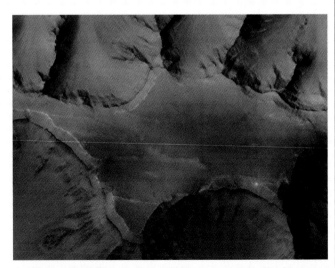

▲ **Figure 6.1.8** The surface of Mars as seen by the Mars *Global Surveyor*, showing a vast canyon 6000 km long and layered rocks indicating a geologically active history.

Jupiter is by far the largest planet in the Solar System. It is a gaseous planet and is noted for its Great Red Spot, Figure 6.1.9, which is a massive swirling storm that has been visible from Earth for over 200 years.

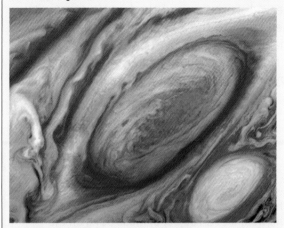

▲ **Figure 6.1.9** Jupiter's Great Red Spot pictured by the *Voyager* 2 space probe

Saturn has spectacular rings made up of ice particles (Figure 6.1.10) which are clearly visible through a telescope. Like Jupiter it has an ever-changing, very turbulent atmosphere of hydrogen, helium, ammonia and methane gas.

▲ **Figure 6.1.10** Saturn's rings as seen by the *Voyager* 1 probe

Uranus and *Neptune* are large, cold and windy planets with small rocky cores surrounded by an icy mass of water, methane and ammonia; they both have rings and many moons. Their atmospheres consist of methane, as well as hydrogen and helium. The four outer 'gas giants' are able to retain these lighter gases in their atmospheres, unlike the Earth, because of the greater gravitational attraction their large masses exert.

The dwarf planet *Pluto* is smaller than our Moon and has an atmosphere of frozen methane and nitrogen. It has five moons, the largest of which is Charon, which are thought to have formed in a collision between two dwarf planets early in the life of the Solar System. The reddish-brown cap around the north pole of Charon was found to contain organic molecules, which together with water, can form chemicals needed for the evolution of life.

Unmanned space missions such as *Voyager 1* and *2* obtained much information about the outer planets and their many moons on flybys. The *Cassini* spacecraft spent 13 years orbiting and collecting data on Saturn and its moons, before plunging into its atmosphere in 2017. Much of the information we have on the dwarf planet Pluto and its moons was obtained from the *New Horizons* spacecraft flyby in 2015; it also collected data on Jupiter's atmosphere, ring system and moons on its outward journey. Mars has been comprehensively mapped by the *Global Surveyor* satellite, while the robotic vehicles *Spirit* and *Opportunity*, which were landed on the planet and roved the surface for many years, obtained extensive data on Martian rocks. Currently the *Curiosity* rover is making its way across the Martian landscape sampling geological features. On 26 November 2018 the InSight spacecraft landed on Mars with a large package of scientific instruments. The seismometers have recorded many Mars quakes, providing information about the planet's interior. The rover *Perseverance*, due to land on Mars in February 2021, will search for signs of past microbial life. It carries the Mars helicopter, *Ingenuity*, designed to test if powered flight is possible for future human exploration of the planet.

Gravity and planetary motion

In Topic 1.5 we saw that to keep a body moving in a circular path requires a centripetal force acting towards the centre of the circle. In the case of the planets orbiting the Sun in near circular paths, it is the force of gravity between the Sun and the planet that provides the necessary centripetal force. The strength of the Sun's gravitational field decreases with distance so the further a planet is away from the Sun, the weaker the centripetal force; this results in a lower orbital speed and longer orbital duration.

In the case of a comet with a large elliptical orbit, its speed increases as it approaches the Sun and decreases as it moves further away. Energy is conserved, with some of the kinetic energy it has when close to the Sun being transferred into potential energy as it moves away.

The Moon is kept in a circular orbit around the Earth by the force of gravity between it and the Earth.

Test yourself

9 From the data given in Table 6.1.2 calculate
  a the circumference of the orbit of the planet Mars about the Sun
  b the speed of Mars in its orbit.

10 From the data given in Table 6.1.2 state
  a the surface gravity on Jupiter
  b the weight of an object of mass 50 kg on the surface of Jupiter.

11 Would you expect the orbital speed of Jupiter to be greater or less than that of Saturn? Explain your answer.

Revision checklist

After studying Topic 6.1 you should know and understand:

✔ that the Earth is a planet which orbits the Sun once in approximately 365 days and spins on its axis once in 24 hours
✔ that the Moon orbits the Earth once in approximately one month and how the motion of the Moon explains the phases of the Moon
✔ how the motion of the Earth explains day and night, the rising and setting of the Sun and the periodic nature of the seasons
✔ that the force that keeps an object in orbit around the Sun is due to the gravitational attraction of the Sun and that the Sun contains over 99% of the mass of the Solar System
✔ the differences between the inner and outer planets in terms of how the Solar System was formed

✔ that the strength of the Sun's gravitational field weakens, and orbital speeds of planets decrease, as the distance from the Sun increases.

After studying Topic 6.1 you should be able to:

✔ define and calculate average orbital speed from
$$v = \frac{2\pi r}{T}$$

✔ recall that planets, dwarf planets and comets orbit the Sun, while moons orbit planets
✔ recall that the four inner planets are rocky and small and the four outer planets are gaseous and large
✔ calculate the time taken for light to travel over a large distance, such as between objects in the Solar System

✔ interpret planetary data and use it to solve problems
✔ recall that the planets, dwarf planets and comets have elliptical orbits and that the Sun is not at the centre of the ellipse
✔ explain in terms of conservation of energy why the speed of an object in orbit is greater when it is closer to the Sun.

Exam-style questions

1 The motion of the Earth and Moon explain many natural events.
State which of the following statements is *not* true.
  A The Earth orbits the Sun once every 365 days.
  B The Moon orbits the Earth in approximately one month.
  C Day and night are due to the Earth spinning on its axis.
  D The Sun rises in the west and sets in the east in the Southern Hemisphere.
  [Total: 1]

2 The appearance of the Moon from Earth can be explained by the motion of the Moon and the Earth. Explain why
  a we never see the other side of the Moon   [2]
  b the Moon has phases   [3]
  c the Moon rises and sets.   [1]
  [Total: 6]

3 Calculate how long it takes light to travel from the Sun to Mars. Take the distance of Mars from the Sun to be 228×10^6 km and the speed of light to be 3×10^8 m/s.
  [Total: 3]

4 The inner and outer planets are very different in size and composition. In terms of their formation explain why
  a the inner planets are small and rocky   [4]
  b the outer planets are large and gaseous.   [4]
  [Total: 8]

5 a When a comet enters the Solar System from beyond Pluto and approaches the Sun its speed changes.
    i State how the strength of the Sun's gravitational field changes as the distance from the Sun increases.   [1]
    ii State how the speed of the comet changes as it approaches the Sun.   [1]
  b Venus is nearer to the Sun than the Earth.
    i Compare the orbital speed of Venus with that of the Earth. Explain your answer.   [2]
    ii Explain why Venus takes less than one Earth year to orbit the Sun.   [3]
  [Total: 7]

6 a Jupiter is further away from the Sun than is the Earth.
    i Compare the surface temperature of Jupiter with that of the Earth. Explain your answer.   [2]
    ii Compare the mass of Jupiter with that of the Earth.   [1]
    iii Compare the orbital speed of Jupiter with that of the Earth.   [1]
  b The orbital time of Mars round the Sun is 1.9 Earth years and the orbital path is 1.5 times longer than that of the Earth.
    i Calculate the ratio of the orbital speed of Mars to the orbital speed of the Earth.   [3]
    ii Compare the speed of Mars with that of the Earth.   [1]
  [Total: 8]

6.2 Stars and the Universe

6.2.1 The Sun as a star

FOCUS POINTS

★ Know that the Sun is a medium-sized star and describe its properties.

★ Know that stars are powered by nuclear reactions that release energy and that these reactions involve the fusion of hydrogen into helium in stable stars.

The Sun is our closest star and is so bright that it banishes the darkness of space with its light. The radiation it emits comes from glowing hydrogen and provides energy for all life on the Earth.

The hydrogen is heated by the energy released in nuclear reactions in the Sun's interior. In stable stars these reactions involve the conversion of hydrogen into helium by nuclear fusion. Stars have several stages in their life cycle and our Sun is currently in its stable phase which will last for a few more billion years.

The Sun

The Sun is a medium-sized star which consists mainly of hydrogen and helium. The radiant energy it emits is mostly in the infrared, visible and ultraviolet regions of the electromagnetic spectrum. This radiation is emitted from glowing hydrogen which is heated by the energy released in nuclear reactions within the Sun.

Nuclear reactions in stars

Stars are powered by nuclear reactions. Stable stars such as our Sun are hot and dense enough in their centre (core) for nuclear fusion (see Topic 5.1) of hydrogen into helium to occur. The high temperature in the core required to sustain the nuclear reactions is maintained by the large amount of energy released in the fusion process; the Sun is powered by nuclear fusion.

Some of the energy generated in the core is transferred to the outer layers of the star. These are cooler and less dense than the core, but are still hot enough for the hydrogen gas to glow, and emit electromagnetic radiation into space.

Stars vary in age, size, mass, surface temperature, colour and brightness. Colour and brightness both depend on surface temperature which in turn increases with the mass of the star. Stars that are white or blue are hotter and brighter (surface temperatures of 6000 to 25 000°C) than those that are red or yellow (surface temperatures 3000 to 6000°C).

> **Test yourself**
>
> 1 Name the main types of radiation emitted by the Sun.
>
> 2 How are stable stars powered?

6.2.2 Stars

FOCUS POINTS

★ Know that galaxies consist of billions of stars.
★ Know that the Sun is a star in the Milky Way and that other stars in the Milky Way are much further away from Earth than the Sun.
★ Know that light-years are used to measure astronomical distances, and understand what is meant by a light-year.

★ Use the fact that one light-year is equal to 9.5×10^{15} m.
★ Describe the life cycle of a star, including the terms protostar, stable star, red giant and supergiant, planetary nebula, white dwarf star, supernova, neutron star and black hole.

Our Sun lies in the Milky Way galaxy that contains billions of stars far enough away that the light from even the nearest takes 4 years to arrive. The most distant events, observable today through ever more powerful telescopes, occurred many millions of years ago.

Observations of events both near and far have enabled astronomers to build up a picture of how stars form from the clouds of dust and gas in interstellar space, how they grow and evolve over millions of years and how they eventually die, sometimes in a violent explosion such as a supernova, in which their constituents are again scattered into space. In this section you will learn about how stars are formed, how they are powered by nuclear fusion and the life cycle of some low and high mass stars.

The Sun is our closest star and is so bright that it banishes the darkness of space with its light.

When the Sun sets, due to the rotation of the Earth, the night sky is revealed. Away from light produced by humans, the night sky presents a magnificent sight with millions of visible stars.

The night sky

The night sky has been an object of wonder and study since the earliest times. On a practical level it provided our ancestors with a calendar, a clock and a compass. On a theoretical level it raised questions about the origin and nature of the Universe and its future. In the last 100 years or so, technological advances have enabled much progress to be made in finding the answers and making sense of what we see.

Although the stars may seem close on a dark night, in fact the distances involved are enormous compared with the distances across our Solar System. So great are they, that we need a new unit of length, the **light-year**. This is the distance travelled in (the vacuum of) space by light in one year.

The star nearest to the Solar System is Alpha Centauri, 4.31 light-years away, which means the light arriving from it at the Earth today left 4.3 years ago. The Pole Star is 142 light-years from Earth. The whole of the night sky we see is past history and we will have to wait a long time to find out what is happening there right now.

Light-years
One light-year is the distance travelled in (the vacuum of) space by light in one year and equals nearly 10 million million kilometres:

1 light-year = 9.5×10^{12} km = 9.5×10^{15} m

Galaxies

A **galaxy** is a large collection of stars; there are billions of stars in a galaxy.

As well as containing stars, galaxies consist of clouds of gas, mostly hydrogen and dust. They move in space, many rotating as spiral discs like huge Catherine wheels with a dense central bulge. The **Milky Way**, the spiral galaxy to which our Solar System belongs, can be seen on dark nights as a narrow band of light spread across the sky, Figure 6.2.1a.

Key definitions

Light-year used to measure astronomical distances; the distance travelled in (the vacuum of) space by light in one year

Galaxy made up of many billions of stars

Milky Way the spiral galaxy to which our Solar System belongs. The Sun is a star in this galaxy and other stars that make up this galaxy are much further away from the Earth than the Sun is from the Earth.

▲ **Figure 6.2.1a** The Milky Way from Earth

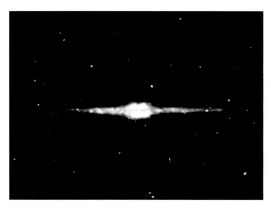

▲ **Figure 6.2.1b** An image of the Milky Way from a space probe (the Cosmic Background Explorer)

We are near the outer edge of the galaxy, so what we see when we look at the Milky Way is the galaxy's central bulge. Figure 6.2.1b is an infrared photograph of the galaxy taken from beyond the Earth's atmosphere, clearly showing the central bulge.

Galaxies vary considerably in size and number of stars. The diameter of a typical galaxy is 30 000 light-years, with a distance to its neighbours of around 3 million light-years.

Galaxies travel in groups. The nearest spiral in our local cluster is the Andromeda galaxy; it is around 2.5 million light-years away but is visible to the naked eye.

> **Test yourself**
>
> 3   The Universe is composed of stars and galaxies. Which of the following statements are *not true*?
>   A   The Universe is a collection of galaxies.
>   B   There are billions of stars in a galaxy.
>   C   The Sun is a star.
>   D   Our Solar System belongs to the Andromeda galaxy.
> 4   How long does it take light to reach the Earth from a galaxy 10 million light-years distant from the Earth? Choose your answer from the following times.
>   A   10 years
>   B   30 years
>   C   300 thousand years
>   D   10 million years

Origin of stars

When interstellar clouds of dust and gas containing hydrogen collapse under the force of gravitational attraction, a *protostar* is formed. As the mass of the protostar increases, its core temperature rises; gravitational potential energy is transferred to kinetic energy as the protostar contracts under internal gravitational forces. When the core is hot enough, nuclear fusion (see Topic 5.1) can start; hydrogen is converted into helium, and a star is born. If the young star has a very large mass, it forms a blue or white star. If it has a smaller mass, such as our Sun, it forms a yellow or red dwarf and this is more common.

Life cycle of stars

In a *stable star,* such as our Sun, the very strong forces of gravity pulling it inwards are balanced by the opposing forces trying to make it expand due to its extremely high temperature. This thermal pressure arises from the kinetic energy of the nuclei in the core. When the forces are balanced, the star is in a stable state which may last for up to 10 000 million years (i.e. 10^{10} = 10 billion years), during which time most of the hydrogen in the core is converted to helium.

When the star starts to run out of hydrogen as a fuel for nuclear reactions, it becomes unstable. There is less energy being produced by nuclear fusion to sustain the outward thermal pressure and the core collapses inward under gravitational attraction; potential energy is transferred to kinetic energy, so that the core becomes hotter. There is a fast burn-up of the remaining hydrogen envelope (in contrast to the core) and a huge expansion and subsequent cooling of the surface gases; the star turns into a *red giant* (or red supergiant if the star is very massive). As the core heats up, its temperature becomes high enough for the nuclear fusion of helium into carbon to occur. This will happen to our Sun (which is halfway through its stable phase) in about 5000 million years' time from now. The inner planet Mercury will then be engulfed and life on Earth destroyed.

The further stages in the life cycle of a star depend on its mass (Figure 6.2.2).

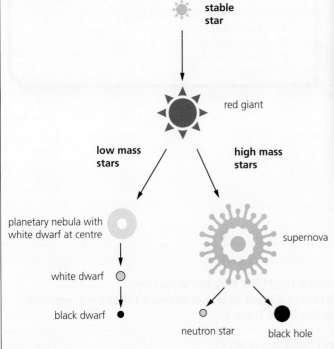

▲ **Figure 6.2.2** Life cycle of stars

Low mass stars

For stars with a mass up to about eight times that of our Sun, when all the helium is used up, the core of the red giant collapses under its own gravity, and enough energy is released to cause the outer layers to be expelled. The small core becomes a *white dwarf* at the centre of a glowing shell of ionised gas known as a *planetary nebula*. Planetary nebulae are thought to play an important role in distributing elements formed in the star into the interstellar medium. They have been photographed from the Hubble Space telescope, but have a short lifetime of about 10 000 years and appear to mark the transition of the red giant into a white dwarf. The white dwarf has a lifetime of about a billion years and eventually cools into a cold *black dwarf* consisting mainly of carbon. This will be the fate of our Sun.

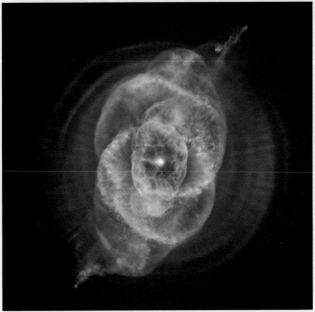

▲ **Figure 6.2.3** White dwarf at centre of a planetary nebula

High mass stars

Stars with a mass greater than about eight times that of our Sun are hotter than the Sun and use up their hydrogen more quickly; their stable stage is shorter and may last for only about 100 million years. The core then collapses into a *red supergiant* and nuclear fusion of helium into carbon occurs. When all the helium has been used up, the core collapses further under gravity and it becomes hot enough for the nuclear fusion of carbon into oxygen, nitrogen and finally iron to occur. Nuclear fusion then stops and the energy of the star is released in a *supernova* explosion.

In the explosion there is a huge increase in the star's brightness and the temperature becomes high enough for fusion of nuclei into many elements heavier than iron to occur. Material, including these heavy elements, is thrown into space as a nebula, and becomes available for the formation of new stars and their associated planetary systems.

The Crab Nebula is the remains of the supernova seen by Chinese astronomers on Earth in 1054. It is visible through a telescope as a hazy glow in the constellation Taurus. Figure 6.2.4 is an image of the Crab Nebula taken by the Hubble Space Telescope.

The centre of the supernova collapses to a very dense *neutron star*, which spins rapidly and acts as a *pulsar*, sending out pulses of radio waves. If the red giant is very massive, the remnant at the centre of the supernova has such a large density that its gravitational field stops anything escaping from its surface, even light; this is a *black hole*. In a black hole matter is packed so densely that the mass of the Earth would only occupy the volume of one cubic centimetre! Since neither matter nor radiation can escape from a black hole, we cannot see it directly. However, if nearby material, such as gas from a neighbouring star, falls towards a black hole, intense X-ray radiation may be emitted which alerts us to its presence. Objects believed to be massive black holes have been identified at the centre of many spiral galaxies.

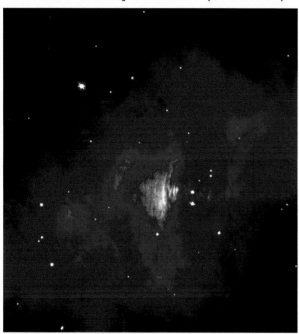

▲ **Figure 6.2.4** The Crab Nebula

> ### Test yourself
>
> 5 When a low mass red giant has consumed most of the helium in its core it may turn into a
>   A yellow dwarf
>   B white dwarf
>   C supernova
>   D neutron star.
> 6 When a high mass red giant has consumed most of the helium in its core it may become a
>   A white dwarf
>   B yellow dwarf
>   C planetary nebula
>   D neutron star.

6.2.3 The Universe

FOCUS POINTS

★ Know that the Universe is made up of billions of galaxies, the Milky Way being just one.
★ Describe redshift and know that light emitted from stars in distant galaxies appears redshifted, which is evidence the Universe is expanding and supports the Big Bang Theory.

★ Describe and explain cosmic microwave background radiation (CMBR).
★ Know and use various equations to calculate the speed at which a galaxy is moving away from Earth and the distance of a far galaxy.
★ Use the Hubble constant in calculations, including estimating the age of the Universe.

Stars cluster together in galaxies. A typical galaxy has a diameter of around 30 000 light-years with neighbouring galaxies around 3 million light-years away. The Universe is composed of billions of galaxies that all appear to be receding from the Earth so that the light from glowing hydrogen takes on a redder colour. Redshift measurements of the light from distant galaxies suggest that the Universe is expanding and support the Big Bang theory of the origin of the Universe.

Cosmic microwave background radiation (CMBR) pervades all of space. It is thought to have been produced in the early stages of the formation of the Universe and has been redshifted into the microwave region as the Universe expanded. Hubble's law, which relates the speed of recession of a distant galaxy to its distance away, and the evaluation of the Hubble constant enable the age of the Universe to be estimated as 14 billion years.

The Universe

The Milky Way is one of the millions of galaxies that make up the Universe. The diameter of the Milky Way galaxy is around 100 000 light-years and it contains 800 billion or more stars.

The expanding Universe

In developing a theory about the origin of the Universe, two discoveries about galaxies have to be taken into account. The first is that light emitted from glowing hydrogen in stars in distant galaxies, is 'shifted' to the red end of the spectrum (longer wavelength) in comparison with the value on Earth. The second is that the further away a galaxy is from us, the greater is this **redshift**. These observations can be explained if other galaxies are moving away from us very rapidly, and the further away they are, the faster is their speed of recession. Evidently the Universe is expanding.

This interpretation is based on the Doppler effect, which occurs when a source of waves is moving. If the source approaches us, waves are crowded into a smaller space and their wavelength seems smaller and their frequency greater. If the source moves away, the wavelength seems larger (Figure 6.2.5). Receding stars and galaxies are therefore shifted to the red end of the spectrum.

The Doppler effect occurs with sound waves and explains the rise and fall of pitch of a siren as the vehicle approaches and passes us. The same effect is shown by light. When the light source is receding, the wavelength seems longer, that is the light is redder. From the size of the redshift of starlight, the speed of recession of the galaxy can be calculated; the most distant ones visible are receding with speeds up to one-third of the speed of light.

The redshift in the light from distant galaxies provides evidence that the Universe is expanding and gives support to the Big Bang theory of the formation of the Universe.

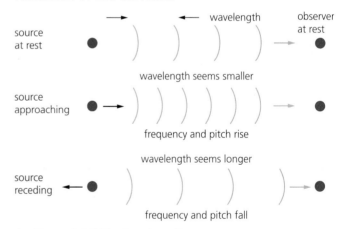

▲ **Figure 6.2.5** The Doppler effect

The Big Bang theory

If the galaxies are receding from each other, it follows that in the past they must have been closer together. It is therefore possible that initially all the matter in the Universe was packed together in an extremely dense state. The Big Bang theory proposes that this was the case, and that the Universe started about 14 billion years ago from one place with a huge explosion – the Big Bang.

The resulting expansion of the Universe continues today, but predictions vary about what will happen in the future. The critical factor appears to be the density of matter in the Universe. This is difficult to calculate because scientists currently believe that as much as 80% of the material in the Universe is invisible, since it does not emit radiation. It may be that this hidden mass exerts sufficient gravitational force to lead the Universe to collapse. The gravitational force between masses not only determines the motion and evolution of planets, stars and galaxies, but will also control the ultimate fate of the Universe.

Microwave background radiation

The Big Bang produced radiation energy which still exists in the universe today in the form of cosmic microwave background radiation (CMBR) of a specific frequency. This was first identified in 1965 by Arno Penzias and Robert Wilson as a background hiss in microwave signals picked up by large radio antennae. It fills the whole Universe with an intensity that is nearly the same in all directions. It was first mapped by NASA's Cosmic Background Explorer (COBE) satellite launched in 1989. More detailed measurements have been obtained from the Wilkinson Microwave Anisotropy Probe (WMAP) launched in 2001, and the European Space Agency's Planck telescope launched in 2010 (Figure 6.2.6).

▲ **Figure 6.2.6** Cosmic microwave background radiation recorded by the Planck telescope

This radiation, left over from the Big Bang, has maximum intensity at a wavelength of 1.1 mm and it has been suggested that slight variations in this value, 'hot' and 'cold' spots, may act as seeding sites for galaxy formation. Since the Big Bang, the Universe has continued to expand, resulting in a redshift of the cosmic background radiation into the microwave region of the electromagnetic spectrum.

The existence of the cosmic background radiation provides strong evidence in support of the Big Bang theory, and gives us an insight into the earliest days of the Universe.

Age of the Universe

Edwin Hubble, an American astronomer working at the Mount Wilson Observatory in California in the 1920s, discovered that the speed of recession v of a galaxy is directly proportional to its distance away d. This is called *Hubble's law* and can be written

$$v = H_0 d$$

The **Hubble constant**, H_0, is defined as the ratio of the speed at which the galaxy is moving away from the Earth to its distance from the Earth, so

$$H_0 = \frac{v}{d}$$

> **Key definitions**
>
> **Hubble constant H_0** the ratio of the speed at which the galaxy is moving away from the Earth to its distance from the Earth
>
> $$H_0 = \frac{v}{d}$$

The Hubble constant represents the rate at which the Universe is expanding at the present time. Its value is found by measuring the speed of recession of large numbers of galaxies whose distances are known. Redshifts are used to find the speed of recession of a galaxy and its distance can be calculated from brightness measurements of a supernova in the same galaxy. The maximum brightness of a particular type of supernova can be calculated (from its mass) and compared with its measured (apparent) value. The apparent brightness decreases as the inverse square of our distance from the supernova, enabling the distance of the galaxy to be determined. Considerable difficulties are involved, but H_0 is estimated to be 2.2×10^{-18} per second. This means that a galaxy 1 million light-years away is receding from us at a speed of

$$2.2 \times 10^{-18} \times 10^6 = 2.2 \times 10^{-12} \text{ light-years/s}$$
$$= 2.2 \times 10^{-12} \times 9.5 \times 10^{12} \text{ km/s}$$
$$= 21 \text{ km/s}$$

The more distant a galaxy is from us, the faster it is receding. Hubble's law provides further evidence in support of the Big Bang theory.

The age of the Universe can be shown to be equal to $\dfrac{1}{H_0}$ and calculated very roughly from the equation $\dfrac{d}{v} = \dfrac{1}{H_0}$ in terms of the distance d and speed of recession v of distant galaxies.

We assume that the time for which two galaxies are close together (when they were formed at the Big Bang) is negligible, compared with their present age. If so, their age is approximately the age of the Universe, which equals the time since expansion began.

For the two galaxies, assuming the speed of recession has not changed, we have

$$\text{age} = \text{distance apart/speed of recession} = \frac{d}{v}$$

From Hubble's law, it follows that

$$\text{age} = \frac{1}{H_0} \approx \frac{1}{2.2 \times 10^{-18}} = 4.5 \times 10^{17}\,\text{s}$$

But 1 year $\approx 3.2 \times 10^7$ s, so the age of the Universe

$$\approx \frac{4.5 \times 10^{17}\,\text{s}}{3.2 \times 10^7\,\text{s/year}}$$

$$\approx 1.4 \times 10^{10}\ \text{years}$$

$$\approx 14\ \text{billion years}$$

▶ Test yourself

7 Explain the evidence in favour of the Big Bang theory of the origin of the Universe.

8 Redshift measurements from a distant galaxy show that it is moving away from the Earth at a speed of 8500 km/s. Assuming $H_0 = 2.2 \times 10^{-18}$ per second, how far away is the galaxy from the Earth?

 A 200 million light years
 B 400 million light years
 C 600 million light years
 D 800 million light years

Revision checklist

After studying Topic 6.2 you should know and understand:

✔ that the Sun is a star, consists of mostly hydrogen and helium and emits radiant energy from glowing hydrogen

✔ that stable stars are powered by the release of energy during nuclear fusion of hydrogen into helium

✔ that the Sun is one of the billion stars in our galaxy, the Milky Way, and that other stars in the Milky Way are much further away from Earth than the Sun

✔ the term light-year and relate it to star distances

✔ that a galaxy is a large collection of stars, that the Milky Way is one of many billions of galaxies making up the Universe and that the diameter of the Milky Way galaxy is around 100 000 light-years

✔ that light from glowing hydrogen in stars in all distant galaxies is redshifted in comparison with light from glowing hydrogen on Earth and this is evidence of an expanding Universe and the Big Bang theory.

After studying Topic 6.2 you should be able to:

✔ recall that a protostar becomes a stable star when the inward gravitational force is balanced by the outward force due to the high temperature in the centre of the star

✔ describe how stars are thought to originate and outline the life cycle of low and high mass stars

✔ recall that the speed at which a galaxy is receding can be found from the change in wavelength of starlight due to redshift

✔ recall the origin and properties of cosmic microwave background radiation and know that it has been redshifted into the microwave region of the electromagnetic spectrum as the Universe expanded

✔ recall the evidence in favour of the Big Bang theory

✔ recall how the distance d of a galaxy can be estimated

✔ define the Hubble constant H_0 and recall that $\dfrac{1}{H_0}$ represents the estimated age of the Universe.

Exam-style questions

1 **a** Explain what is meant by the redshift of starlight. [3]

  b Explain what the redshift of starlight tells us about the motion of distant galaxies. [2]
  [Total: 5]

2 State which of the following provides evidence in support of the Big Bang theory.

  A Gravitational attraction

  B Supernova explosions

  C Redshift of starlight from distant galaxies

  D Fusion of hydrogen into helium
  [Total: 1]

3 **a** Describe how a stable star is formed and how it is powered. [5]

  b When does a stable star turn into a red giant? [2]
  [Total: 7]

4 **a** State the forces, and give their direction, which are balanced when a star is in a stable state. [4]

  b Write down the sequence of stages in the life cycle of a star like the Sun. [5]
  [Total: 9]

5 **a** Calculate Hubble's constant if a galaxy 500 million light-years from Earth is receding at 11 000 km/s. [5]

  b Explain the significance of $\dfrac{1}{H_0}$. [2]
  [Total: 7]

Mathematics for physics

Use this section for reference

Solving physics problems

When tackling physics problems using mathematical equations it is suggested that you *do not substitute numerical values until you have obtained the expression in symbols which gives the answer.* That is, work in symbols until you have solved the problem and only then insert the numbers in the expression to get the final result.

This has two advantages. First, it reduces the chance of errors in the arithmetic (and in copying down). Second, you write less since a symbol is usually a single letter whereas a numerical value is often a string of figures.

Adopting this 'symbolic' procedure frequently requires you to change round an equation first. The next two sections and the questions that follow them are intended to give you practice in doing this and then substituting numerical values to get the answer.

Equations – type 1

In the equation $x = a/b$, the subject is x. To change it we *multiply or divide both* sides of the equation by the same quantity.

To change the subject to a, we have

$$x = \frac{a}{b}$$

If we multiply both sides by b, the equation will still be true.

$$\therefore x \times b = \frac{a}{b} \times b$$

The bs on the right-hand side cancel

$$\therefore b \times x = \frac{a}{b} \times b = a$$

and

$$a = b \times x$$

To change the subject to b, we have

$$x = \frac{a}{b}$$

Multiplying both sides by b as before, we get

$$a = b \times x$$

Dividing both sides by x:

$$\frac{a}{x} = \frac{b \times x}{x} = b$$

$$\therefore b = \frac{a}{x}$$

Note that the *reciprocal* of x is $1/x$.

Can you show that

$$\frac{1}{x} = \frac{b}{a}?$$

Now try the following questions using these ideas.

> ## Questions
>
> 1 What is the value of x if
>
>   a $\quad 2x = 6$ b $\quad 3x = 15$ c $\quad 3x = 8$
>
>   d $\quad \dfrac{x}{2} = 10$ e $\quad \dfrac{x}{3} = 4$ f $\quad \dfrac{2x}{3} = 4$
>
>   g $\quad \dfrac{4}{x} = 2$ h $\quad \dfrac{9}{x} = 3$ i $\quad \dfrac{x}{6} = \dfrac{4}{3}$
>
> 2 Change the subject to
>
>   a $\quad f$ in $v = f\lambda$ b $\quad \lambda$ in $v = f\lambda$
>
>   c $\quad I$ in $V = IR$ d $\quad R$ in $V = IR$
>
>   e $\quad m$ in $d = \dfrac{m}{V}$ f $\quad V$ in $d = \dfrac{m}{V}$
>
>   g $\quad s$ in $v = \dfrac{s}{t}$ h $\quad t$ in $v = \dfrac{s}{t}$
>
> 3 Change the subject to
>
>   a $\quad I^2$ in $P = I^2R$ b $\quad I$ in $P = I^2R$
>
>   c $\quad a$ in $s = \dfrac{1}{2}at^2$ d $\quad t^2$ in $s = \dfrac{1}{2}at^2$
>
>   e $\quad t$ in $s = \dfrac{1}{2}at^2$ f $\quad v$ in $\dfrac{1}{2}mv^2 = mgh$
>
>   g $\quad y$ in $\lambda = \dfrac{ay}{D}$ h $\quad p$ in $R = \dfrac{pl}{A}$

4 By replacing (substituting) find the value of $v = f\lambda$ if
  a $f = 5$ and $\lambda = 2$ **b** $f = 3.4$ and $\lambda = 10$
  c $f = 1/4$ and $\lambda = 8/3$ **d** $f = 3/5$ and $\lambda = 1/6$
  e $f = 100$ and $\lambda = 0.1$ **f** $f = 3 \times 10^5$ and $\lambda = 10^3$

5 By changing the subject and replacing find
  a f in $v = f\lambda$, if $v = 3.0 \times 10^8$ and $\lambda = 1.5 \times 10^3$
  b h in $p = 10hd$, if $p = 10^5$ and $d = 10^3$
  c a in $n = a/b$, if $n = 4/3$ and $b = 6$
  d b in $n = a/b$, if $n = 1.5$ and $a = 3.0 \times 10^8$
  e F in $p = F/A$ if $p = 100$ and $A = 0.2$
  f s in $v = s/t$, if $v = 1500$ and $t = 0.2$

Equations – type 2

To change the subject in the equation $x = a + by$ *we add or subtract the same quantity from each side.* We may also have to divide or multiply as in type 1. Suppose we wish to change the subject to y in

$$x = a + by$$

Subtracting a from both sides,

$$x - a = a + by - a = by$$

Dividing both sides by b,

$$\frac{x-a}{b} = \frac{by}{b} = y$$

$$\therefore \; y = \frac{x-a}{b}$$

Questions

6 What is the value of x if

  a $x + 1 = 5$ **b** $2x + 3 = 7$ **c** $x - 2 = 3$

  d $2(x - 3) = 10$ **e** $\dfrac{x}{2} - \dfrac{1}{3} = 0$ **f** $\dfrac{x}{3} + \dfrac{1}{4} = 0$

  g $2x + \dfrac{5}{3} = 6$ **h** $7 - \dfrac{x}{4} = 11$ **i** $\dfrac{3}{x} + 2 = 5$

7 By changing the subject and replacing, find the value of a in $v = u + at$ if
  a $v = 20$, $u = 10$ and $t = 2$
  b $v = 50$, $u = 20$ and $t = 0.5$
  c $v = 5/0.2$, $u = 2/0.2$ and $t = 0.2$

8 Change the subject in $v^2 = u^2 + 2as$ to a.

Proportion (or variation)

One of the most important mathematical operations in physics is finding the relation between two sets of measurements.

Direct proportion

Suppose that in an experiment two sets of readings are obtained for the quantities x and y as in Table M1 (units omitted).

▼ **Table M1**

x	1	2	3	4
y	2	4	6	8

We see that when x is doubled, y doubles; when x is trebled, y trebles; when x is halved, y halves; and so on. There is a one-to-one correspondence between each value of x and the corresponding value of y.

  We say that y is *directly proportional* to x, or y *varies directly* as x. In symbols

$$y \propto x$$

Also, the ratio of one to the other, e.g. y to x, is always the same, i.e. it has a constant value which in this case is 2. Hence

$$\frac{y}{x} = \text{a constant} = 2$$

The constant, called the *constant of proportionality* or *constant of variation*, is given a symbol, e.g. k, and the relation (or law) between y and x is then summed up by the equation

$$\frac{y}{x} = k \;\text{ or }\; y = kx$$

Notes

1 In practice, because of inevitable experimental errors, the readings seldom show the relation so clearly as here.
2 If instead of using numerical values for x and y we use letters, e.g. x_1, x_2, x_3, etc., and y_1, y_2, y_3, etc., then we can also say

$$\frac{y_1}{x_1} = \frac{y_2}{x_2} = \frac{y_3}{x_3} = \ldots = k$$

or

$$y_1 = kx_1, \; y_2 = kx_2, \; y_3 = kx_3, \; \ldots$$

Inverse proportion

Two sets of readings for the quantities p and V are given in Table M2 (units omitted).

▼ **Table M2**

p	3	4	6	12
V	4	3	2	1

There is again a one-to-one correspondence between each value of p and the corresponding value of V, but when p is doubled, V is halved, when p is trebled, V has one-third its previous value, and so on.

We say that V is *inversely proportional* to p, or V *varies inversely* as p, i.e.

$$V \propto \frac{1}{p}$$

Also, the *product* $p \times V$ is always the same and we write

$$V = \frac{k}{p} \text{ or } pV = k$$

where k is the constant of proportionality or variation and equals 12 in this case.

Using letters for values of p and V we can also say

$$p_1 V_1 = p_2 V_2 = p_3 V_3 = \dots = k$$

Graphs

Another useful way of finding the relation between two quantities is by a graph.

Straight-line graphs

When the readings in Table M1 are used to plot a graph of y against x, a *continuous* line joining the points is a *straight line passing through the origin* as in Figure M1. Such a graph shows there is direct proportionality between the quantities plotted, i.e. $y \propto x$. But note that the line must go through the origin.

A graph of p against V using the readings in Table M2 is a curve, as in Figure M2. However, if we plot p against $1/V$ (Table M3) (or V against $1/p$) we get a straight line through the origin, showing that $p \propto 1/V$, as in Figure M3 (or $V \propto 1/p$).

▼ **Table M3**

p	V	$1/V$
3	4	0.25
4	3	0.33
6	2	0.50
12	1	1.00

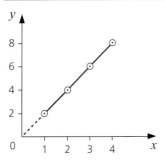

▲ **Figure M1**

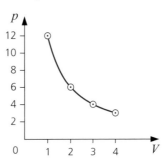

▲ **Figure M2**

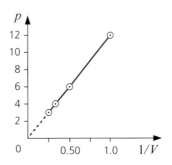

▲ **Figure M3**

Gradient or slope

The gradient or slope of a straight-line graph equals the constant of proportionality. In Figure M1, the slope is $y/x = 2$; in Figure M3 it is $p/(1/V) = 12$.

In practice, points plotted from actual measurements may not lie exactly on a straight line due to experimental errors. The best straight line is then drawn through them so that they are equally distributed about it. This automatically averages the results. Any points that are well off the line stand out and may be investigated further.

Variables

As we have seen, graphs are used to show the relationship between two physical quantities. In an experiment to investigate how potential difference, V, varies with the current, I, a graph can be drawn of V/V values plotted against the values of I/A. This will reveal how the potential difference depends upon the current (see Figure M4).

In the experiment there are two *variables*. The quantity I is varied and the value for V is dependent upon the value for I. So V is called the **dependent variable** and I is called the **independent variable**.

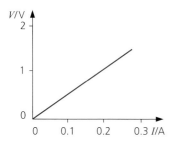

▲ Figure M4

Note that in Figure M4 each axis is labelled with the quantity *and* the unit. Also note that there is a scale along each axis. The statement V/V *against* I/A means that V/V, the dependent variable, is plotted along the y-axis and the independent variable I is plotted along the x-axis (see Figure M5).

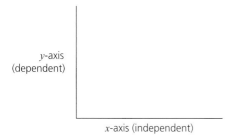

▲ Figure M5

Practical points

i   The axes should be labelled giving the quantities being plotted and their units, e.g. I/A meaning current in amperes.

ii   If possible, the origin of both scales should be on the paper and the scales chosen so that the points are spread out along the graph. It is good practice to draw a large graph.

iii The scale should be easy to use. A scale based on multiples of 10 or 5 is ideal. Do not use a scale based on a multiple of 3; such scales are very difficult to use.

iv   Mark the points ⊙ or ×.

▶ Questions

9 In an experiment different masses were hung from the end of a spring held in a stand and the extensions produced were as shown below.

Mass/g	100	150	200	300	350	500	600
Extension/cm	1.9	3.1	4.0	6.1	6.9	10.0	12.2

  a Plot a graph of extension along the vertical (y) axis against mass along the horizontal (x) axis.

  b What is the relation between extension and mass? Give a reason for your answer.

10 Pairs of readings of the quantities m and v are given below.

m	0.25	1.5	2.5	3.5
v	20	40	56	72

  a Plot a graph of m along the vertical axis and v along the horizontal axis.

  b Is m directly proportional to v? Explain your answer.

  c Use the graph to find v when $m = 1$.

11 The distances s (in metres) travelled by a car at various times t (in seconds) are shown below.

s/m	0	2	8	18	32	50
t/s	0	1	2	3	4	5

Draw graphs of
a s against t
b s against t^2.
What can you conclude?

Pythagoras' theorem

The square of the hypotenuse of a right-angled triangle (see Figure M6) is equal to the sum of the squares of the other two sides: $c^2 = a^2 + b^2$. This is a useful equation for determining the unknown length of one side of a right-angled triangle. Also the angle θ between sides b and c is given by: $\sin \theta = a/c$ or $\cos \theta = b/c$ or $\tan \theta = a/b$.

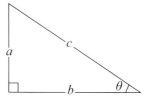

▲ Figure M6

Additional exam-style questions

1 Motion, forces and energy

Physical quantities and measurement techniques, Motion, Mass and weight and Density

1 Which are the basic SI units of mass, length and time?
   A kilogram, kilometre, second
   B gram, centimetre, minute
   C kilogram, centimetre, second
   D kilogram, metre, second

   [Total: 1]

2 a A girl cycles along a level road for a distance of 1.6 km in 8 minutes. Calculate her average speed. [3]
   b The road then goes downhill and the girl begins to accelerate. What happens to her average speed? [1]

   [Total: 4]

3 An astronaut has a mass of 90 kg on Earth. The gravitational force on an object on the Moon is 1/6th of the value on the Earth. Taking $g = 9.8\,\text{m/s}^2$ on the Earth, state
   a the weight of the astronaut on the Earth [3]
   b the mass of the astronaut on the Moon [1]
   c the weight of the astronaut on the Moon. [2]

   [Total: 6]

4 Density can be calculated from which of the following expressions?
   A mass/volume
   B mass × volume
   C volume/mass
   D weight/area

   [Total: 1]

5 a The smallest division marked on a metre ruler is 1 mm. A student measures a length with the ruler and records it as 0.835 m. Is he justified in giving three significant figures? [1]
   b State the SI unit of density. [2]
   c The mass of an object is 120 g and its volume is 15 cm³. Calculate the density of the object. [3]

   [Total: 6]

Forces, Momentum, Energy, work and power and Pressure

6 A 3.1 kg mass falls with its terminal velocity. Which of the combinations **A** to **D** gives its weight, the air resistance and the resultant force acting on it?

	Weight	Air resistance	Resultant force
A	0.3 N down	zero	zero
B	3 N down	3 N up	3 N up
C	10 N down	10 N up	10 N down
D	30 N down	30 N up	zero

[Total: 1]

7 A boy whirls a ball at the end of a string round his head in a horizontal circle, centre O. He lets go of the string when the ball is at X in the diagram. In which direction does the ball fly off?
   A 1   B 2   C 3   D 4

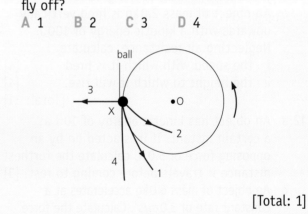

[Total: 1]

8 a Write down an expression for the mechanical work done by a force. [2]
   b Give the units of work. [1]
   c A force of 3.0 N is used to move a box a distance of 5.0 m. Calculate the work done by the force. [2]
   d How much energy is transferred? [1]

   [Total: 6]

9 a State how energy is transferred from the following devices:
  i an electric lamp [1]
  ii a battery [1]
  iii an electric motor [1]
  iv the generator in a power station. [1]
 b The efficiency of a certain coal power station is 30%.
  Explain the meaning of efficiency. [2]
  [Total: 6]

10 Which one of the following statements is *not* true?
  A Pressure is the force acting on unit area.
  B Pressure is calculated from force/area.
  C The SI unit of pressure is the pascal (Pa) which equals 1 newton per square metre ($1\,N/m^2$).
  D The greater the area over which a force acts the greater is the pressure.
  [Total: 1]

11 a A stone of mass 2.0 kg is dropped from a height of 4.0 m. Neglecting air resistance, calculate the kinetic energy of the stone just before it reaches the ground. [3]
 b An object of mass 2.0 kg is fired vertically upwards with a kinetic energy of 100 J. Neglecting air resistance, calculate
  i the speed with which it is fired [4]
  ii the height to which it will rise. [4]
  [Total: 11]

12 a An object has kinetic energy of 10 J at a certain instant. If it is acted on by an opposing force of 5.0 N, calculate the furthest distance it travels before coming to rest. [3]
 b An object of mass 6.0 kg accelerates at a constant rate of $3.0\,m/s^2$. Calculate the force acting on it in the direction of acceleration. [3]
 c An object is brought to rest by a force of 18 N acting on it for 3.0 s. Calculate
  i the impulse of the force [2]
  ii the change of momentum of the object. [1]
  [Total: 9]

2 Thermal physics

13 a Explain the action and use of a bimetallic strip. [4]
 b i State the relationship between pressure and volume for a fixed mass of gas at constant temperature. [1]
  ii If the volume of a fixed mass of gas at constant temperature doubles, what happens to the pressure of the gas? [2]
 c Explain in terms of the molecular kinetic theory how the pressure of a gas is affected by a rise in temperature if the volume remains constant. [3]
  [Total: 10]

14 Explain why
 a in cold weather the wooden handle of a saucepan feels warmer than the metal pan [2]
 b convection occurs when there is a change of density in parts of a fluid [4]
 c conduction and convection cannot occur in a vacuum. [2]
  [Total: 8]

3 Waves

15 a The wave travelling along the spring in the diagram is produced by someone moving end X of the spring back and forth in the directions shown by the arrows.
  i Is the wave longitudinal or transverse? [1]
  ii What is the region called where the coils of the spring are closer together than normal? [1]
  ii What is the region called where the coils of the spring are further apart than normal? [1]

 b When the straight water waves in the diagram pass through the narrow gap in the barrier they are diffracted. What changes (if any) occur in
  i the shape of the waves [1]
  ii the speed of the waves [1]
  iii the wavelength? [1]
  [Total: 6]

16 In the diagram a ray of light is shown reflected at a plane mirror.
   a State the angle of incidence. [1]
   b State the angle the reflected ray makes with the mirror. [1]
   c List the characteristics of the image formed in a plane mirror. [5]
   [Total: 7]

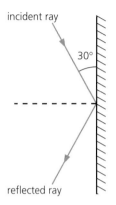

incident ray
30°
reflected ray

17 In the diagram, which of the rays **A** to **D** is most likely to represent the ray emerging from the parallel-sided sheet of glass?

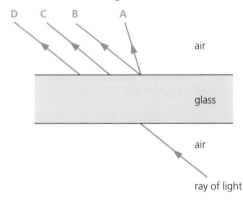

D C B A
air
glass
air
ray of light

[Total: 1]

18 A narrow beam of white light is shown passing through a glass prism and forming a spectrum on a screen.
   a Name the effect. [1]
   b Which colour of light appears at
      i A [1]
      ii B? [1]

c State the seven colours of the visible spectrum in order of increasing wavelength. [3]

white light
prism
screen
A
B

[Total: 6]

19 a Explain what is meant by the terms
      i principal axis of a lens [2]
      ii principal focus of a converging lens. [2]
   b A magnifying glass is used to view a small object.
      i   How far from the lens should the object be? [1]
      ii Is the image upright or inverted? [1]
      iii Is the image real or virtual? [1]
   [Total: 7]

20 The diagram below shows the complete electromagnetic spectrum.

radio waves	microwaves	A	visible light	ultraviolet	B	γ-rays

   a Name the radiation found at
      i A [1]
      ii B. [1]
   b State which of the radiations marked on the diagram would have
      i the lowest frequency [1]
      ii the shortest wavelength. [1]
   c Are electromagnetic waves longitudinal or transverse? [1]
   d Name two uses of microwaves. [2]
   [Total: 7]

21 If a note played on a piano has the same pitch as one played on a guitar, state which of the following is the same
   A frequency
   B amplitude
   C quality
   D loudness.
   [Total: 1]

22 The waveforms of two notes P and Q are shown below. Which one of the statements **A** to **D** is true?
A P has a higher pitch than Q and is not so loud.
B P has a higher pitch than Q and is louder.
C P and Q have the same pitch and loudness.
D P has a lower pitch than Q and is not so loud.

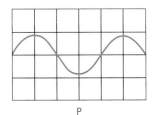

P

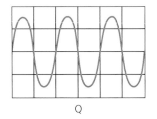

Q

[Total: 1]

23 a Name **two** examples of waves that can be modelled as transverse. [2]
b Name **two** examples of waves that can be modelled as longitudinal. [2]
c Name **three** properties of waves. [3]
[Total: 7]

4 Electricity and magnetism

24 Which one of the following statements about the diagram below is *not* true?

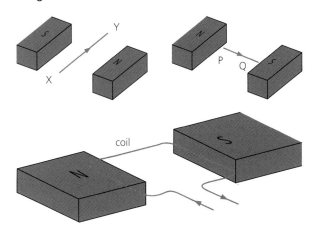

coil

A If a current is passed through the wire XY, a vertically upwards force acts on it.
B If a current is passed through the wire PQ, it does not experience a force.
C If a current is passed through the coil, it rotates clockwise.
D If the coil had more turns and carried a larger current, the turning effect would be greater.
[Total: 1]

25 For the circuit below calculate
a the total resistance [3]
b the current in each resistor [4]
c the p.d. across each resistor. [3]

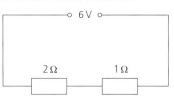

6 V

2 Ω   1 Ω

[Total: 10]

26 For the circuit below calculate
a the total resistance [4]
b the current in each resistor [4]
c the p.d. across each resistor. [1]

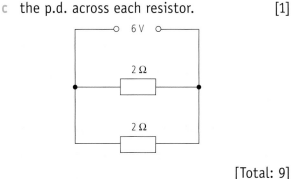

6 V

2 Ω

2 Ω

[Total: 9]

27 a An electric kettle for use on a 230 V supply is rated at 3000 W.
 i Calculate the current required by the kettle. [3]
 ii For safe working, the cable supplying it should be able to carry at least
  A 2 A   B 5 A   C 10 A   D 15 A [1]
b Calculate the cost of operating three 100 W lamps for 10 hours. Take the cost of 1 kWh to be 10 cents. [3]
[Total: 7]

28 Which one of the following statements is *not* true?
A In a house circuit, lamps are wired in parallel.
B Switches, fuses and circuit breakers should be placed in the neutral wire.
C An electric fire has its earth wire connected to the metal case to prevent the user receiving a shock.
D When connecting a three-core cable to a 13 A three-pin plug the brown wire goes to the live pin.
[Total: 1]

29 State the units for the following quantities
  a electric charge   [1]
  b electric current   [1]
  c p.d.   [1]
  d energy   [1]
  e power.   [1]
      [Total: 5]

5 Nuclear physics

30 a A radioactive source which has a half-life of 1 hour gives a count-rate of 100 counts per second at the start of an experiment and 25 counts per second at the end. Calculate the time taken for the experiment. [3]
  b i State the relative penetrating powers of α-particles, β-particles and γ-radiation. [2]
    ii State the relative ionising powers of α-particles, β-particles and γ-radiation. [2]
      [Total: 7]

6 Space physics

31 The Earth is tilted on its axis as it orbits the Sun.
  a In December is the northern hemisphere tilted towards or away from the Sun? [1]
  b State when the Sun is highest on the horizon in the northern hemisphere. [2]
  c State when the longest hours of daylight occur in the southern hemisphere. [2]
  d State the months in which the hours of day and night are equal. [2]
      [Total: 7]

32 The Moon is kept in a circular orbit around the Earth by the force of gravity between it and the Earth. If the average radius of the orbit is 385 000 km and the orbital speed is approximately 1 km/s, calculate
  a the distance travelled by the Moon in one orbit of the Earth [2]
  b the orbital period of the Moon in days. [4]
      [Total: 6]

33 a One light-year is the distance travelled by light in one year and is equal to 9.5×10^{12} km. Calculate the distance in light years of a star that is 2.85×10^{14} km away from the Earth. [2]
  b Choose the approximate distance from the Sun of the following bodies.
    i Earth   [1]
    ii Pluto   [1]
    iii the nearest star outside the Solar System   [1]
    iv the Andromeda galaxy.   [1]
      A 4 light-years
      B 150 million km
      C 6000 million km
      D 2 million light-years
      [Total: 6]

34 a Outline two pieces of evidence which support the Big Bang theory of the origin of the Universe.   [5]
  b Redshift measurements show that a galaxy is receding from Earth at a speed of 16 000 km/s. Use Hubble's Law to find how many light-years the galaxy is distant from the Solar System. (Take Hubble's constant to be 2.2×10^{-18} s^{-1} and 1 light-year $= 9.5 \times 10^{12}$ km.)   [5]
      [Total: 10]

Theory past paper questions

Please note that from 2023, Cambridge Theory, Practical Test and Alternative to Practical examination questions will expect students to use $g = 9.8\,\text{N/kg}$ or $9.8\,\text{m/s}^2$ in calculations. However, in these past paper questions $g = 10\,\text{N/kg}$ or $10\,\text{m/s}^2$ is used.

1 Motion, forces and energy

Physical quantities and measurement techniques, Motion, Mass and weight and Density

1 A student investigates water dripping from a tap (faucet).
Fig. T1 shows the dripping tap and a rule next to a container collecting the drops of water.

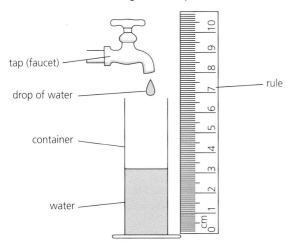

▲ Fig.T1

a Name the quantity that the student is measuring with the rule. [1]

b The student uses a digital stopwatch to measure the time between the drops of water. She repeats her measurement.
Fig. T2 shows the reading on the stopwatch for all her measurements.

▲ Fig. T2

i Record the time, in seconds, measured by the student. [1]
ii Calculate the average time between drops of water. Show your working. [2]

c The student collects drops of water for 15.5 minutes.
Calculate how many drops leave the tap in 15.5 minutes. Use your answer to part b ii. [3]

[Total: 7]

Cambridge IGCSE Physics 0625 Paper 32 Q1 May/June 2016

2 Fig. T3 shows a simple pendulum swinging backwards and forwards between P and Q.
One complete oscillation of the pendulum is when the bob swings from P to Q and then back to P.

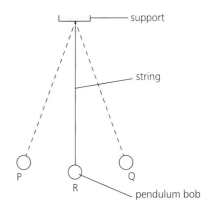

▲ Fig. T3

a A student starts two stopwatches at the same time while the pendulum bob is swinging. The student stops one stopwatch when the pendulum bob is at P. He stops the other stopwatch when the pendulum bob next is at Q. Fig. T4 shows the readings on the stopwatches.

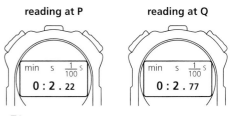

▲ Fig. T4

i   Use readings from Fig. T4 to determine the time for one complete oscillation of the pendulum.
    time = _____ s   [2]
ii The method described in **a** does not give an accurate value for one complete oscillation of the pendulum.
    Describe how the student could obtain an accurate value for one complete oscillation of the pendulum.   [4]

b As the pendulum bob moves from R to Q it gains 0.4 J of gravitational potential energy. Air resistance can be ignored.
   State the value of kinetic energy of the pendulum bob at
   1. R
   2. Q.   [2]
                                          [Total: 8]

Cambridge IGCSE Physics 0625 Paper 32 Q3
May/June 2019

3 A person on roller skates makes a journey. Fig. T5 shows the speed–time graph for the journey.

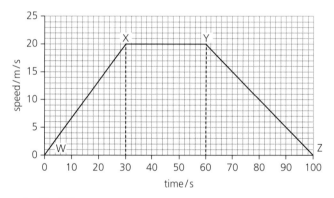

▲ Fig. T5

a The graph shows three types of motion. Copy and complete the table to show when each type of motion occurs. Use the letters shown on Fig. T5. Add a letter to each of the blank spaces.
   The first row is done for you.

motion	start of motion	end of motion
acceleration	W	X
deceleration		
constant speed		

[2]

b Calculate the distance travelled between 60 s and 100 s.   [3]
c The size of the acceleration is greater than the deceleration.
   Describe how Fig. T5 shows this.   [1]
                                          [Total: 6]

Cambridge IGCSE Physics 0625 Paper 32 Q1
Oct/Nov 2018

4 A student moves a model car along a bench. Fig. T6 is the speed–time graph for the motion of the model car.

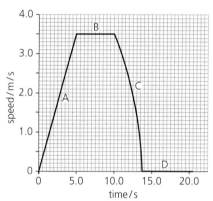

▲ Fig. T6

a Describe the motion of the car in each of the sections A, B, C and D.   [4]
b Determine the distance moved by the model car in the first five seconds.
   distance = _____ m   [3]
                                          [Total: 7]

Cambridge IGCSE Physics 0625 Paper 32 Q1
May/June 2019

5 a Fig. T7 shows the axes of a distance–time graph for an object moving in a straight line.

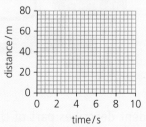

▲ Fig. T7

   i 1. On a copy of Fig. T7, draw between time = 0 and time = 10 s, the graph for an object moving with a constant speed of 5.0 m/s. Start your graph at distance = 0 m.

2. State the property of the graph that represents speed. [2]

ii Between time = 10 s and time = 20 s the object accelerates. The speed at time = 20 s is 9.0 m/s.
Calculate the average acceleration between time = 10 s and time = 20 s. [2]

b Fig. T8 shows the axes of a speed–time graph for a different object.

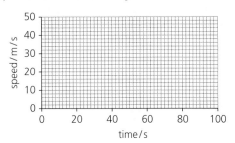

▲ Fig. T8

i The object has an initial speed of 50 m/s and decelerates uniformly at 0.35 m/s² for 100 s.
On a copy of Fig. T8, draw the graph to represent the motion of the object. [2]

ii Calculate the distance travelled by the object from time = 0 to time = 100 s. [3]
[Total: 9]

Cambridge IGCSE Physics 0625 Paper 42 Q1
May/June 2018

6 Fig. T9 shows students getting on to a school bus.

▲ Fig. T9

a A student describes part of the journey.
The bus accelerates from rest at a constant rate for 10 s. It reaches a maximum speed of 10 m/s.
The bus maintains a constant speed of 10 m/s for 60 s.
The bus then decelerates at a constant rate for 15 s, until it stops.

On a copy of Fig. T10, draw the speed–time graph for this part of the journey made by the bus.

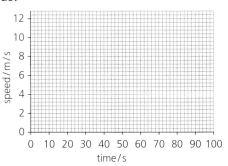

▲ Fig. T10 [5]

b On another part of the journey, the average speed of the bus is 7.5 m/s.
Calculate the distance the bus travels in 150 s. [3]
[Total: 8]

Cambridge IGCSE Physics 0625 Paper 32 Q2
Feb/March 2019

7 a Define *acceleration*. [1]

b Fig. T11 shows the distance–time graph for the journey of a cyclist.

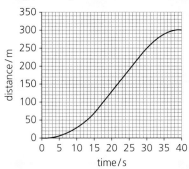

▲ Fig. T11

i Describe the motion of the cyclist in the time between:
1. time = 0 and time = 15 s
2. time = 15 s and time = 30 s
3. time = 30 s and time = 40 s. [3]

ii Calculate, for the 40 s journey:
1. the average speed [2]
2. the maximum speed. [2]
[Total: 8]

Cambridge IGCSE Physics 0625 Paper 42 Q1
Feb/March 2019

8 Fig. T12 shows a set of masses made from the same material.

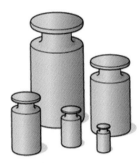

▲ Fig. T12

a Identify the quantity that is the same for all the masses.
   Choose **one** box.
   ☐ density
   ☐ volume
   ☐ weight   [1]
b The largest mass is 2.5 kg.
   State the number of grams in 2.5 kg.   [1]
c The three largest masses are 2.5 kg, 1.0 kg and 0.5 kg.
   Calculate the combined **weight** of these three masses. Include the unit. [4]
                                            [Total: 6]

   Cambridge IGCSE Physics 0625 Paper 32 Q1
                          Feb/March 2019

9 Fig. T13 shows a wooden raft. The raft is made from 8 logs.
   The logs are all of the same type of wood.

   log of wood

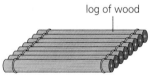

▲ Fig. T13

a The average mass of each log is 65.0 kg.
   Calculate the total weight of the raft.   [3]
b i The mass of one of the logs is 66.0 kg. It is 3.0 m long and has a cross sectional area of 0.040 m^2.
      Calculate the density of the wood in the log. [3]
   ii Explain why the log in b i floats on water. [1]
                                            [Total: 7]

   Cambridge IGCSE Physics 0625 Paper 32 Q2
                            May/June 2018

Forces, Momentum, Energy, work and power and Pressure

10 A load is attached to a spring, as shown in Fig. T14. Two arrows indicate the vertical forces acting on the load. The spring and the load are stationary.

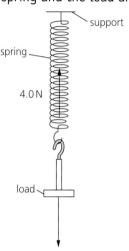

support

spring

4.0 N

load

▲ Fig. T14

a i State the name of the force acting vertically downwards. [1]
   ii The vertical force that acts upwards is 4.0 N.
      State the value of the force acting vertically downwards. [1]
b The load is pulled downwards and then released. The load moves up and down.
   Fig. T15 represents the vertical forces acting on the load at some time after it is released.

7.6 N

2.8 N

▲ Fig. T15

Calculate the resultant force on the load and state its direction. [2]
c i State the principle of conservation of energy. [1]
   ii Eventually the load stops moving up and down.
      Describe and explain why the load stops moving. Use your ideas about conservation of energy. [2]
                                            [Total: 7]

   Cambridge IGCSE Physics 0625 Paper 32 Q3
                          Feb/March 2019

11 a An object is moving in a straight line at constant speed.
State **three** ways in which a force may change the motion of the object. [2]

b Fig. T16 shows an object suspended from two ropes. The weight of the object is 360 N. The magnitude of the tension in each rope is *T*.

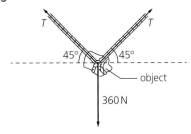

▲ Fig. T16

Determine the tension *T* by drawing a vector diagram of the forces acting on the object. State the scale you have used. [5]
[Total: 7]

Cambridge IGCSE Physics 0625 Paper 42 Q3 Feb/March 2019

12 Fig. T17 shows a tyre hanging from the branch of a tree.

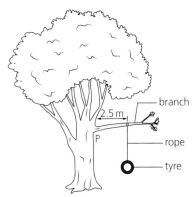

▲ Fig. T17

a The mass of the tyre is 15 kg. Calculate its weight. [2]

b The weight of the tyre exerts a moment on the branch, about point P where the branch joins the tree.

i   Explain what is meant by the term *moment*. [1]

ii A child sits on the tyre. The weight of the child and tyre together is 425 N. Calculate the moment of this force about point P. Use information given in Fig. T17. Include the unit. [4]

iii A heavier child wants to sit on the tyre. Describe how the tyre position should be adjusted so that the moment is the same as in b ii. [1]
[Total: 8]

Cambridge IGCSE Physics 0625 Paper 31 Q3 May/June 2017

13 Fig. T18 shows the speed–time graph for a student cycling along a straight, flat road.

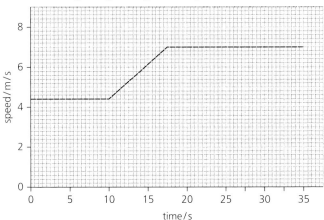

▲ Fig. T18

a Calculate the distance he travels in the first 10 s. [3]

b Fig. T19 shows three pairs of forces A, B and C.

▲ Fig. T19

Identify which pair of forces, A, B or C, acts on the cyclist between 11 s and 16 s. Explain your choice. [3]

c The cyclist pushes on one pedal with a force of 120 N. The area of his shoe in contact with the pedal is 16 cm².

Calculate the pressure on the pedal. Include the unit. [4]

[Total: 10]

Cambridge IGCSE Physics 0625 Paper 32 Q2
May/June 2016

14 a The velocity of an object of mass *m* increases from *u* to *v*.
State, in terms of *m*, *u* and *v*, the change of momentum of the object. [1]

b In a game of tennis, a player hits a stationary ball with his racquet.

i The racquet is in contact with the ball for 6.0 ms. The average force on the ball during this time is 400 N.
Calculate the impulse on the tennis ball. [2]

ii The mass of the ball is 0.056 kg.
Calculate the speed with which the ball leaves the racquet. [2]

iii State the energy transfer that takes place:

1. as the ball changes shape during the contact between the racquet and the ball

2. as the ball leaves the racquet. [2]

[Total: 7]

Cambridge IGCSE Physics 0625 Paper 42 Q3
Oct/Nov 2018

15 Fig. T20 shows a model fire engine. Its brakes are applied.

▲ Fig. T20

0.80 kg of water is emitted in the jet every 6.0 s at a velocity of 0.72 m/s relative to the model.

a Calculate the change in momentum of the water that is ejected in 6.0 s. [2]

b Calculate the magnitude of the force acting on the model because of the jet of water. [2]

c The brakes of the model are released.
State and explain the direction of the acceleration of the model. [2]

d In c the model contains a water tank, which is initially full.

State and explain any change in the magnitude of the initial acceleration if the brakes are first released when the tank is nearly empty. [3]

[Total: 9]

Cambridge IGCSE Physics 0625 Paper 42 Q2
May/June 2019

16 a A force is used to move an object from the Earth's surface to a greater height.
Explain why the gravitational potential energy (g.p.e.) of the object increases. [1]

b Fig. T21 shows a train moving up towards the top of a mountain.

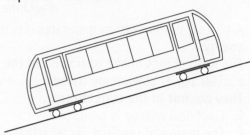

▲ Fig. T21

The train transports 80 passengers, each of average mass 65 kg, through a vertical height of 1600 m.
Calculate the increase in the total gravitational potential energy (g.p.e.) of the passengers.
increase in g.p.e. = _____ [2]

c The engine of the train has a power of 1500 kW. The time taken to reach the top of the mountain is 30 minutes.
Calculate the efficiency of the engine in raising the 80 passengers 1600 m to the top of the mountain. [4]

[Total: 7]

Cambridge IGCSE Physics 0625 Paper 42 Q2
Feb/March 2018

17 Fig. T22 shows an aircraft on the deck of an aircraft carrier.

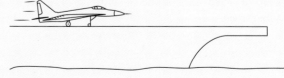

▲ Fig. T22

The aircraft accelerates from rest along the deck. At take-off, the aircraft has a speed of 75 m/s.

The mass of the aircraft is 9500 kg.
a Calculate the kinetic energy of the aircraft at take-off. [3]
b On an aircraft carrier, a catapult provides an accelerating force on the aircraft. The catapult provides a constant force for a distance of 150 m along the deck.
Calculate the resultant force on the aircraft as it accelerates. Assume that all of the kinetic energy at take-off is from the work done on the aircraft by the catapult. [2]
[Total: 5]

Cambridge IGCSE Physics 0625 Paper 42 Q3
May/June 2018

18 a A nuclear power station generates electrical energy.
The main stages in the operation of the nuclear power station are listed.
They are **not** in the correct order.
E Electrical energy is produced.
F The fission of uranium nuclei releases thermal energy.
G A turbine drives a generator.
H Thermal energy heats water to produce steam.

Copy and complete the flow chart to describe how a nuclear power station works.
In each empty box, insert the letter for the correct statement.
The nuclear power station uses uranium as a fuel.

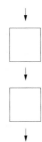

The steam drives a turbine.

Electrical energy is transmitted. [2]

b Electrical energy from the power station is used to power two different lamps. Fig. T23 shows how the light outputs from two types of lamp vary with the power input.

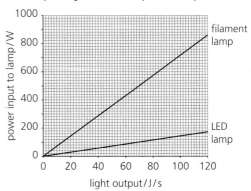

▲ Fig. T23

i An experiment requires a lamp with a light output of 70 J/s.
For the LED lamp and for the filament lamp determine the input power required to give a light output of 70 J/s. Use information from Fig. T23.
1. For the LED lamp, input power = _____ W
2. For the filament lamp, input power = _____ W [2]
ii Explain why using LED lamps is better for the environment. Use information from Fig. T23 in your answer. [2]
[Total: 6]

Cambridge IGCSE Physics 0625 Paper 32 Q5
May/June 2019

19 An archaeologist is investigating a shipwreck and discovers a wooden box on the seabed.

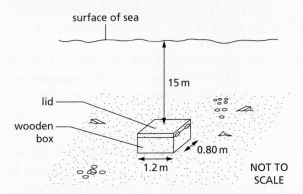

▲ Fig. T24

The dimensions of the lid of the box are 1.2 m by 0.80 m and the pressure of the atmosphere is 1.0×10^5 Pa. The lid is 15 m below the surface of the sea.

a The density of sea-water is 1020 kg/m³.
  Calculate
    i   the pressure on the lid of the box due
        to the sea-water,   [2]
    ii the total pressure on the lid,   [1]
    iii the downward force that the total
        pressure produces on the lid.   [2]
b The force needed to open the lid is **not**
  equal to the value calculated in a iii.
  Suggest **two** reasons for this.   [2]
                                        [Total: 7]

Cambridge IGCSE Physics 0625 Paper 42 Q4
May/June 2016

2 Thermal physics

20 Fig. T25 shows workers pouring liquid metal.

▲ Fig. T25

a The metal changes from hot liquid to cool solid.
  Describe what happens to the arrangement, separation and motion of the atoms as the metal changes from hot liquid to cool solid.   [3]
b The workers cool their tools in water. They spill some water onto the floor but later the floor is dry.
  Explain what happens to the water. State the name of the process.   [3]
                                        [Total: 6]

Cambridge IGCSE Physics 0625 Paper 31 Q6
May/June 2017

21 a Fig. T26 shows the apparatus used to observe the motion of smoke particles that are in the air in a box.

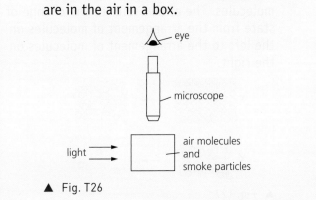

▲ Fig. T26

Light from a lamp enters the box through a window in one side of the box. The smoke particles are observed using a microscope fixed above a window in the top of the box.

i   The motion of a single smoke particle is observed through the microscope.
Draw a circle as shown, and sketch the path of this smoke particle.

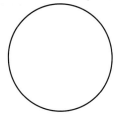

[1]

ii Explain why the smoke particle follows the path that is observed. [3]

b A tennis player is practising by hitting a ball many times against a wall.
The ball hits the wall 20 times in 60 s.
The average change in momentum for each collision with the wall is 4.2 kg m/s.
Calculate the average force that the ball exerts on the wall. [3]

[Total: 7]

Cambridge IGCSE Physics 0625 Paper 42 Q5
Feb/March 2018

22 a In Fig. T27, the small circles represent molecules. The arrows refer to the change of state from the arrangement of molecules on the left to the arrangement of molecules on the right.

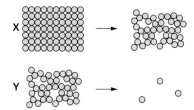

▲ Fig. T27

Complete the following by writing solid, liquid or gas in each of the blank spaces.
1. Change of state **X** is from _____ to _____
2. Change of state **Y** is from _____ to _____
[2]

b Explain, in terms of the forces between their molecules, why gases expand more than solids when they undergo the same rise in temperature. [2]

c A cylinder of volume 0.012 m^3 contains a compressed gas at a pressure of 1.8 × 10^6 Pa. A valve is opened and all the compressed gas escapes from the cylinder into the atmosphere.
The temperature of the gas does not change. Calculate the volume that the **escaped** gas occupies at the atmospheric pressure of 1.0 × 10^5 Pa. [3]

[Total: 7]

Cambridge IGCSE Physics 0625 Paper 42 Q7
Feb/March 2019

23 a State and explain, in terms of molecules, any change in the pressure of a gas when the volume is reduced at a constant temperature. [3]

b Copy and complete Table T1 to give the relative order of magnitude of the expansion of gases, liquids and solids for the same increase of temperature.
Write one of these words in each blank space:

gas   **liquid**   **solid**

▼ Table T1

expands most	
expands least	

[2]

[Total: 5]

Cambridge IGCSE Physics 0625 Paper 42 Q4
May/June 2019

24 A thermometer is used to measure the temperature inside a room in a house.
a State a physical property that varies with temperature and can be used in a thermometer. [1]
b Fig. T28 shows how the temperature of the room changes between 6:00 pm and 11:00 pm.

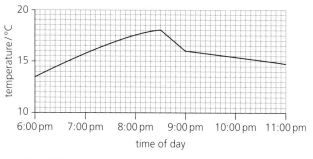

▲ Fig. T28

A heater in the room is switched on at 6 pm. The room has a large window. A large amount of thermal energy is transferred through the window. The window in the room has thick curtains. Closing the curtains reduces the loss of thermal energy from the room.

  i Suggest the time at which the heater is switched off. [1]

  ii Suggest the time at which the curtains were closed and explain your answer. Use information from the graph. [2]

c In cool climates, people use mineral wool to reduce heat loss from houses. Mineral wool is made of fibres and trapped air, as shown in Fig. T29.

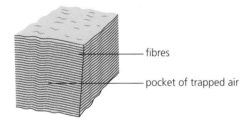

▲ Fig. T29

Use words from the box to complete the sentences. Each word may be used once, more than once, or not at all.

conductor	conduction	convection
emitter	insulator	radiation
radiator		

Air is a good _____.
When air is trapped between fibres, it reduces heat loss by _____ and by _____. [3]

[Total: 7]

Cambridge IGCSE Physics 0625 Paper 32 Q6
Feb/March 2018

25 Fig. T30 shows apparatus used by a student to measure the specific heat capacity of iron.

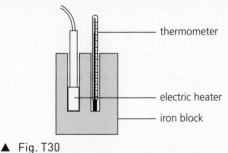

▲ Fig. T30

a The student improves the accuracy of the experiment by placing material around the block, as shown in Fig. T31.

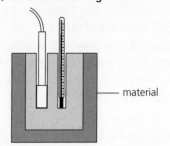

▲ Fig. T31

  i Suggest the name of a possible material the student could use and explain how it improves the accuracy of the experiment. [3]

  ii State how the student could further improve the accuracy of the experiment by using more of the material used in Fig. T31. [1]

b The current in the heater is 3.8 A and the potential difference (p.d.) across it is 12 V. The iron block has a mass of 2.0 kg. When the heater is switched on for 10 minutes, the temperature of the block rises from 25°C to 55°C.
Calculate the specific heat capacity of iron. [4]

[Total: 8]

Cambridge IGCSE 0625 Paper 42 Q4
Oct/Nov 2018

26 a i A liquid is heated so that bubbles of its vapour rise to the surface and molecules escape to the atmosphere.
State the name of this process. [1]

  ii At a lower temperature than in **a i**, molecules escape from the surface to the atmosphere.
State the name of this process. [1]

b i Fig. T32 shows apparatus used to determine the power output of a heater.

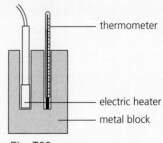

▲ Fig. T32

The metal block has a mass of 2.7 kg. The metal of the block has a specific heat capacity of 900 J/(kg °C). In 2 min 30 s, the temperature of the block increases from 21°C to 39°C. Calculate the power of the heater. [4]

ii State and explain a precaution that can be taken to improve the accuracy of the experiment. [2]

[Total: 8]

Cambridge IGCSE Physics 0625 Paper 42 Q5
May/June 2019

3 Waves

27 a A ray of light refracts as it travels from air into glass, as shown in Fig. T33.

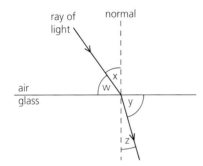

▲ Fig. T33

i State which angle w, x, y or z, is the angle of refraction. [1]

ii Light is a transverse wave.
State another example of a transverse wave. [1]

b Fig. T34 represents some wavefronts approaching a barrier with a narrow gap.

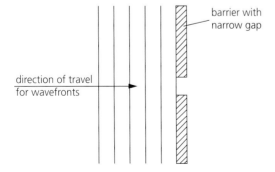

▲ Fig. T34

i On a copy of Fig. T34, draw three wavefronts that have passed through the gap. [2]

ii State the name of the effect in bi. [1]

[Total: 5]

Cambridge IGCSE Physics 0625 Paper 32 Q7
Feb/March 2018

28 Fig. T35 shows a ray of light that is reflected by a mirror.

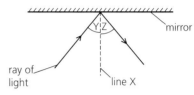

▲ Fig. T35

a i State the name of line X shown on Fig. T35. [1]

ii State the name of angle Y shown on Fig. T35. [1]

iii A student moves the ray of light and doubles the size of angle Y. State the effect on angle Z. [1]

b Fig. T36 shows a converging lens used to form an image I of an object O.

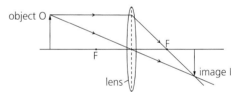

▲ Fig. T36

i State the name of the points labelled F on Fig. T36. [1]

ii Describe the nature of the image I. [2]

[Total: 6]

Cambridge IGCSE Physics 0625 Paper 32 Q6
May/June 2019

29 Fig. T37 represents an object positioned on the principal axis of a thin lens.

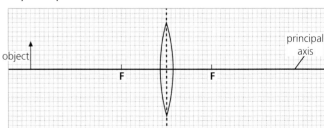

▲ Fig. T37

Each small square of the grid represents 0.5 cm. Each principal focus of the lens is labelled F.

a Use the grid to determine the focal length of the lens. [1]

b i On a copy of Fig. T37, draw a ray from the top of the object that passes through a principal focus, then through the lens and beyond it. [1]

ii Draw a second ray from the top of the object that passes through the centre of the lens. Continue the path of this ray to the edge of the grid. [1]

iii Draw an arrow to show the position and nature of the image produced by the lens. [2]

[Total: 5]

Cambridge IGCSE Physics 0625 Paper 32 Q5
May/June 2018

30 a Fig. T38 shows a visible spectrum focused on a screen by passing light from a source of white light through a lens and a prism.

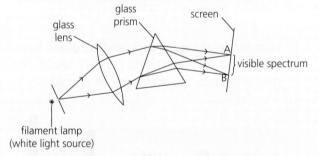

▲ Fig. T38

i State the name of the process that separates the colours in white light. [1]

ii State the colour of the light on the screen at:
point A _____
point B _____ [1]

iii State the property of the glass of the prism that causes white light to be split into the different colours of the spectrum. [1]

b Fig. T39 shows a section of an optical fibre in air. A ray of light is incident on the fibre wall at X.

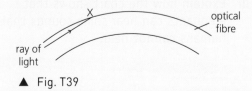

▲ Fig. T39

i Copy Fig. T39 and continue the path of the ray of light up to the end of the fibre. [1]

ii The refractive index of the material of the fibre is 1.46. Calculate the critical angle of the material of the fibre. [2]

iii State **two** uses of optical fibres. [2]

[Total: 8]

Cambridge IGCSE Physics 0625 Paper 42 Q5
Oct/Nov 2018

31 a A ray of light in glass is incident on a boundary with air.
State what happens to the ray when the angle of incidence of the ray is

i less than the critical angle of the glass, [1]

ii greater than the critical angle of the glass. [1]

b Fig. T40 shows a ray of light incident on a glass block at A. The critical angle of the glass is 41°.

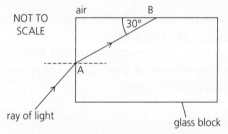

▲ Fig. T40

i Copy Fig. T40 and, without calculation, continue the ray from point B until it leaves the glass block. [2]

ii Calculate the refractive index of the glass. [2]

[Total: 6]

Cambridge IGCSE Physics 0625 Paper 41 Q6
May/June 2017

32 a Choose **two** of the following that apply to an ultrasound wave travelling in air.
frequency 3.5 Hz   frequency 350 Hz
frequency 35 000 Hz   longitudinal
transverse   speed 1.5 m/s
speed 1.5×10^3 m/s   speed 1.5×10^6 m/s [2]

b Calculate the wavelength in a vacuum of X-rays of frequency 1.3×10^{17} Hz. [3]

c A dentist takes an X-ray photograph of a patient's teeth. Explain why it is safe for the patient to be close to the source of X-rays, but the dentist must stand away from the source. [2]

d State, with a reason, why microwave ovens are designed only to work with the door closed. [2]

[Total: 9]

Cambridge IGCSE Physics 0625 Paper 42 Q6
May/June 2018

33 a A ray of white light is incident on a glass prism. It forms a spectrum that is visible on the screen. Fig. T41 shows the arrangement.

▲ Fig. T41

Two of the colours in the visible spectrum are listed in the box below.
Copy and complete the box. List the five missing colours of the visible spectrum, in the correct order.

| red | | | | | | violet |

[2]

b Electromagnetic radiation has many uses.
i Draw a line from each use to the type of radiation it requires.

use	type of radiation
	radio waves
detecting an intruder at night	microwaves
	infra-red
communicating by satellite for a telephone	visible light
	ultraviolet
detecting broken bones in the body	X-rays
	gamma rays

[3]

ii The types of radiation listed in bi form the electromagnetic spectrum.

| amplitude | frequency | velocity |

Complete the sentence. Choose a word from the box.
The position of each type of radiation in the electromagnetic spectrum depends on its _____. [1]

[Total: 6]

Cambridge IGCSE Physics 0625 Paper 32 Q7
Oct/Nov 2018

34 a Complete the sentences about sound.
Use words from the box above each sentence.

i

| glows | reflects | refracts | vibrates |

Sound is produced when a source _____. [1]

ii

| electromagnetic | longitudinal | transverse |

Sound waves are _____ waves. [1]

iii

| metal | vacuum | liquid |

Sound waves cannot travel through a _____. [1]

b Humans, elephants, mice and dolphins have different hearing ranges. Fig. T42 shows the hearing range for each type of animal.

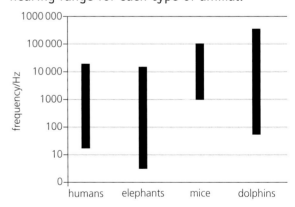

▲ Fig. T42

i State the lowest frequency of sound that can be heard by mice. [1]

ii State the highest frequency of sound that can be heard by elephants. [1]

iii Explain how the chart shows that elephants can hear some sounds that humans **cannot** hear. [2]

iv State the term given to the high frequencies that dolphins can hear but humans **cannot** hear. [1]
[Total: 8]

Cambridge IGCSE Physics 0625 Paper 32 Q8
May/June 2018

35 a A healthy human ear can hear a range of frequencies.
Three frequency ranges are shown.
Choose the range for a healthy human ear.
0 Hz–20 Hz   10 Hz–10 000 Hz   20 Hz–20 000 Hz [1]

b Explain the meaning of the term *ultrasound*. [2]

c A student listens to two different sounds, P and Q.
The two different sounds are represented on a computer screen on the same scale.
Fig. T43 shows the screens.

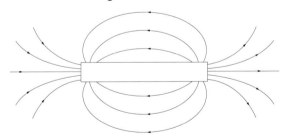

▲ Fig. T43

State and explain how sound P is different from sound Q. [3]
[Total: 6]

Cambridge IGCSE Physics 0625 Paper 32 Q8
Oct/Nov 2018

4 Electricity and magnetism

36 a Fig. T44 shows the magnetic field pattern around a bar magnet.

▲ Fig. T44

i Copy Fig. T44 and mark the North and South poles of the magnet. Use the letter N for the North pole and S for the South pole. [1]

ii A small bar of unmagnetised iron is placed next to a bar magnet, as shown in Fig. T45.

▲ Fig. T45

The iron bar moves towards the magnet. Explain why the iron bar moves. [2]

b Fig. T46 shows a coil of wire wrapped around an iron core. A student uses these to make an electromagnet.

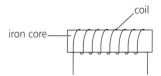

▲ Fig. T46

i Complete the diagram in Fig. T46 to show how it could be used to make an electromagnet. [1]

ii State **one** advantage of an electromagnet compared to a permanent magnet. [1]
[Total: 5]

Cambridge IGCSE Physics 0625 Paper 32 Q8
May/June 2019

37 a Describe the movement of charge that causes an object to become positively charged. [1]

b Fig. T47 shows a negatively charged rod held over an uncharged metal sphere.

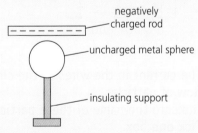

▲ Fig. T47

i On a copy of Fig. T47, add + and – signs to represent the results of the movement of charge within the sphere. [2]

ii Describe the actions that must be taken to obtain an even distribution of positive charge on the surface of the sphere. [2]
[Total: 5]

Cambridge IGCSE Physics 0625 Paper 41 Q10
May/June 2017

38 a A student rubs a polythene rod with a dry cloth. The polythene rod becomes negatively charged. Describe and explain how the rod becomes negatively charged. [3]

b The negatively charged polythene rod hangs from a nylon thread so that it is free to turn. The student charges a second polythene rod and brings it close to the first rod, as shown in Fig. T48.

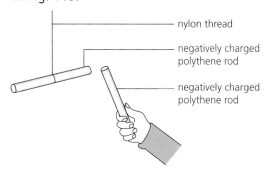

▲ Fig. T48

Describe and explain what happens when the negatively charged rods are close to each other. [2]

[Total: 5]

Cambridge IGCSE Physics 0625 Paper 32 Q11
Feb/March 2018

39 a Fig. T49 shows a simple circuit.

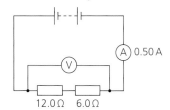

▲ Fig. T49

i The current in the wires of the circuit is a flow of particles.
Indicate the name of these particles.
Tick **one** box.
☐ electrons
☐ atoms
☐ protons [1]

ii Calculate the combined resistance of the two resistors. [1]

iii Calculate the potential difference (p.d.) reading that would be shown on the voltmeter. [3]

b The circuit is changed.
The two resistors are connected in parallel.
Explain what happens, if anything, to the current reading on the ammeter. [2]

[Total: 7]

Cambridge IGCSE Physics 0625 Paper 32 Q9
May/June 2018

40 a The lamp of a car headlight is rated at 12 V, 50 W.
Calculate the current in the lamp when operating normally. [2]

b A car is driven at night.
In a journey, the total charge that passes through the 12 V battery is 270 kC.
i Calculate the electrical energy transferred. [3]

ii The fuel used by the car provides $3.6 \times 10^4 \, \text{J/cm}^3$.
Calculate the volume of fuel used to provide the energy calculated in **b i**. [2]

[Total: 7]

Cambridge IGCSE Physics 0625 Paper 42 Q8
Feb/March 2018

41 Fig. T50 shows current–potential difference graphs for a resistor and for a lamp.

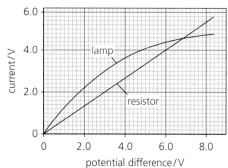

▲ Fig. T50

a i The potential difference (p.d.) applied to the resistor is increased. Choose the box that indicates the effect on the resistance of the resistor.
☐ resistance increases
☐ resistance is constant
☐ resistance decreases [1]

ii The potential difference (p.d.) applied to the lamp is increased. Choose the box that indicates the effect on the resistance of the lamp.
☐ resistance increases
☐ resistance is constant
☐ resistance decreases [1]

b The p.d. across the lamp is 6.0 V. Calculate the resistance of the lamp. [2]

c The lamp and the resistor are connected in **parallel** to a 6.0 V supply.
Calculate the current from the supply. [2]

d The lamp and the resistor are connected in **series** to another power supply. The current in the circuit is 4.0 A.
Calculate the total p.d. across the lamp and the resistor. [2]
[Total: 8]

Cambridge IGCSE Physics 0625 Paper 42 Q9
Feb/March 2018

42 Fig. T51 shows a circuit used by a student to test a metal wire made of nichrome.

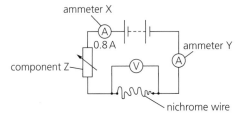

▲ Fig. T51

a State the name of component Z. [1]

b The current reading on ammeter X is 0.8 A.
State the reading on ammeter Y. [1]

c The current in the nichrome wire is 0.8 A. The potential difference (p.d.) across the nichrome wire is 4.5 V.
Calculate the resistance of the nichrome wire. [3]

d The student tests a different nichrome wire, which is thicker than the wire in c, but of the same length. When testing this wire, the current in the wire is different from the value given in c.
State and explain the difference in current. [2]
[Total: 7]

Cambridge IGCSE 0625 Physics Paper 32 Q10
Oct/Nov 2018

43 Fig. T52 shows a circuit containing a filament lamp of resistance 0.30 Ω and two resistors, each of resistance 0.20 Ω.

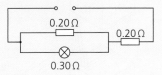

▲ Fig. T52

a Calculate the combined resistance of the lamp and the two resistors. [3]

b The potential difference (p.d.) of the supply is increased so that the current in the lamp increases.
State and explain any change in the resistance of the lamp. [2]
[Total: 5]

Cambridge IGCSE Physics 0625 Paper 42 Q10
May/June 2019

44 a The resistance of a long piece of wire is 6.0 Ω. The potential difference across the wire is 2.0 V.
Calculate the current in the wire. [3]

b A force acts on a wire carrying a current in a magnetic field.
Fig. T53 shows the direction of the current in the wire and the direction of the force acting on the wire.

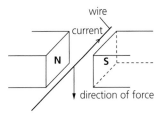

▲ Fig. T53

i Copy Fig. T53 and draw arrows to indicate the direction of the magnetic field. [1]

ii The magnetic field is reversed.
State what happens, if anything, to the direction of the force on the wire. [1]

c Fig. T54 shows a current-carrying coil in a magnetic field.

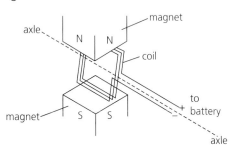

▲ Fig. T54

The coil starts to turn about its axle.
i State **two** ways of increasing the turning effect on the coil. [2]
ii Describe and explain the effect of reversing the connections to the battery. [2]
[Total: 9]

Cambridge IGCSE Physics 0625 Paper 32 Q9
May/June 2016

45 Fig. T55 shows two circuits, A and B.

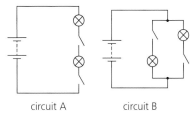

circuit A   circuit B

▲ Fig. T55

Both circuits contain a 6 V power supply and two 6 V lamps.
a State **two** advantages of circuit B compared to circuit A. [2]
b Fig. T56 shows the energy input and outputs, in one second, for one electric lamp.

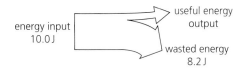

useful energy output

energy input 10.0 J

wasted energy 8.2 J

▲ Fig. T56

i Calculate the useful energy output, in one second, of the lamp. [1]
ii Draw a labelled diagram, similar to Fig. T56, for a more efficient lamp. [1]
c Electricity can be generated using wind turbines.
Fig. T57 shows two wind turbines.

▲ Fig. T57

State **two** advantages and **two** disadvantages of using wind turbines, rather than fossil fuels, to generate electricity. [4]
[Total: 8]

Cambridge IGCSE Physics 0625 Paper 32 Q5
May/June 2016

46 a The resistance of a circuit component varies with the brightness of the light falling on its surface.
i State the name of the component. [1]
ii Draw the circuit symbol for this component. [1]
b Fig. T58 shows a 6.0 V battery connected in series with a 1.2 kΩ resistor and a thermistor.

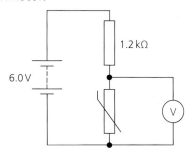

1.2 kΩ

6.0 V

V

▲ Fig. T58

i At a certain temperature, the resistance of the thermistor is 2.4 kΩ.
Calculate the reading on the voltmeter. [4]
ii The battery connected to the circuit in Fig. T58 is not changed.
Suggest a change that would cause the reading of the voltmeter to decrease. [1]
[Total: 7]

Cambridge IGCSE Physics 0625 Paper 41 Q9
May/June 2017

47 A student demonstrates electromagnetic induction.
  a Describe how to demonstrate electromagnetic induction using a magnet, a coil of wire and a sensitive ammeter. You may include a diagram. [3]
  b State **two** factors that affect the size of an induced electromotive force (e.m.f.). [2]

[Total: 5]

Cambridge IGCSE Physics 0625 Paper 31, Q12
May/June 2017

48 a The electrical energy produced by a power station is transmitted over long distances at a very high voltage.
  Explain why a very high voltage is used. [3]
  b Fig. T59 represents a transformer.

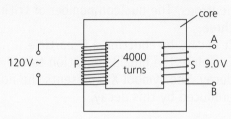

▲ Fig. T59

  i The primary coil P has 4000 turns and an input of 120 V. The secondary coil S has an output of 9.0 V.
  Calculate the number of turns in the secondary coil. [2]
  ii State a suitable material for the core of the transformer. [1]

[Total: 6]

Cambridge IGCSE Physics 0625 Paper 42 Q10
Feb/March 2019

5 Nuclear physics

49 Fig. T60 shows a radioactive source placed close to a radiation detector and counter. The detector can detect α, β and γ radiation.

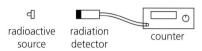

▲ Fig. T60

The radioactive source emits β-particles only. Describe how you could show that the source emits β-particles only.
As part of your answer, you may draw on a copy of Fig. T60 and add any other apparatus you may need.

[Total: 4]

Cambridge IGCSE Physics 0625 Paper 32 Q12
Feb/March 2018

50 a A detector of ionising radiation measures the background count rate in a classroom where there are no radioactive samples present. The readings, in counts/minute, taken over a period of time are shown in this chart.

counts/minute	16	12	14	16	15	17

  i State **two** possible sources of this background radiation. [2]
  ii Explain why the readings are not the same. [1]
  b With no radioactive sample present, a scientist records a background radiation count of 40 counts/minute.
  He brings a radioactive sample close to the detector. The count rate increases to 200 counts/minute.
  After 24 days the count rate is 50 counts/minute.
  Calculate the half-life of the radioactive sample. [4]
  c On a copy of this chart, draw a line between each type of ionising radiation and its property and another line between the property and its use. One has been done for you.

Name of ionising radiation	Property	Use
X-ray	It is the most ionising radiation and is most easily absorbed by very small amounts of substance	Remotely detecting leaks in underground water pipes
α-particle	Penetration is affected by small changes in the amount of solid it is passing through	Detecting fractures in bones
β-particle	It is highly penetrating and is poorly ionising	Detecting smoke in a fire alarm system
γ-ray	Can pass easily through soft living tissue. Calcium absorbs more than soft tissue	Detecting a change in the thickness of aluminium foil during its manufacture

[3]
[Total: 10]

Cambridge IGCSE Physics 0625 Paper 42 Q10
Oct/Nov 2018

51 a A radioactive nucleus of uranium-235 decays to a nucleus of thorium and emits an α-particle. Complete the equation.

$$^{235}_{92}\text{U} \rightarrow \;^{\cdots}_{\cdots}\text{Th} + {}^{4}_{2}\alpha$$

[2]

  b A nucleus of uranium-235 undergoes nuclear fission in a reactor.
  i State what is meant by *nuclear fission*. [1]
  ii Suggest why a nuclear reactor is surrounded by thick concrete walls. [2]
  iii State one environmental advantage and one environmental disadvantage of using a fission reactor to generate electrical energy in a power station. [2]
  c The thorium produced by the decay in a is also radioactive and has a half-life of 26 hours. At a certain time, a pure sample of this isotope initially contains 4.8 × 10⁹ atoms. Calculate the number of atoms of this sample that decay in the following 52 hours. [3]
[Total: 10]

Cambridge IGCSE Physics 0625 Paper 42, Q11
May/June 2018

52 a A radioactive source is tested over a number of hours with a radiation detector. The readings are shown in Table T2.

▼ Table T2

time/ hours	0	1	2	3	4	5	6	7	8	9	10
detector reading/ (counts/s)	324	96	39	23	21	17	21	20	19	20	18

Use the readings to suggest a value for the background count-rate during the test, and to determine the half-life of the sample. [4]

  b Hydrogen-3 (tritium) has one proton and two neutrons. The nucleon number of tritium is three.
It decays by emitting a β-particle.
Complete the nuclide equation to show this decay. The symbol X represents the nuclide produced by this decay.

$$^{\cdots}_{\cdots}\text{H} \rightarrow {}^{\cdots}_{\cdots}\beta + {}^{\cdots}_{\cdots}\text{X}$$

[3]

  c The arrows in Fig. T61 show the paths of three α-particles moving towards gold nuclei in a thin foil.

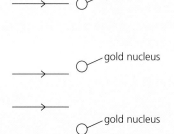

▲ Fig. T61

On Fig. T61, complete the paths of the three α-particles. [3]
[Total: 10]

Cambridge IGCSE Physics 0625 Paper 41 Q11
May/June 2017

53 Radon-220 is a radioactive isotope.
The nuclide notation for radon-220 is $^{220}_{86}\text{Rn}$.
  a Describe the composition of a neutral atom of radon-220. [3]
  b A nucleus of radon-220 decays to an isotope of polonium (Po) by emitting an alpha particle.

Complete the nuclide equation for the decay of radon-220.

$$^{220}_{86}\text{Rn} \rightarrow \text{....}\alpha + \text{.....Po}$$ [3]

c A detector of radiation is placed near a sample of radon-220 and gives a reading of 720 counts/s. The half-life of radon-220 is 55 s. Calculate the reading after 220 s. Ignore background radiation. [2]

[Total: 8]

Cambridge IGCSE Physics 0625 Paper 42 Q11
May/June 2016

c A radioactive substance decays by emitting an α-particle.
An α-particle can be represented as $^{4}_{2}\alpha$.
Draw a labelled diagram showing the composition of an α-particle. [3]

[Total: 7]

Cambridge IGCSE Physics 0625 Paper 32 Q12
May/June 2019

54 a Radioactive emission is a random process. Explain the meaning of the word *random*. [1]

b The table compares three types of radioactive emission.

emission	relative ionising ability	relative penetrating ability
alpha		
beta		
gamma		

Complete the table by choosing words from the box.

| high | low | medium | [3]

Practical Test past paper questions

Please note that from 2023, Cambridge Theory, Practical Test and Alternative to Practical examination questions will expect students to use $g = 9.8$ N/kg or 9.8 m/s^2 in calculations. However, in these past paper questions $g = 10$ N/kg or 10 m/s^2 is used.

1 In this experiment, you will compare the oscillations of two pendulums. Carry out the following instructions, referring to Figs. P1 and P2.

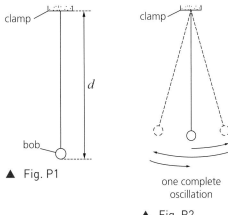

▲ Fig. P1

one complete oscillation

▲ Fig. P2

A pendulum has been set up for you as shown in Fig. P1. This pendulum has a spherical pendulum bob.

a Adjust the pendulum until the distance d measured to the **bottom** of the bob is 50.0 cm. Explain briefly how to use the set-square to avoid a parallax (line of sight) error when measuring the distance d. You may draw a diagram. [1]

b Move the bob slightly to the side and release it so that it swings. Fig. P2 shows one complete oscillation of the pendulum.
  i Measure the time t_1 for 20 complete oscillations. [1]
  ii Calculate the period T_1 of the pendulum. The period is the time for one complete oscillation. [2]

c Remove the pendulum from the clamp and attach the second pendulum provided to the clamp. This has a long, thin bob. Adjust the pendulum until the distance d measured to the **bottom** of the bob is 50.0 cm.

Displace the bob slightly and release it so that it swings.
  i Measure the time t_2 for 20 complete oscillations. [1]
  ii Calculate the period T_2 of the pendulum. [1]

d A student suggests that the periods T_1 and T_2 should be equal.
State whether your results support this suggestion. Justify your answer by reference to the results. [2]

e The period T of a pendulum can be determined by measuring the time t for 20 complete oscillations and then calculating the period. Some students are asked to explain the reason for this method being more accurate than measuring the time taken for one oscillation.
Choose which of these sentences gives the best explanation.
☐ The method eliminates errors from the measurements.
☐ The method is more accurate because the experiment is repeated.
☐ The method includes more readings so there is less chance for errors.
☐ The method reduces the effect of errors when starting and stopping the stopwatch. [1]

f A student plans to carry out more pendulum experiments. He considers possible variables and precautions to improve accuracy.
On a copy of the following list, mark the possible variables with the letter **V** and the precautions with the letter **P**.
☐ amplitude of swing
☐ length of pendulum
☐ mass of pendulum bob
☐ shape of pendulum bob
☐ use of a reference point to aid counting
☐ viewing the rule at right angles when measuring the length [2]

[Total: 11]

Cambridge IGCSE Physics 0625 Paper 52 Q1 Oct/Nov 2017

2 In this experiment, you will determine the density of water.
Carry out the following instructions, referring to Fig. P3.
You are provided with a plastic drinks cup.

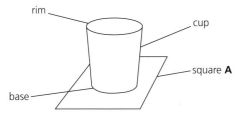

▲ Fig. P3

a i Place the cup, with the base at the bottom, in a copy of square **A**. Draw carefully around the base of the cup. Remove the cup from the paper. Take measurements from your drawing to determine an accurate value for the diameter D_B of the base of the cup.

square **A**

[2]

ii Place the cup, with the rim at the bottom, in a copy of square **B**. Draw carefully around the rim of the cup.
Remove the cup from the paper.
Take measurements from your drawing to determine the diameter D_T of the rim of the cup. [1]

square **B**

iii Calculate the average diameter D of the cup using the equation

$$D = \frac{1}{2}(D_B + D_T).$$ [1]

b i Measure the vertical height h of the cup. [1]

ii 1. Calculate the volume V of the cup using the equation $V = 0.785D^2h$.
2. Calculate $V/2$. [1]

c You are provided with water in beaker W. Pour a volume $V/2$ of water into the measuring cylinder. Pour this water into the cup.

i Use the balance provided to measure the mass m of the cup containing the water. [1]

ii Determine the density ρ of water using the equation $\rho = \frac{2m}{V}$.

Give your answer to a suitable number of significant figures for this experiment. Include the unit. [3]

d A student carries out all the instructions for this experiment with care, but his value for the density of water ρ is not equal to the expected value.
Suggest, with a reason, a part of the procedure, a, b or c that could give an unreliable result. [1]

[Total: 11]

Cambridge IGCSE Physics 0625 Paper 52 Q1
May/June 2018

3   In this experiment, you will investigate the stretching of a spring.
Carry out the following instructions, referring to Fig. P4.

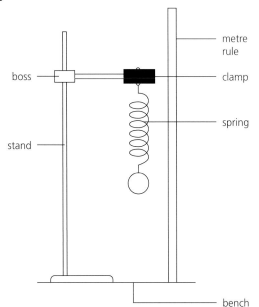

▲ Fig. P4

a   •   Do **not** remove the spring from the clamp. Use the metre rule to measure the length l_0 of the coiled part of the spring.
    •   Record l_0, in a copy of Table P1 at load $L = 0.0\,\text{N}$.
        On a copy of Fig. P4, show clearly the length l_0. [1]
b   •   Place a load $L = 1.0\,\text{N}$ on the spring. Record, in Table P1, the length l of the coiled part of the spring.
    •   Repeat this procedure using loads $L = 2.0\,\text{N}$, $3.0\,\text{N}$, $4.0\,\text{N}$ and $5.0\,\text{N}$.

▼ Table P1

L/N	0.0	1.0	2.0	3.0	4.0	5.0
l/mm						

[2]

c   Describe **one** precaution that you took in order to obtain reliable readings. [1]
d   On graph paper, plot a graph of l/mm (y-axis) against L/N (x-axis). [4]
e   A student suggests that the length l of the spring is directly proportional to the load L.

State whether your readings support this suggestion. Justify your answer by reference to the graph line. [1]
f   Use your results to predict the load L that would give a length l twice the value of l_0. Show clearly how you obtained your answer. [2]

[Total: 11]

Cambridge IGCSE Physics 0625 Paper 51 Q1
May/June 2017

4   A student is investigating whether the distance that a toy truck will travel along a horizontal floor, before stopping, depends on its mass.
The following apparatus is available to the student:
a ramp
blocks to support the ramp as shown in Fig. P5
toy truck
a selection of masses
other standard apparatus from the physics laboratory.

Plan an experiment to investigate whether the distance that the toy truck will travel along a horizontal floor, before stopping, depends on its mass.
You are **not** required to carry out this investigation.
In your plan, you should:
•   explain briefly how you would carry out the investigation
•   state any apparatus that you would use that is not included in the list above
•   state the key variables that you would control
•   draw a table, or tables, with column headings to show how you would display your readings (you are **not** required to enter any readings in the table).
You may add to a copy of the diagram in Fig. P5 to help your description.

▲ Fig. P5

[Total: 7]

Cambridge IGCSE Physics 0625 Paper 52 Q4
May/June 2018

5 In this experiment, you will investigate how the volume of water affects the rate at which water in a beaker cools.
Carry out the following instructions, referring to Fig. P6.
The thermometer must remain in the clamp throughout the experiment.

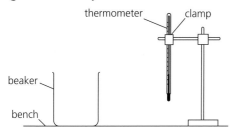

▲ Fig. P6

a Pour 200 cm³ of hot water into the beaker. Place the thermometer in the water.
Copy Table P2 and in the first row, record the maximum temperature θ of the water and immediately start the stopclock.
Record, in the table, the temperature θ of the water at times $t = 30\,s$, $60\,s$, $90\,s$, $120\,s$, $150\,s$ and $180\,s$.
Remove the thermometer from the beaker and empty the beaker. [1]

b i Repeat a, using 100 cm³ of hot water in the beaker. [1]
   ii Complete the headings and the time column in the table. [2]

▼ Table P2

	beaker with 200 cm³ of hot water	beaker with 100 cm³ of hot water
$t/$	$\theta/$	$\theta/$
0		

c Write a conclusion stating how the volume of water in the beaker affects the rate of cooling of the water. Justify your answer by reference to your results. [2]

d i Using your results for 100 cm³ of water, calculate the average rate of cooling x_1 for the **first** 90 s of the experiment. Use your readings from the table and the equation

$$x_1 = \frac{\theta_0 - \theta_{90}}{t},$$

where $t = 90\,s$ and θ_0 and θ_{90} are the temperatures at 0 s and 90 s.
Include the unit for the rate of cooling. [1]

   ii Using your results for 100 cm³ of water, calculate the average rate of cooling x_2 in the **last** 90 s of the experiment. Use your readings from the table and the equation

$$x_2 = \frac{\theta_{90} - \theta_{180}}{t},$$

where $t = 90\,s$ and θ_{90} and θ_{180} are the temperatures at 90 s and 180 s.
Include the unit for the rate of cooling. [1]

e A student suggests that it is important that the experiments with the two volumes of water should have the same starting temperatures.
State whether your values for x_1 and x_2 support this suggestion. Justify your statement with reference to your results. [1]

f Another student wants to investigate whether more thermal energy is lost from the water surface than from the sides of the beakers.
Describe an experiment that could be done to investigate this.
You are **not** required to carry out the experiment.
You may draw a diagram to help your description. [2]

[Total: 11]

Cambridge IGCSE Physics 0625 Paper 52 Q3
Feb/March 2018

6 In this experiment, you will investigate the reflection of light by a plane mirror.
Carry out the following instructions, using the separate ray-trace sheet provided.

You may refer to Fig. P7 for guidance.

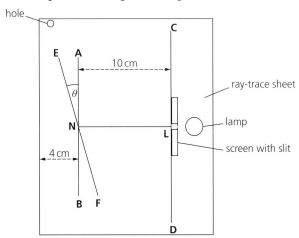

▲ Fig. P7

a • Draw a line **AB** 4 cm from the edge of the ray-trace sheet and in the middle of the paper, as shown in Fig. P7.
   • Draw a line **CD** parallel to line **AB** and 10 cm from it.
   • Draw a normal to line **AB** at a point **N** in the centre of line **AB**. Point **N** must be an equal distance from the top and bottom of the sheet.
   • Extend the normal to line **CD** and label the point at which it crosses line **CD** with the letter **L**. [1]

b Draw a line **EF**, through point **N**, as shown in Fig. P7 and at an angle $\theta = 5°$. [1]

c • Place the plane mirror on line **EF** with the reflecting surface facing to the right.
   • Place the screen with a slit on line **CD** and arrange the lamp so a ray of light shines along line **LN**.
   • Mark the ray that is reflected from the mirror, using a small cross at a suitable distance from point **N**. Label this cross **G**.
   • Remove the mirror, screen and lamp from the ray-trace sheet. [1]

d Draw a line joining point **N** and point **G**. Extend this line until it meets line **CD**.
   • Label the point at which line **NG** meets line **CD** with the letter **H**.
   • Measure, and record in a copy of Table P3, the length a of line **LH**. [1]

e Repeat **b**, **c** and **d** for values of $\theta = 10°$, 15°, 20° and 25°.

▼ Table P3

$\theta/°$	$a/$cm
5	
10	
15	
20	
25	

[1]

f On graph paper, plot a graph of $a/$cm (y-axis) against $\theta/°$ (x-axis). [4]

g Suggest a possible source of inaccuracy in this experiment, even if it is carried out carefully. [1]

h A student wishes to check if his values for a are reliable.
   Suggest how he could extend the experiment, using the same apparatus, to check the reliability of his results.
   You are **not** required to carry out this extended experiment. [1]

[Total: 11]

Cambridge IGCSE Physics 0625 Paper 52 Q1
Feb/March 2019

7 In this experiment, you will determine the focal length f of a lens.
   Carry out the following instructions referring to Fig. P8.

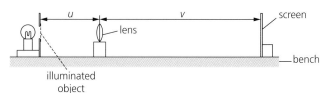

▲ Fig. P8

a i Place the screen 100 cm from the illuminated object.
   • Place the lens between the object and the screen so that the centre of the lens is at a distance $u = 20.0$ cm from the object.
   • Adjust the position of the screen until a clearly focused image is formed on the screen.
   • Measure the distance v between the centre of the lens and the screen. Record the values of u and v in Table P4.

▼ Table P4

u/cm	v/cm

ii Repeat the procedure using values for u of 22.0 cm, 25.0 cm, 30.0 cm and 35.0 cm. [3]

b On graph paper, plot a graph of v/cm (y-axis) against u/cm (x-axis). You do not need to start your axes at the origin (0, 0). Draw the best-fit curve. [4]

c i • Mark, with a cross, the point on the graph grid where $u = 25.0$ cm and $v = 25.0$ cm.
  • Mark, with a cross, the point on the graph grid where $u = 35.0$ cm and $v = 35.0$ cm.
  • Join these two points with a straight line. [1]

ii • Record u_1, the value of u at the point where the straight line crosses your graph line.
  • Record v_1, the value of v at the point where the straight line crosses your graph line. [1]

iii Calculate the focal length f of the lens using the equation

$$f = \frac{(u_1 + v_1)}{4}.$$ [2]

[Total: 11]

Cambridge IGCSE Physics 0625 Paper 52 Q3
Oct/Nov 2017

8 A student is investigating whether the resistance of a wire depends on the material from which the wire is made.

Resistance R is given by the equation $R = \dfrac{V}{I}$.

The following apparatus is available to the student:
ammeter
voltmeter
power supply (0–3 V)

micrometer screw gauge
variable resistor
switch
connecting leads
wires made of different materials.
Plan an experiment to investigate whether the resistance of a wire depends on the material from which the wire is made. You are **not** required to carry out this investigation.
You should:
• draw a diagram of the circuit you would use to determine the resistance of each wire
• explain briefly how you would carry out the investigation, including the measurements you would take
• state the key variables that you would control
• draw a suitable table, with column headings, to show how you would display your readings (you are **not** required to enter any readings in the table).

[Total: 7]

Cambridge IGCSE Physics 0625 Paper 51 Q4
May/June 2017

9 In this experiment, you will investigate a circuit containing different lamps.
The circuit has been set up for you.
Carry out the following instructions, referring to Fig. P9.

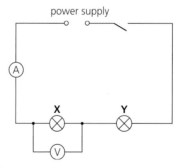

▲ Fig. P9

a i Switch on. Record the current I_S in the circuit. [1]

ii Record the potential difference (p.d.) V_X across lamp **X**.
Disconnect the voltmeter.
Connect the voltmeter to measure the p.d. V_Y across lamp **Y**.
Record V_Y. [1]

iii Disconnect the voltmeter.
Connect the voltmeter to measure the p.d. V_S across both lamps **X** and **Y** connected in series.
Record V_S. Switch off. [1]

iv A student suggests that V_S should be equal to $(V_X + V_Y)$.
State whether your readings support this suggestion. Justify your statement with reference to your results. [2]

b Calculate the resistance R_1 of lamp **X**. Use your readings from **a i** and **a ii** and the equation

$$R_1 = \frac{V_X}{I_S}.$$

Record your answer to a suitable number of significant figures for your experiment. [2]

c The circuit components are to be rearranged so that
• lamps **X** and **Y** are connected in parallel
• the ammeter measures the current in lamp **X** only
• the voltmeter measures the p.d. across the lamps.
Draw a circuit diagram of this arrangement. [2]

d i Set up the circuit as described in **c**.
Switch on. Measure and record the current I_P in lamp **X** and the p.d. V_P across the lamps. [1]
Switch off.

ii Calculate the new resistance R_2 of lamp **X**.
Use your readings from **d i** and the equation

$$R_2 = \frac{V_P}{I_P}.$$ [1]

[Total: 11]

Cambridge IGCSE Physics 0625 Paper 52 Q2
Feb/March 2018

Alternative to Practical past paper questions

Please note that from 2023, Cambridge Theory, Practical Test and Alternative to Practical examination questions will expect students to use $g = 9.8\,\text{N/kg}$ or $9.8\,\text{m/s}^2$ in calculations. However, in these past paper questions $g = 10\,\text{N/kg}$ or $10\,\text{m/s}^2$ is used.

1 The class is investigating the motion of a pendulum.
   Fig. P10 shows the apparatus.

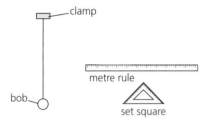

▲ Fig. P10

a i On a copy of Fig. P10, show clearly the length l of the pendulum. [1]

   ii Use Fig. P11 to explain how you would measure the length l accurately. You may draw on a copy of the diagram.

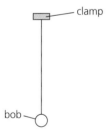

▲ Fig. P11

[2]

b A student determines the period T of the pendulum. The period is the time taken for one complete oscillation. The student measures the time t for 20 oscillations. Fig. P12 shows the time t.

▲ Fig. P12

i Calculate the period T of the pendulum. [1]

ii Explain how measuring the time for 20 oscillations rather than one oscillation helps the student to obtain a more reliable value for the period. [2]

c The student wants to determine a value for the acceleration of free fall from his results. He needs the value of T^2 to do this. Calculate T^2.
   Give your answer to a suitable number of significant figures and include the unit. [2]

[Total: 8]

Cambridge IGCSE Physics 0625 Paper 61 Q4
May/June 2017

2 A student is determining the density of water. She is provided with a plastic cup, shown in Fig. P13.

▲ Fig. P13

a She draws around the base of the cup. Her drawing is shown in Fig. P14.

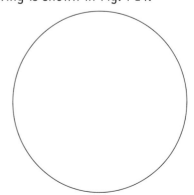

▲ Fig. P14

i From Fig. P14, take and record measurements to determine an accurate value for the diameter D_B of the base of the cup. [2]

ii The student places the cup upside down and draws around the rim of the cup. She determines the diameter D_T of the rim of the cup.

$D_T =$ .... 7.2 cm

Calculate the average diameter D of the cup using the equation $D = \dfrac{D_B + D_T}{2}$. [1]

b i On Fig. P15, measure the vertical height h of the cup.

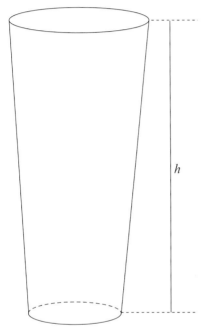

▲ Fig. P15

ii Calculate the volume V of the cup using the equation $V = 0.785D^2h$. [1]

c The student fills the cup with water. The mass of the cup with the water is shown in Fig. P16.

232g

▲ Fig. P16

Determine the density ρ of water using the equation $\rho = \dfrac{m}{V}$ and your value from b ii.

Give your answer to a suitable number of significant figures for this experiment. Include the unit. [3]

d Suggest, with a reason, a part of the procedure a, b or c that could give an unreliable result for the density of water. [1]

e The student pours the water from the cup into a measuring cylinder.
Draw a diagram to show water in a measuring cylinder. Show clearly the meniscus and the line of sight the student should use to obtain an accurate value for the volume of the water. [2]

[Total: 10]

Cambridge IGCSE Physics 0625 Paper 62 Q1
May/June 2018

3 A student has a selection of rubber bands of different widths. He is investigating the extension produced by adding loads. Fig. P17 shows the set-up used.

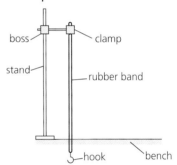

boss   clamp

stand   rubber band

hook   bench

▲ Fig. P17

In addition to the apparatus shown in Fig. P17, the following apparatus is available to the student:
a metre rule
a selection of different rubber bands
a selection of loads.
Plan an experiment to investigate how strips of rubber of different widths stretch when loaded. You should
• explain briefly how you would carry out the investigation
• state the key variables that you would control
• draw a table, or tables, with column headings to show how you would display your readings (You are **not** required to enter any readings in the table.)
• explain briefly how you would use your readings to reach a conclusion.

[Total: 7]

Cambridge IGCSE Physics 0625 Paper 62 Q4
Oct/Nov 2017

4 Students are investigating how the use of a lid or insulation affects the rate of cooling of hot water in a beaker. They use the apparatus shown in Fig. P18.

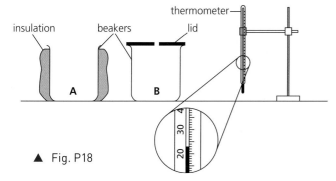

▲ Fig. P18

a Record the room temperature θ_R shown on the thermometer in Fig. P18. [1]

b • 100 cm³ of hot water is poured into beaker **A** and the initial temperature θ is recorded in Table P5.
  • The temperature θ of the water at times $t = 30\,s, 60\,s, 90\,s, 120\,s, 150\,s$ and $180\,s$ are shown in Table P5.
  • This process is repeated for beaker **B**.
  Copy Table P5 and complete the headings and the time column. [2]

▼ Table P5

t/	beaker **A** with insulation	beaker **B** with a lid
	θ/	θ/
0	83.0	86.0
	79.0	84.0
	75.5	82.5
	73.0	81.0
	71.0	80.0
	69.5	79.0
	68.5	78.5

c Write a conclusion stating whether the insulation or the lid is more effective in reducing the cooling rate of the water in the beakers in this experiment.
Justify your answer by reference to the results. [2]

d One student thinks that the experiment does not show how effective insulation is on its own or how effective a lid is on its own.
Suggest an additional experiment which could be used to show how effective a lid or insulation is.
Explain how the additional results could be used. [2]

e i Calculate x_A, the average cooling rate for beaker **A** over the whole experiment. Use the readings for beaker **A** from Table P5 and the equation

$$x_A = \frac{\theta_0 - \theta_{180}}{T}$$

where $T = 180\,s$ and θ_0 and θ_{180} are the temperatures at time $t = 0$ and time $t = 180\,s$.
Include the unit for the cooling rate. [2]

ii Students in another school are carrying out this experiment using identical equipment. State why they should make the initial temperature of the water the same as in this experiment if they are to obtain average cooling rates that are the same as in Table P5. Assume that the room temperature is the same in each case.
Use the results from beaker **A** to explain why this factor should be controlled. [2]

[Total: 11]

Cambridge IGCSE Physics 0625 Paper 62 Q2 Feb/March 2019

5 The class is investigating the refraction of light passing through a transparent block. A student is using optics pins to trace the paths of rays of light. Fig. P19 shows the student's ray-trace sheet.

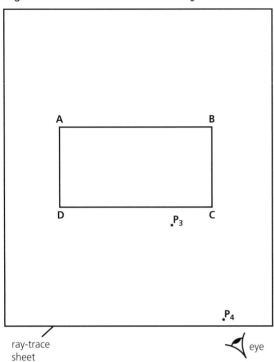

ray-trace sheet — eye

▲ Fig. P19

a • On a copy of Fig. P19, draw a normal at the centre of side **AB**. Label this line **NL**. Label the point **E** where the normal crosses **AB**. Label the point **M** where the normal crosses **CD**.
   • Draw a line above **AB** to the left of the normal and at an angle of incidence $i = 30°$ to the normal. Label this line **FE**.
   • Label the positions of two pins P_1 and P_2 placed a suitable distance apart on **FE** for accurate ray tracing.   [2]

b The student observes the images of P_1 and P_2 through side **CD** of the block so that the images of P_1 and P_2 appear one behind the other. He places two pins P_3 and P_4 between his eye and the block so that P_3 and P_4, and the images of P_1 and P_2 seen through the block, appear one behind the other. The positions of P_3 and P_4 are marked on Fig. P19. Draw a line joining the positions of P_3 and P_4. Continue the line until it meets the normal. Label this point **K**.   [1]

c • Measure and record the angle α between the line joining the positions of P_3 and P_4 and the line **KM**.
   • Measure and record the length x between points **M** and **K**.   [2]

d The student repeats the procedure with the angle of incidence $i = 50°$.
   His readings for α and x are shown.
   $\alpha = 52°$
   $x = 19\,mm$
   A student suggests that the angle α should always be equal to the angle of incidence i. State whether the results support this suggestion. Justify your answer by reference to the values of α for $i = 30°$ and $i = 50°$.   [2]

e Suggest **one** precaution that you would take with this experiment to obtain reliable results.   [1]

[Total: 8]

Cambridge IGCSE Physics 0625 Paper 61 Q2
May/June 2017

6 A student is investigating a circuit containing different lamps.
She is using the circuit shown in Fig. P20.

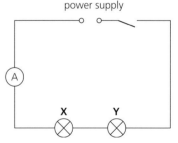

▲ Fig. P20

a On a copy of Fig. P20, draw a voltmeter connected so that it measures the potential difference (p.d.) across lamp **X**.   [1]

b The student uses the ammeter to measure the current in the circuit.

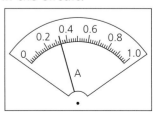

▲ Fig. P21

Record the current I_s in the circuit, as shown in Fig. P21.

I_s ———————   [1]

c i The student uses the voltmeter to measure the p.d. V_X across lamp **X** and then reconnects the voltmeter to measure the p.d. V_Y across lamp **Y**.

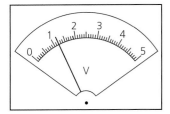

▲ Fig. P22a   ▲ Fig. P22b

Record the value of the p.d. V_X across lamp **X**, shown in Fig. P22a.
Record the value of the p.d. V_Y across lamp **Y**, shown in Fig. P22b. [1]

ii She then measures the p.d. V_S across both lamps in series.

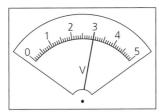

▲ Fig. P23

Record the value of the p.d. V_S across both lamps in series, shown in Fig. P23. [1]

iii A student suggests that V_S should be equal to $(V_X + V_Y)$.
State whether the readings support this suggestion. Justify your statement with reference to the results. [2]

d Calculate the resistance R_1 of lamp **X**.
Use the readings from **b** and **c i** and the equation $R_1 = \dfrac{V_X}{I_S}$. [1]

e i The circuit components are to be rearranged so that
   • lamps **X** and **Y** are connected in parallel
   • the ammeter measures the current in lamp **X** only
   • the voltmeter measures the p.d. across the lamps.
   Draw a circuit diagram of this arrangement. [2]

ii The student sets up the circuit as described in **e i**.
She measures and records the current in lamp **X** and the p.d. across the lamps.
She then calculates a new resistance R_2 for lamp **X** in this parallel circuit.
$R_2 = 8.3\,\Omega$
The student notices that lamp **X** is very bright in this parallel circuit, but it was dim in the series circuit in **a**.
Suggest how temperature affects the resistance of a lamp.
Justify your suggestion by reference to the value of R_1 from **d** and the value of R_2. [2]
[Total: 11]

Cambridge IGCSE Physics 0625 Paper 62 Q3
Feb/March 2018

7 A student is investigating the power of two lamps. The circuit is shown in Fig. P24.

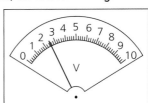

▲ Fig. P24

a i Record the potential difference (p.d.) V_T across the lamps and the current I_T in the circuit, as shown in Fig. P25 and Fig. P26.

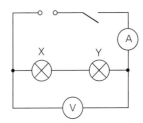

▲ Fig. P25

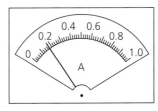

▲ Fig. P26 [2]

ii Calculate the power P_T produced by the lamp filaments, using the equation $P_T = V_T I_T$. [1]

b The student connects the voltmeter across lamp X only. She records the p.d. V_X across lamp X and the current I_X in the circuit.

$V_X =$ ......1.3 V......
$I_X =$ ......0.18 A......

She repeats the procedure with the voltmeter connected across lamp Y only.

$V_Y =$ ......1.2 V......
$I_Y =$ ......0.18 A......

i Calculate the power P_X produced by the lamp filament X using the equation $P_X = V_X I_X$, and calculate the power P_Y produced by the lamp filament Y using the equation $P_Y = V_Y I_Y$. [1]

ii State and explain briefly whether the two values for power P_X and P_Y are the same within the limits of experimental accuracy. [2]

c The student repeats the experiment using two other lamps. She notices that one lamp is dimly lit, but the other lamp does not light at all.

The p.d. V_T across the lamps is the same as in b, but the current I_T in the circuit is approximately half of the original value.

The student concludes that the filament of one of the lamps is broken.

State whether you agree with the student and give a reason for your answer. [2]

d Draw a circuit diagram to show the circuit in Fig. P24 rearranged so that:
  • the lamps are connected in parallel
  • a variable resistor is connected to control the total current in the circuit
  • the ammeter will measure the total current in the circuit
  • the voltmeter will measure the p.d. across the lamps. [3]

[Total: 11]

Cambridge IGCSE Physics 0625 Paper 62 Q2 Oct/Nov 2018

8 A student is investigating the resistance of three wires **A**, **B** and **C**. He is using the circuit shown in Fig. P27.

The circuit is set up to test wire **A**. The length, l, of each wire is measured and recorded.

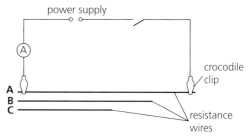

▲ Fig. P27

a On a copy of Fig. P27, draw a voltmeter connected so that it will measure the potential difference across wire **A**. [1]

b In the first line of a copy of Table P6, record the potential difference V and current I for wire **A**, as shown in Figs. P28 and P29. [2]

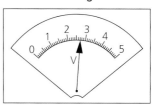

▲ Fig. P28

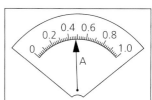

▲ Fig. P29

▼ Table P6

wire	l/m	V/V	I/A	R/Ω
A	0.900			
B	0.500	2.4	0.75	
C	0.400	2.2	0.85	

c The student connects the crocodile clips to wire **B** and then wire **C** in turn. His readings of potential difference and current are shown in Table P6.

Calculate, and record in a copy of Table P6, the resistance R of each wire.

Use the equation $R = \dfrac{V}{L}$. [2]

d i Calculate the resistance per unit length r of each wire using the equation $r = \dfrac{R}{l}$.

Include the unit.

r for wire **A** = _____.

r for wire **B** = _____

r for wire **C** = _____   [2]

ii Another student suggests that r should be the same for each wire.

State whether your results support this suggestion. Justify your statement with reference to your results.   [2]

e The student measures the length of each wire to be tested.

On a copy of Fig. P30, draw an arrow (↔) to indicate **precisely** between which two points he should measure l.

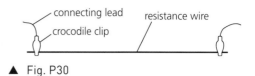

▲ Fig. P30

[1]

f One possible problem with this type of experiment is heating of the resistance wires. Suggest a precaution that could be taken to reduce this.   [1]

[Total: 11]

Cambridge IGCSE Physics 0625 Paper 62 Q2
Feb/March 2017

List of equations

$$v = \frac{s}{t}$$

$$\text{average speed} = \frac{\text{total distance travelled}}{\text{total time taken}}$$

$$a = \frac{\Delta v}{\Delta t}$$

$$g = \frac{W}{m}$$

$$\rho = \frac{m}{V}$$

$$k = \frac{F}{x}$$

$$F = ma$$

$$\text{moment} = \text{force} \times \text{perpendicular distance from the pivot}$$

$$p = mv$$

$$\text{impulse} = F\Delta t = \Delta(mv)$$

$$F = \frac{\Delta p}{\Delta t}$$

$$E_k = \frac{1}{2} mv^2$$

$$\Delta E_p = mg\Delta h$$

$$W = Fd = \Delta E$$

$$(\%) \text{ efficiency} = \frac{(\text{useful energy output})}{(\text{total energy input})} (\times 100\%)$$

$$(\%) \text{ efficiency} = \frac{(\text{useful power output})}{(\text{total power input})} (\times 100\%)$$

$$P = \frac{W}{t}$$

$$P = \frac{\Delta E}{t}$$

$$p = \frac{F}{A}$$

$$\Delta p = \rho g \Delta h$$

$$T \text{ (in K)} = \theta \text{ (in °C)} + 273$$

$$pV = \text{constant}$$

$$c = \Delta E / m\Delta\theta$$

$$v = f\lambda$$

$$n = \frac{\sin i}{\sin r}$$

$$n = \frac{1}{\sin c}$$

$$I = \frac{Q}{t}$$

$$E = \frac{W}{Q}$$

$$V = \frac{W}{Q}$$

$$R = \frac{V}{I}$$

$$P = IV$$

$$E = IVt$$

$$\frac{R_1}{R_2} = \frac{V_1}{V_2}$$

$$\frac{V_p}{V_s} = \frac{N_p}{N_s}$$

$$I_p V_p = I_s V_s$$

$$P = I^2 R$$

$$v = \frac{2\pi r}{T}$$

$$H_0 = \frac{v}{d}$$

$$\frac{d}{v} = \frac{1}{H_0}$$

Symbols and units for physical quantities

	Core			Supplement	
Quantity	Usual symbol	Usual unit	Quantity	Usual symbol	Usual unit
length	l, h, d, s, x	km, m, cm, mm			
area	A	m^2, cm^2			
volume	V	m^3, cm^3, dm^3			
weight	W	N			
mass	m, M	kg, g	mass	m, M	mg
time	t	h, min, s	time	t	ms, μs
density	ρ	g/cm^3, kg/m^3			
speed	u, v	km/h, m/s, cm/s			
acceleration	a	m/s^2			
acceleration of free fall	g	m/s^2			
force	F	N			
gravitational field strength	g	N/kg			
			spring constant	k	N/m, N/cm
			momentum	p	kg m/s
			impulse		N s
moment of a force		N m			
work done	W	J, kJ, MJ			
energy	E	J, kJ, MJ, kWh			
power	P	W, kW, MW			
pressure	p	N/m^2, N/cm^2	pressure	p	Pa
temperature	θ, T	°C, K			
			specific heat capacity	c	J/(g °C), J/(kg °C)
frequency	f	Hz, kHz			
wavelength	λ	m, cm	wavelength	λ	nm
focal length	f	m, cm			
angle of incidence	i	degree (°)			
angle of reflection	r	degree (°)			
angle of refraction	r	degree (°)			
critical angle	c	degree (°)			

Core			Supplement		
Quantity	**Usual symbol**	**Usual unit**	**Quantity**	**Usual symbol**	**Usual unit**
			refractive index	n	
potential difference/ voltage	V	V, mV, kV			
current	I	A, mA			
e.m.f.	E	V			
resistance	R	Ω			
charge	Q	C			
count-rate		counts/s, counts/ minute			
half-life		s, minutes, h, days, weeks, years			
			Hubble constant	H_0	s^{-1}

Glossary

Terms marked with an * indicate Supplement material.

absolute zero lowest possible temperature: −273°C or 0 K

absorber takes in heat or radiation

* **acceleration** change of velocity per unit time

acceleration of free fall g; for an object near to the surface of the Earth, this is approximately constant and is approximately $9.8 \, \text{m/s}^2$

accretion model build up of mass due to gravitational attraction between smaller particles leading to the formation of the Solar System

activity the average number of atoms which decay per second of a sample of radioactive material

advantages of high-voltage transmission of electricity (i) lower power loss in transmission cables and (ii) lower currents in cables so thinner/cheaper cables can be used

air resistance frictional force opposing the motion of a body moving in air

alpha particles α-particles; radiation consisting of helium ions with a double positive charge $(_{2}^{4}\text{He})$

alternating current a.c.; the direction of current flow reverses repeatedly

* **alternator** a.c. generator

ammeter instrument used to measure electric current

ampere A; unit of current

amplitude height of crest (or depth of trough) of a wave, measured from the undisturbed position of the medium carrying the wave

* **analogue** continuously varying

angle of incidence angle between incident ray and the normal to a surface

angle of reflection angle between reflected ray and the normal to a surface

angle of refraction angle between refracted ray and the normal to a surface

anode positively charged electrode

area of a circle πr^2 where r is the radius of the circle

area of a square or rectangle length × breadth

area of a triangle base × height/2

* **armature** coil and core of an electric motor

atom tiny constituent of matter

* **atomic bomb** produces an uncontrolled chain reaction

availability able to be used at all times

* **average orbital speed** $v = \dfrac{2\pi r}{T}$

average speed distance moved/time taken

background radiation ever-present radiation resulting from cosmic rays from outer space and radioactive materials in rocks, the air, buildings

balance instrument used for measuring mass

battery consists of two or more electric cells

beam (1) many rays of light (2) rectangular-shaped rod

beta particles β-particles; radiation consisting of high-speed electrons $(_{-1}^{0}\text{e})$

bimetallic strip equal lengths of two different metals riveted together

biofuels fuel obtained from plants or animal dung and sewage

biogas a mixture of methane and carbon dioxide produced from animal and human waste

boiling temperature temperature at which bubbles of vapour form in a liquid

Brownian motion random motion due to molecular collisions, experienced for example by smoke particles suspended in air

* **brushes** in a motor or generator; carbon blocks which act as electrical contacts

bubble chamber a device which uses bubbles in liquid hydrogen to make the paths of α-, β- and γ-radiation visible

cathode negatively charged electrode

cathode rays beams of high-speed electrons

centimetre 10^{-2} m

centre of gravity the point at which all of an object's weight can be considered to be concentrated

* **centripetal force** force acting towards the centre which keeps a body moving in a circular path

* **chain reaction** a set of nuclear reactions which are started by a single fission

chemical energy energy stored (for example in food and fuel) which can be released by chemical reactions

circuit breaker used instead of a fuse to break a circuit when the current exceeds a certain value

combined resistance of two resistors in parallel less than that of either resistor by itself

condensation change of a gas or vapour to a liquid

conduction flow of thermal energy through matter from places of higher temperature to places of lower temperature without movement of the matter as a whole

conductor material which allows thermal energy or electrons to flow easily

constant having the same value

constant of proportionality the ratio of two variables which are directly proportional to each other

continuous ripples a wave

convection flow of thermal energy through a fluid from places of higher temperature to places of lower temperature by movement of the fluid itself

convection currents streams of warm moving fluids

convector warms air using convection currents

conventional current direction in which a positive charge would flow in a circuit

converging light bends inwards

* coulomb C; unit of charge

count-rate counts/second

crest of a wave maximum amplitude of a wave

* critical a chain reaction becomes critical when on average each fission results in another fission event, so that the reaction is sustained

critical angle angle of incidence which produces an angle of refraction of 90°

crumple zones front and rear sections of a vehicle, designed to collapse on impact so that kinetic energy is absorbed gradually

cubic centimetre unit used to measure volume (cm^3)

cubic metre SI unit of volume (m^3)

current flow of electric charges; symbol I, measured in ampere (A)

* deceleration a negative acceleration; velocity decreases as time increases

degrees Celsius °C; unit of temperature

density mass per unit volume

dependent variable quantity whose value depends on that of another

deuterium ^2_1H; isotope of hydrogen with one proton and one neutron in the nucleus

diffraction spreading of a wave at the edges of an obstacle

diffuse reflection a parallel beam of light is reflected in many different directions at a rough surface

diffusion cloud chamber a device which makes the paths of α-, β- and γ-radiation visible

* digital using discrete values only

direct current d.c.; electrons flow in one direction only

direction of a magnetic field at a point the direction of the force on the N pole of a magnet at that point

direction of an electric field at a point the direction of the force on a positive charge at that point

dispersion separation of white light into its component colours

displacement distance moved in a stated direction

displacement–distance graph graph of the displacement from their undisturbed position of the particles transmitting a wave, plotted against their distance from the source, at a particular instant of time

distance–time graph graph of distance on the vertical axis plotted against time on the horizontal axis

diverging light spreads out

dynamic friction frictional force acting on a body moving at a constant speed

* dynamo d.c. generator

echo reflection of a sound wave

* eddy currents currents induced in a metal which is in a changing magnetic field; they cause energy loss in the core of a transformer

* efficiency (useful energy output/energy input) × 100%; (useful power output/power input) × 100%

effort force required to lift a weight

* elastic limit for extensions beyond the elastic limit, a material will not return to its original length when unloaded; it is permanently stretched and the law of proportionality no longer applies

* electric current the charge passing a point per unit time (current $I = Q/t$ where Q is the charge flowing past a particular point in time t)

electrical energy energy transferred by an electric current

electric cell changes chemical energy into electrical energy

* electric field a region of space where an electric charge experiences a force due to other charges

electric force force between electric charges

electrode emitter or collector of electric charges

electromagnet temporary magnet produced by passing an electric current through a coil of wire wound on a soft iron core

electromagnetic induction the production of a p.d. across a conductor when it moves through a magnetic field or is at rest in a changing magnetic field

electromagnetic radiation radiation resulting from electrons in an atom undergoing an energy change; all types travel in a vacuum at 3×10^8 m/s (the speed of light), obey the wave equation and exhibit interference, diffraction and polarisation

electromotive force (e.m.f.); the electrical work done by a source in moving unit charge around a complete circuit

electron $_{-1}^{\;0}e$; negatively charged elementary particle

electrostatic induction production of charge on a conductor when a charge is brought close to it

emitter gives out heat or radiation

energy may be stored as kinetic, gravitational potential, chemical, elastic (strain), nuclear, electrostatic and internal (thermal)

energy density energy/volume

energy sources materials and resources from which energy can be produced

energy transfer change of store or location of energy

equilibrium when there is no resultant force and no resultant moment on an object

evaporation loss of vapour from a liquid surface at a temperature below the boiling temperature of the liquid

* exciter supplies d.c. to the electromagnets in the rotor of an alternator

expansion increase in size

extension change in length of a body being stretched

factors affecting the magnitude of an induced e.m.f. e.m.f increases with increases of: (i) the speed of motion of the magnet or coil, (ii) the number of turns on the coil, (iii) the strength of the magnet

field line line which shows the direction of an electric, magnetic or gravitational field at each point

filament thin coil of wire which can transfer electrical energy to heat and light (as in a lamp)

* fission the break-up of a large nucleus into smaller parts

* Fleming's left-hand (motor) rule when the thumb and first two fingers of the left hand are held at right angles to each other with the first finger pointing in the direction of the magnetic field and the second finger in the direction of the current then the thumb points in the direction of the thrust on a wire

* Fleming's right-hand (dynamo) rule used to show the relative directions of force, field and induced current: when the thumb and first two fingers of the right hand are held at right angles to each other with the first finger pointing in the direction of the magnetic field and the thumb in the direction of the motion of the wire, then the second finger points in the direction of the induced current

fluid a liquid or gas

fluorescent emitting light when struck by ultraviolet radiation or electrons

focal length distance between the optical centre and the principal focus of a lens

focus bring beams to a point

force a push or pull on a body

fossil fuels coal, oil and natural gas formed from the remains of plants and animals which lived millions of years ago

* forward bias p.d. connected across a diode such that it has a low resistance and conducts

freezing temperature temperature at which a liquid changes to a solid

frequency number of complete oscillations per second

friction force which opposes one surface moving, or trying to move, over another surface

fulcrum pivot

fuse a short length of wire which melts when the current in the circuit exceeds a certain value; it protects the circuit from carrying a large current

* fusion the union of light nuclei into a heavier one

galaxy made up of many billions of stars

galvanometer an instrument used to detect small currents and p.d.s

gamma radiation γ-radiation; high-frequency, very penetrating electromagnetic waves

gas turbine gas is used to turn the blades of a rotor

Geiger–Müller tube GM tube; detects radiation

generator electricity-producing machine

geostationary satellite satellite which travels at the same speed as that at which the Earth rotates and so appears to be stationary at a particular point above the Earth's surface

geothermal energy energy obtained from hot rocks below the Earth's surface is used to heat water to steam which is then used to drive a turbine and generate electricity

gigametre (Gm) = 10^9 m = 1 billion metres

* gravitational field a region where the Earth exerts a force on a body

gravitational field strength the force per unit mass

gravitational potential energy energy a body has due to its height above the Earth's surface (*mgh*)

half-life of a particular isotope the time taken for half the nuclei of that isotope in any sample to decay

hard in magnetic terms, a permanent magnet made from a material such as steel

hazards associated with using mains electricity supply include damaged insulation, overheated cables, damp conditions, excess current from overloaded plugs, extension leads, single and multiple sockets

heat exchanger transfers heat from one fluid to another

hertz units of frequency (cycles per second)

* Hubble constant, H_0 the ratio of the speed at which the galaxy is moving away from the Earth to its distance from the Earth; $H_0 = v/d$

hydraulic car brakes oil is used to transmit pressure from the brake pedal to the car brakes

hydraulic jack machine in which a liquid is used to transmit pressure and lift an object

hydroelectric energy flow of water from a high to a low level is used to drive a water turbine connected to an electricity generator

image likeness of an object

* impulse force × time for which force acts

independent variable quantity whose value does not depend on that of another

induction motor a.c. motor

infrared radiation electromagnetic waves emitted by hot, but not glowing, bodies

insulator material which does not allow thermal energy or electrons to flow easily

intensity measure of the magnitude of a quantity such as sound, light or electric field

internal energy energy of the molecules in a body (both potential and kinetic)

inversely proportional two variables are inversely proportional if their product is a constant

ionosphere electrically charged particles in the upper atmosphere

ionisation process by which an atom or molecule becomes an ion

ions charged atoms or molecules which have lost or gained one or more electrons so that they are no longer neutral

isotopes of an element, are atoms which have the same number of protons but different numbers of neutrons in the nucleus

joule J; SI unit of energy

joulemeter used to measure the electrical energy transferred by an appliance

kelvin K; SI unit of temperature; a kelvin has the same size as a degree Celsius but 0°C = 273 K

kilogram kg; SI unit for mass

kilometre km; 10^3 m

kilowatt kW; 10^3 W

kilowatt-hour kWh; the electrical energy used by a 1 kW appliance in 1 hour; 1 kWh = 1000 J/s × 3600 s = 3 600 000 J = 3.6 MJ

kinetic energy E_k; energy a body has due to its motion ($mv^2/2$)

laminated layered

law of moments when a body is in equilibrium, the sum of the clockwise moments about any point equals the sum of the anticlockwise moments about the same point

law of reflection the angle of incidence is equal to the angle of reflection

length greatest dimension of an object; SI unit is the metre (m)

light-dependent resistor LDR; semiconductor device in which the electrical resistance decreases when the intensity of light falling on it increases

* light emitting diode LED; semiconductor device which emits light when it is forward biased but not when it is reverse biased

light-year used to measure astronomical distances; the distance travelled in (the vacuum of) space by light in one year

* limit of proportionality the point at which the load-extension graph becomes non-linear

limits of audibility the approximate range of frequencies audible to humans, 20 Hz to 20 000 Hz

* linear conductor conductor which obeys Ohm's law

linearly values lie along a line

line of force *see* field line

live wire connected to a high p.d.

load weight

longitudinal wave direction of vibration of particles of the transmitting medium is parallel to the direction of travel of the wave

loudspeaker converts an electric current into a sound wave

luminous object which makes its own light

magnetic field a region of space where a magnet experiences a force due to other magnets or an electric current

magnetic force force between magnets

magnetic materials materials that can be magnetised by a magnet; in their unmagnetised state they are attracted by a magnet

mass a measure of the quantity of matter in an object at rest relative to an observer

* mass defect mass lost in a reaction; it can be large in nuclear reactions such as fission or fusion and become a source of energy ($E = mc^2$)

mass number A; number of protons and neutrons (nucleons) in the nucleus of an atom

mechanical waves to and fro vibration of the particles of a transmitting medium

* medical ultrasound imaging technique used to image internal organs of the body using the reflection of ultrasonic pulses

megawatt MW; 10^6 W

melting temperature temperature at which a solid changes to a liquid

meniscus curved liquid surface

metre SI unit for length (m)

micrometre µm; 10^{-6} m

microphone converts sound waves to an electric current

microwaves radio waves with a wavelength of a few cm

Milky Way the spiral galaxy to which our Solar System belongs. The Sun is a star in this galaxy and other stars that make up this galaxy are much further away from the Earth than the Sun is from the Earth

millimetre 10^{-3} m

* moderator graphite core of a nuclear reactor which slows down fission neutrons

molecule combination of atoms

moment of a force moment = force × perpendicular distance from pivot

* momentum mass × velocity

* monochromatic single colour (frequency) of light

* motor rule *see* Fleming's left-hand rule

multimeter an instrument which can be used to measure a.c or d.c. currents and voltages and also resistances

multiplying factor number of times a force is increased, for example in a hydraulic system

musical notes produced by regular vibrations

mutual induction occurs when a changing current in one coil produces a changing current in a nearby coil as a result of electromagnetic induction

nanometre 10^{-9} m

National Grid network of electricity transmission lines

negative charges repel other negative charges, but negative charges attract positive charges

neutral equilibrium occurs when a body stays in its new position when slightly displaced and then released

neutral point point at which magnetic fields cancel

neutral wire connected to earth

neutron uncharged particle found in the nucleus of an atom (except that of hydrogen)

newton SI unit of force; N

Newton's first law of motion a body stays at rest, or if moving continues to move with uniform velocity, unless an external force makes it behave differently

* Newton's second law of motion force = mass × acceleration

noise produced by irregular vibrations

non-luminous object which does not make its own light; it may reflect light from a luminous source

non-magnetic materials materials that cannot be magnetised and are not attracted by a magnet

non-ohmic device which does not obey Ohm's law

non-renewable cannot be replaced when used up

normal line which is perpendicular to a surface

nuclear energy energy derived from nuclei

nuclear fuels radioactive materials such as uranium, used in the core of a nuclear reactor to produce heat

nucleon proton or neutron

nucleon number A; number of protons and neutrons in the nucleus

nucleus dense core of an atom containing protons and neutrons

nuclide atom of an element characterised by the mass number A and the proton number Z

octave notes are an octave apart if the frequency of one note is twice that of the other

ohm Ω; unit of resistance

* ohmic conductor conductor which obeys Ohm's law

Ohm's law the current through a metallic conductor is directly proportional to the p.d. across its ends if the temperature and other conditions are constant

optical centre centre of a lens

orbit curved path, such as that taken by a moon around a planet

parallel lines with the same direction

parallel circuit components are connected side by side and the current splits into alternative paths and then recombines; current from the source is larger than the current in each branch

parallelogram law if two forces acting at a point are represented in size and direction by the sides of a parallelogram drawn from the point, their resultant is represented in size and direction by the diagonal of the parallelogram drawn from the point

pascal Pa; SI unit of pressure $1\,Pa = 1\,N/m^2$

period time for one complete oscillation (1/frequency)

permanent magnets made of steel, retain their magnetism

perpendicular at 90°

phase vibrating particles transmitting a wave are in phase if they are moving in the same direction and have the same displacement; if this is not the case, they are out of phase

pitch frequency of a sound wave

polar near the north or south pole of the Earth

positive charges repel other positive charges, but positive charges attract negative charges

potential difference p.d.; the work done by a unit charge passing through a component

potential divider variable resistor connected so that the p.d. applied to a device can be changed

potential energy energy a body has because of its position or condition (mgh)

potentiometer resistor whose resistance can be varied

power the work done per unit time and the energy transferred per unit time

powers of ten way of writing numbers; index gives number of times number must be multiplied by 10

pressure the force per unit area

primary main, most important

principal axis line through the optical centre of a lens at right angles to the lens

principal focus (focal point) point on the principal axis of a lens to which light rays parallel to the principal axis converge, or appear to diverge from

principle of conservation of energy energy cannot be created or destroyed; it is always conserved

progressive wave travelling wave carrying energy from one place to another

proportional two variables are proportional if their ratio is a constant

proton positively charged particle found in the nucleus of an atom

proton number Z; number of protons in the nucleus

pulse a few cycles of a wave

pumped storage electricity generated at off-peak periods is used to pump water from a low-level reservoir to a high-level one

quality or timbre of a sound is determined by the number and strength of the overtones present

radar system which enables the position of a distant object to be found by the use of microwaves directed at and reflected from the object

radial along a radius

radiant emits radiation

radiation transfer of thermal energy from one place to another by electromagnetic waves

radio waves electromagnetic waves with the longest wavelength

radioactive material which emits α-, β- or γ-radiation

radioactive decay the emission of α-, β- or γ-radiation from unstable nuclei

radioisotope isotope of an element which is radioactive

radionuclide *see* radioisotope

radiotherapy use of ionising radiation to treat cancer

random irregular, erratic or haphazard

range spread of values

ratemeter instrument which counts the number of current pulses/second (often used with a GM tube)

ray direction of the path in which light travels

reaction force exerted by a support on a body

real image an image which can be formed on a screen

* **rectifier** changes a.c. to d.c.

redshift shift in the wavelengths of light from distant galaxies towards longer wavelengths (the red end of the spectrum)

reed switch electromagnetic switch

refraction bending of rays when they pass from one medium to another

* **refractive index** *n*; the ratio of the speeds of a wave in two different regions

regular reflection a parallel beam of light is reflected from a smooth surface such as a mirror, in a parallel beam

relative charges of protons, neutrons and electrons +1, 0, –1, respectively

relative motion motion of one object with respect to another

relay electromagnetic switch

renewable can be replaced; cannot be used up

resistance opposition of a conductor to the flow of electric current; symbol *R*, measured in ohms (Ω)

resistance of a metallic wire directly proportional to its length and inversely proportional to its cross-sectional area

resistor conductor designed to have resistance

resultant force the rate of change in momentum per unit time

* **retardation** a negative acceleration; velocity decreases as time increases

reverberation combination of a sound wave and its echo which acts to prolong the sound

* **reverse bias** p.d. connected across a diode such that it has a high resistance and does not conduct

rheostat variable resistor connected so that the current in a circuit can be changed

right angle 90°

right-hand grip rule if the fingers of the right hand grip a solenoid in the direction of the conventional current, the thumb points to the N pole

right-hand screw rule if a right-handed screw moves forwards in the direction of the conventional current, the direction of rotation of the screw gives the direction of the magnetic field

rotor blades or electromagnets on a rotating shaft

* **scalar** a quantity which has magnitude only

scaler instrument which counts current pulses (often used with a GM tube)

second SI unit of time (s)

secondary coming later; of less importance

* **semiconductor diode** electronic device which allows current to flow in one direction but not the other

sensitivity response of a device to a change in input; precision

series circuit components connected one after the other; the current is the same at each point in a series circuit

significant figures number of figures to which a value is given; they indicate the precision of a measurement

sliding friction frictional force acting on a body moving at a constant speed

* **slip rings** rings attached to the ends of the coil of an a.c. generator which rotate with the coil and to which electrical contact is made

soft in magnetic terms, material such as iron which is easily magnetised and demagnetised

solar cells convert sunlight into electricity

solar energy energy from the Sun

solar panels convert sunlight into thermal energy

solenoid long cylindrical coil of wire

* **sonar** an echo technique which enables the depth of an object to be found using ultrasonic waves

spectrum band of colours produced when white light is separated into its component parts

* **specific heat capacity** the energy required per unit mass per unit temperature increase

speed distance travelled per unit time

speed–time graph graph of speed on the vertical axis plotted against time on the horizontal axis

* **split-ring commutator** split ring of copper which rotates with the coil of an electric motor; it enables the current through the coil to be reversed every half-turn

* **spring constant** force per unit extension

stable equilibrium occurs when a body returns to its original position when slightly displaced and then released

standard notation writing numbers using powers of ten

starting friction maximum value of a frictional force which occurs just as a body starts to move

static friction maximum value of a frictional force which occurs just as a body starts to move

* **stator (or diaphragm)** set of fixed blades (or fixed coils in an alternator)

steam turbine steam is used to turn the blades of a rotor

strain energy energy stored in a compressed spring or elastic material

stretching force force causing a body to change shape

systematic error error introduced by the measuring device (the system)

tangent a line touching, but not intersecting, a curve

temperature determines the direction in which thermal energy flows; kinetic theory regards temperature as a measure of the average kinetic energy of the molecules of the body

temporary magnets made of soft iron, lose their magnetism easily

* **terminal velocity** constant velocity reached when the air resistance upwards equals the downward weight of a falling body

thermal energy energy of the molecules in a body

thermal power station thermal energy is used to turn water into steam to drive a turbine and generate electricity

thermistor semiconductor device in which the electrical resistance decreases when the temperature increases

thermometer device for measuring temperature

* **thermonuclear fusion** the combination of light nuclei into a heavier nucleus in a reaction which takes place at a very high temperature (for example in the Sun)

thermostat a device which keeps the temperature of a room or appliance constant

* **thoron** radioactive gas which emits α-particles

tidal energy flow of tidal water from a high to a low level used to generate electricity

time duration; SI unit is the second (s)

total internal reflection occurs when a light ray does not cross the boundary between two media; it is totally reflected at the boundary

* **tracer** radioisotope injected into a system; the motion or concentration of the isotope can be monitored, for example with a GM tube

transformer two coils (primary and secondary) wound on a soft iron core which allow an alternating p.d. to be changed from one value to another

transverse wave direction of vibration is perpendicular to the direction of travel of the wave

tritium $_1^3$H; isotope of hydrogen with one proton and two neutrons in the nucleus

ultrasound sound wave with a frequency greater than 20 kHz

ultraviolet radiation electromagnetic waves having shorter wavelengths than light

uniform having the same value

uniform acceleration constant acceleration

uniform velocity constant velocity

unstable equilibrium occurs when a body moves further away from its original position when slightly displaced and then released

vacuum a space from which all air has been removed

vaporisation change of a liquid to a vapour

variable quantity that can be changed

variable resistor resistor whose resistance can be changed

* **variation of magnetic field strength** the magnetic field decreases with distance from a current-carrying wire and varies around a solenoid

* **vector** a quantity which has both magnitude and direction

velocity speed in a given direction; change in displacement per unit time

virtual image an image which cannot be formed on a screen

volt V; unit of p.d.

voltage p.d.; measured in volts (V)

voltmeter instrument used to measure p.d.

volume of a cylinder $\pi r^2 \times h$, where r is the radius and h is the height of the cylinder

watt SI unit of power; 1 W = 1 J/s

wave energy rise and fall of sea waves used to generate electricity

wave equation $v = f\lambda$

waveform shape of a wave

wavefront in two dimensions, it is a line on which the particles transmitting the wave are vibrating in phase; the crests of waves in a ripple tank are wavefronts. A line drawn perpendicular to a wavefront is a ray

wavelength distance between successive crests of a wave

weight a gravitational force on an object that has mass

wind turbines convert wind energy into electrical energy

work measure of amount of energy transferred. Work done = force × distance moved in the direction of the force. SI unit is the joule (J)

X-rays electromagnetic waves with a shorter wavelength than ultraviolet radiation

Acknowledgements

The publishers would like to thank the following who have given permission to reproduce the following material in this book:

p.1 © Agence DPPI/Rex Features; **p.2** © Somatuscani/stock.adobe.com; **p.7** © nirutft – Fotolia; **p.12** © Images-USA/Alamy Stock Photo; **p.18** © Images&Stories/Alamy; **p.21** © Berenice Abbott/Contributor/Getty Images; **p.22** © Agence DPPI/Rex Features; **p.25** © David J. Green – studio/Alamy; **p.33** *l* © Paul/stock.adobe.com; *r* © Paul/stock.adobe.com; **p.36** © Arnulf Husmo/Stone/Getty Images; **p.37** © Arbortech Pty Ltd; **p.41** © Albaimages/Alamy; **p.42** © ESA; **p.48** © Kerstgens/SIPA Press/Rex Features; **p.49** *t* © Prairie Agricultural Machinery Institute; *b* Crown Copyright material has been reproduced by permission of the Driver and Vehicle Standards Agency, which does not accept any responsibility for the accuracy of the reproduction; **p.58** *l* © Stu Forster/Getty Images; *t* © Duif du Toit/Gallo Images/Getty Images; *b* © Sergii Kumer/Alamy Stock Photo; **p.60** © Wolfgang Zwanzger/stock.adobe.com; **p.62** *tl* © Julia Druzhkova/stock.adobe.com; *tr* © Naoki Nishimura/AFLO Sport/Nippon News/Alamy Stock Photo; *bl* © Freer/stock.adobe.com; *br* © SSE Renewables; **p.63** © Charles Ommanney/Rex Features; **p.66** *l* © Ilkercelik/stock.adobe.com; *r* © TRL Ltd./Science Photo Library; **p.70** *tl* © Tim gartside/Alamy Stock Photo; *tr* © Martin Bond/Science Photo Library; *bl* © Aurora Solar Car Team member Courtney Black; **p.71** *tl* © Hemis /Alamy; *tr* © Mark Edwards/Still Pictures/Robert Harding; *br* © Joerg Boethling/Alamy Stock Photo; **p.72** © GE Steam Power; **p.78** © Nelly Ovchinnikova/stock.adobe.com; **p.81** © Esa Hiltula/Alamy; **p.82** © Orientka/stock.adobe.com; **p.85 & 86** © Dr Linda Stannard, UCT/Science Photo Library; **p.94** © Philippe Plailly/Eurelios/Science Photo Library; **p.101** © Sergey Ryzhov/Alamy Stock Photo; **p.102** © Chris Mattison/Alamy; **p.103** © Martyn F. Chillmaid/Science Photo Library; **p.111** © Mark Sykes/Alamy; **p.117** *l & r* © sciencephotos/Alamy; **p.118** © Don B. Stevenson/Alamy; **p.119** © Sandor Jackal – Fotolia.com; **p.121** © James R. Sheppard; **p.122** *t* © Zoonar RF/Thinkstock; *b* © Glenn Bo/IStockphoto/Thinkstock; **p.123** © Vitaliymateha/stock.adobe.com; **p.127** © Tom Tracy Photography/Alamy; **p.131** *t & b* © Andrew Lambert/Science Photo Library; **p.132** *tr* © HR Wallingford; *c & b* © Andrew Lambert/Science Photo Library; **p.134** © BRUCE COLEMAN INC./Alamy; **p.136** © Alexander Tsiaras/Science Photo Library; **p.137** © Tom Tracy Photography/Alamy; **p.139** © Owen Franken/Corbis; **p.141** *l* © Colin Underhill/Alamy; *r* © Sciencephotos/Alamy Stock Photo; **p.148** *l* © Last Resort; *c* © vario images GmbH & Co.KG/Alamy; *b* © CNRI/Science Photo Library; **p.151** *l* © S.T. Yiap Selection/Alamy; *r* © Last Resort; **p.156** © Alfred Pasieka/Science Photo Library; **p.162** © US Geological Survey/Science Photo Library; **p.163** © Image Source White/Image Source/Thinkstock; **p.167** © Jonathan Watts/Science Photo Library; **p.170** © Andrew Drysale/Rex Features; **p.172** © Science Photo Library; **p.173** © Digitalglobe/Shutterstock; **p.175** © Keith Kent/Science Photo Library; **p.180** *t & b* Andrew Lambert/Science Photo Library; **p.182** © Alex Bartel/Science Photo Library; **p.184** © Keith Kent/Science Photo Library; **p.191** *l* © NaMaKuKi/stock.adobe.com; *r* © Martyn F. Chillmaid/Science Photo Library; **p.198** *t* © Roman Ivaschenko/stock.adobe.com; *b* © David J. Green – electrical/Alamy Stock Photo; **p.204** © Pat Canova/Alamy Stock Photo; **p.211** © Giuliano Del Moretto/Shutterstock.com; **p.226** © Maksym Yemelyanov/stock.adobe.com; **p.232** *t* © GE Steam Power; *b* © Kirill Kukhmar/ITAR-TASS News Agency/Alamy Stock Photo; **p.240** © Patboon/stock.adobe.com; **p.245** © RJH_IMAGES/Alamy Stock Photo; **p.251** © CERN/Photo Science Library; **p.254** *t & b* Courtesy of the Physics Department, University of Surrey; **p.264** *t & b* © The Royal Society, Plate 16, Fig 1 from CTR Wilson , Proc. Roy. Soc. Lond. A104, pp.1–24 (1923); **p.265** © Lawrence Berkeley Laboratory/Science Photo Library; **p.270** © Martin Bond/Science Photo Library; **p.271** © Museum of Cultural Heritage – University of Oslo, Norway (photo Eirik Irgens Johnsen); **p.275** © NASA/Science Photo Library; **p.278** © John Sanford/Science Photo Library; **p.280** © Royal Greenwich Observatory/Science Photo Library; **p.282** © NASA/Science Photo Library; **p.283** *t* © NASA/Science Photo Library; *b* © NASA/JPL/USGS; **p.288** *t* © Allan Morton/Dennis Milon/Science Photo Library; *b* © NASA/Science Photo Library; **p.289** © NG Images/Alamy Stock Photo; **p.290** © NASA/ESA/STSCI/J.Hester & P.Scowen, ASU/Science Photo Library; **p.292** © Photo Researchers/Science History Images/Alamy Stock Photo

b = bottom, *c* = centre, *l* = left, *r* = right, *t* = top
Every effort has been made to trace and acknowledge ownership of copyright. The publishers will be glad to make suitable arrangements with any copyright holders whom it has not been possible to contact.

Index

velocity 13
   circular motion 41
   and kinetic energy 63–4
   terminal 22
ventilation 123
Venus 279
   facts and figures 282
vernier scales 7–8
VHF (very high frequency)
   radio waves 163
virtual images 141
visible spectrum 156
voltage *see* potential difference
voltmeters 195–6
volts (V) 193, 194, 202
volume 4–5
   effect on gas pressure 92–3, 96
   of an irregularly shaped solid 30

W

water
   conduction of heat 116
   specific heat capacity 106–7

specific latent heat of fusion
   for ice 110
specific latent heat of
   vaporisation 111
   thermal expansion 103
water supply systems 80
water waves 128
   ripple tank experiments 130–2
watt (W) 75, 202
wave energy 70
wave equation 129–30, 161
waveforms of musical notes 171
wavefronts 131
wavelength (λ) 129
   effect on diffraction 132
waves 128
   description of 129
   diffraction 132, 134
   electromagnetic 161
   longitudinal 129
   reflection 131, 133
   refraction 131, 134
   ripple tank experiments 130–2

seismic 173
sound 167–72
transverse 128
tsunami 173
wave speed (v) 129
wave theory 133–4
weight 25–6
   relationship to mass 27
wet suits 123
white dwarf stars 289
wind turbines 70, 73
wires, resistance of 200
wood fires 124
work done 62, 67–8

X

X-rays 163
   dangers of 164

Z

zero error 8